$\rho = \frac{be^{-a}}{r^2}$

r contain $\frac{1}{2}$ Total charge

A cylinder = $2\pi rh$

11205 Jones

INTERMEDIATE ELECTROMAGNETIC THEORY

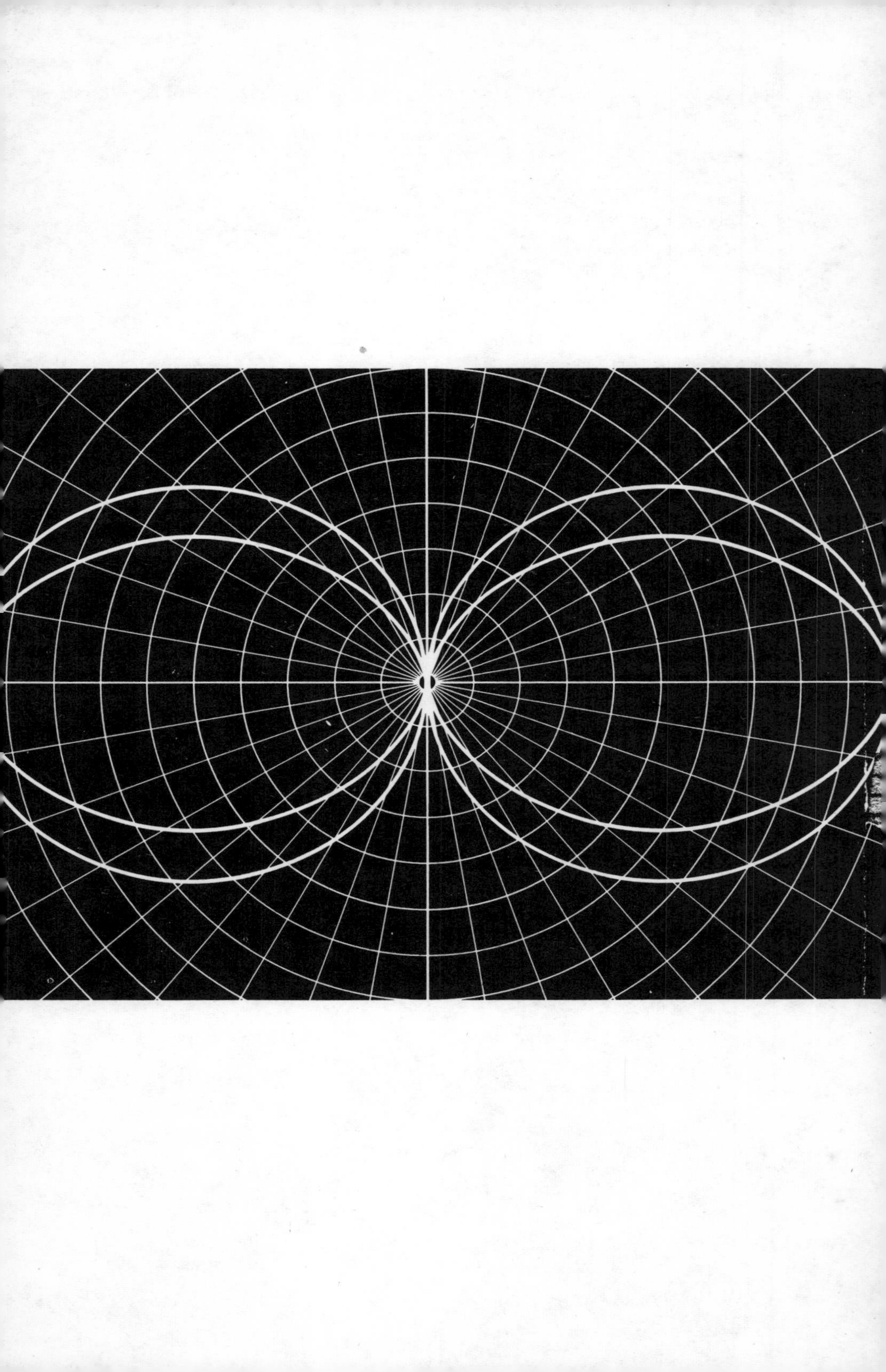

INTERMEDIATE ELECTROMAGNETIC THEORY

W. M. SCHWARZ

Professor of Physics, Union College, Schenectady, New York

ROBERT E. KRIEGER PUBLISHING COMPANY
HUNTINGTON, NEW YORK
1973

ORIGINAL EDITION 1964

REPRINT 1973, (with corrections)

Printed and Published by

ROBERT E. KRIEGER PUBLISHING CO., INC.
BOX 542, HUNTINGTON, NEW YORK. 11743.

Reprinted by arrangement

Library of Congress Card Number: 64-11504
SBN 0-88275-093-3

Printed in the United States of America

Preface

This book has been written for an advanced undergraduate course in electricity and magnetism designed primarily, but not exclusively, for physics majors or advanced electrical engineering students. It reflects the many changes that have been made in the teaching of electricity and magnetism during the past two decades, changes designed to meet the need for more rigorous treatment and to make possible the inclusion of topics that were once reserved for graduate courses.

The traditional one-year course in electricity and magnetism covered Coulomb's law, Ohm's law and d-c circuits, Ampère's law, Faraday's law, a-c currents and circuits, and the associated electrical instrumentation and measurements. In a few concluding lectures Maxwell's field equations and some of their simplier implications were presented.

In order to meet the demands of graduate schools for a more thorough grounding in electromagnetic theory, the course for physics majors at Union College has for the past fifteen years consisted of a first semester covering the basic laws and their implications and a second semester dealing with the Maxwell field equations and their consequences in electromagnetic radiation. Considerable satisfaction has been expressed with this arrangement, and it seemed desirable to support it with a textbook.

Limitations of space and time suggested that this book should be devoted primarily to theory with only limited references to applications of electromagnetism. Therefore, it has been assumed that applications, instruments, and the techniques of measurements are to be taught in laboratory sessions, for which good textbooks in electrical measurements and laboratory techniques are available to the instructor and to the student. The latter should be encouraged to use them as references for his laboratory problems. Where instruments or applications are discussed in this book, such discussion is primarily for the purpose of illustrating the theory.

The mathematical preparation of the student should include, of course, the differential and the integral calculus and at least some knowledge of what a differential equation is. Facility in the techniques of solving differential equations is, however, not expected. The ability to manipulate complex numbers is desirable, as is some acquaintance with Fourier series, although both of these topics are given a brief exposition in the Appendix. Not all students will have the necessary foundation in vector algebra and vector calculus required by electromagnetic theory, and I have chosen to develop the parts of this mathematical tool in course, because my experience suggests that the student finds this method more satisfying than the presentation *in toto* of vector operations in an introductory chapter or in an appendix. However, the sections devoted to vector mathematics appear in a smaller type face so that they may be readily identified and skipped by the student for whom they are unnecessary.

MKS units are used throughout the book, but an appendix is devoted to a rather thorough development of the interrelations among the common systems of units.

Chapters 1, 2, and 3 discuss the consequences of the inverse square law of force between electric charges at rest, including Gauss's law in its integral and differential forms, Poisson's equation, and the modifications required by material media. Chapter 4 introduces the concept of current and develops d-c circuit theory.

In Chapter 5 the decision to develop Ampère's law first from the force between filamentary currents, and later to apply it to moving point charges, was a pedagogical judgment, the basis for which is largely intuitive. In this chapter, Ampère's circuital law and, in Chapter 6, Faraday's law are evolved in both the integral and the differential forms.

Chapter 7 completes the "basic topics" with a consideration of the modifications in the treatment of the magnetic field relations required by the presence of a material medium. The parallelism with dielectrics is pointed out.

In Chapter 8 a discussion of transient currents in circuits containing inductance, capacitance, and resistance leads into the treatment of steady-state alternating currents. Alternating-current bridges are used as illustrations and exercises in the use of complex numbers for solving a-c circuit problems.

Although frequency filters and transmission lines are engineering devices of considerable importance, this fact alone would not justify their inclusion in this book. Simple filter circuits are introduced in Chapter 9, first as an illustration and further exercise in the use of a-c circuit methods. Then the student acquires additional experience in the interpretation of phase relations and meets for the first time the concept of characteristic impedance. Filter circuits are a pedagogical transition from systems employing lumped elements at low frequencies to continuous systems at high frequencies, where delay times cannot be neglected. In filter theory, moreover, there is the beginning of the traveling wave concept. This concept becomes fully developed in the theory of the transmission line, which carries on the transition from lumped circuits to continuous media. The development of transmission line theory in terms of concepts that derive from lumped circuit theory produces much, perhaps most, of what is significant and useful in transmission line behavior; but it is pointed out that this analysis is incomplete, and a more general treatment is discussed later in Chapter 12.

With transmission line theory as a background, the theory of plane electromagnetic waves and their properties in infinite space is developed in Chapter 10, and the behavior of these waves in absorptive and nonuniform media, in Chapter 11. In these chapters further use is made of the characteristic impedance, and analogies with circuit and transmission line behavior are pointed out.

Guided electromagnetic waves follow logically after a study of the interaction of waves at discontinuities in the medium. In Chapter 12 the primary attention is on the behavior of waves in pipes of rectangular cross section. In addition, resonant cavities are discussed and it is shown how the more general theory of the transmission line follows from waveguide theory.

Finally, Chapter 13 on relativity is introduced to integrate the fundamental electromagnetic concepts into a more coherent picture. This chapter is not intended to provide an exhaustive treatment of relativity, but it is hoped that as an introduction the chapter will appeal to the more eager and interested student, whetting his appetite for further investigation.

Discussion questions and problems appear at the end of each chapter, the former being largely qualitative and occasionally of the sort to which there is not just one right answer. Many of the problems serve primarily as exercises to fix the principles of the text in the student's mind, but others are extensions of chapter topics, which it is hoped will sustain the undergraduate student's sense of discovery and encourage him to go beyond the book on projects of his own.

Happily the strains associated with the writing of a textbook are seldom evident in the completed work, but they have been only too evident to my wife, my colleagues, and even on occasion to my students. I am indeed grateful to those around me whose patience and tolerance have helped make this book possible. I must express my appreciation to Union College, whose support during a sabbatical year provided the opportunity to begin the writing of this book; to the readers of the manuscript, whose comments have been of very great value; to my colleagues for profitable discussions; and to the Physics Institute of the University of Vienna for library privileges during my sabbatical year.

Schenectady, N.Y. W. M. Schwarz
November, 1963

"I very much appreciate the many comments and suggestions I have received, and I especially thank those who took the time to point out errors. In this new printing these errors have been corrected, and I shall be chagrined but grateful to be informed of any that have been overlooked. (October 1972). W.M.S."

Contents

1

Electrostatic fields

Whereas classical mechanics deals with the relations of forces to the motions of masses and is not essentially concerned with the origin or nature of those forces, electromagnetic theory is primarily concerned with a particular group of forces—those associated with electric charges—and only on occasion deals with the motions of masses reacting to these forces. The "field" as a device for describing these forces, the potential energy associated with this field, the influence of the motion of *sources* of this field, and the propagation of the field through space are the subjects of primary interest. It must always be remembered however, that the reactions of material bodies are what ultimately give meaning to this field concept.

Inasmuch as the study of the electromagnetic field led to a new understanding of the nature of light, it is interesting to think of electromagnetic theory as a kind of connecting link between the science of mechanics and the science of optics. It is also interesting to note that whereas this theory rests firmly on the concepts of classical mechanics, its depicting of the properties of the electromagnetic field stimulated the development of the Theory of Relativity which, in turn, effected a fundamental reformulation of classical mechanics.

Although some of the simplest manifestations of electric charge and of magnetism were known in ancient times, very little was done in experimenting or creating theory in electromagnetism until about the Eighteenth Century, when interest grew rapidly in investigating the more qualitative features of electrical phenomena. The electrostatic machine came into use, the Leyden jar as a capacitor was developed, the distinction between insulators and conductors and the polarization properties of dielectrics were investigated, and toward the end of this century the law for force between charges was determined. There was considerable discussion about the nature of electricity itself—"electric fire," it was called—and although there seemed to be general agreement that it was a fluid (the word "fluid" was not clearly defined), there was debate as to whether there existed one or two kinds of fluid. The fact that matter could be put into two different electrical states was recognized, but opinion differed as to whether these two states were evidence of two different kinds of fluid, or whether one state represent an excess and the other a deficiency of one universal fluid. Benjamin Franklin was of the latter persuasion, and it is to him that we owe the present designations of positive charge, which he took to be an excess of electrical fire, and negative charge, a deficiency. This debate over the one- and two-fluid theories is historically interesting, but no longer of any consequence since we now have much more explicit theories of the electrical constitution of matter. Negative charges are associated with electrons (but note also the antiproton and the negatively charged particles of intermediate mass); and the positive charges with protons (but note also the positron and the positively charged particles of intermediate mass). We should recognize, however, that although it is sometimes convenient to refer to this modern picture of the constitution of matter, electromagnetic theory is a macroscopic theory involving the properties of bulk matter, and its conclusions do not fundamentally depend on any particular picture of the microscopic constitution of matter.

With one exception, really quantitative work in electricity started with the investigations of Coulomb who measured with a torsion balance the force between small charged bodies. The exception was Henry Cavendish, a brilliant investigator of the latter decades of the Eighteenth Century, who was one of the most curious figures in the history of science. Reputed to be the wealthiest man in England, he was absorbed in scientific investigation to the exclusion of every other interest, was intensely shy to the point of being almost a recluse, and made little effort to communicate his findings to others. Although he made significant discoveries in both chemistry and physics, only a few were known to his contemporaries, and many had to be rediscovered later. The bulk of his work was unknown until his private papers were edited by James Clerk Maxwell in 1879. Then it became known that, among other things, Cavendish had developed the concept of capacitance, the effect of dielectrics on capacitance, and the inverse square law of force between charges, years before they were subsequently investigated by others because Cavendish had not made public his results during his lifetime.

Charles Augustin Coulomb, on the other hand, published an account of his investigation of the law of force between electric charges in 1785, and the law is by common consent known by his name. Although his experimental procedure was simple and direct, it was not capable of high precision. Cavendish's method, though more subtle, was capable of producing very precise results. The Coulomb experiment will be discussed first, and the more elegant Cavendish experiment somewhat later.

It is worthwhile noting that both Cavendish and Coulomb demonstrated only the fact that the force varies with the inverse second power of the separation of the charges. That the force equation also includes the product of the charges was assumed without proof, probably by analogy with Newton's law of gravitation. It was Gauss who pointed out that this assumption actually constitutes a definition of what we mean by quantity of charge. For the earlier workers, quantity of charge had been a somewhat vague and intuitive concept.*

1. Coulomb's Law

Coulomb measured the force between charged bodies by means of the torsion balance, an instrument that Cavendish later used in one of his rare published experiments to measure the gravitational constant in Newton's law of gravitation. Coulomb described with considerable care his experiments to measure the torque required to twist fibers of various cross sections and lengths, and concluded that the torque is proportional to the angle of twist. His torsion balance (Figure 1) consisted of a light

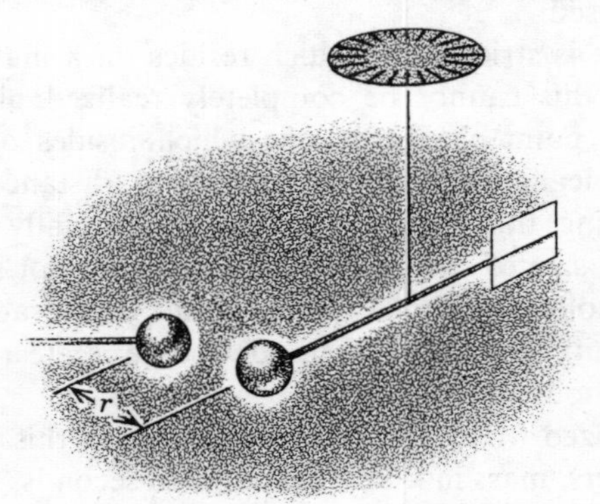

Figure 1. *Coulomb's torsion balance.*

* Von Laue, *History of Physics*, Chapter 5.

cross-arm, hung at its center by a fine fiber, which was hung from a knob calibrated in degrees. One end of the cross-arm carried a light, conducting sphere which could be charged and, on the other end, there was a paper vane to damp out oscillations. When another similarly charged sphere was brought near the first, the torque produced by the force of repulsion was measured in terms of the twist in the fiber required to oppose this torque. Coulomb was thus able to determine that the force of repulsion was inversely proportional to the square of the separation of the charged bodies, provided that the physical size of the bodies was small.

It is more difficult, although not theoretically impossible, to measure the force of *attraction* between opposite charges by this method. But the reader may find it interesting to devise a different method, still using this torsion balance, whereby the attractive force as a function of separation of the charges can be determined (at least to a good approximation). He may then wish to look up Coulomb's original paper for comparison. (See References 3 and 4 at the end of this chapter.)

The relation known today as Coulomb's law states that the magnitude of the force, either a repulsion or an attraction, of one "point charge" on another separated from it by a distance r is directly proportional to the product of the magnitudes of the charges and inversely proportional to the square of the distance r. In symbols,

$$\text{(1-1)} \qquad \mathrm{F} = \frac{kq_1q_2}{r^2}$$

where F is the force which charge q_1 exerts on charge q_2, (or q_2 on q_1), r is the separation distance, and k is a constant of proportionality whose value depends on the medium in which the charges reside and on the system of units used.

A point charge is strictly one which resides on a material body of no size. Obviously this cannot be completely realized physically, but we shall speak of a point charge as one which resides on a body whose dimensions are negligibly small compared to distances used in computations involving this charge. It may occasionally be necessary to stipulate that the size of such a body is small, but not so small as to be comparable to molecular dimensions in order to average out the fundamental discontinuity of matter. By this device we create a quasi-continuum to which the methods of the calculus can be applied.

In the rationalized MKS system of units used in this book, distance is measured in meters, mass in kilograms, time in seconds, force in newtons, and charge in coulombs. In this system the proportionality constant k in Equation 1-1 has the value

$$k = 8.98776 \times 10^9 \text{ newton-meters}^2/\text{coulomb}^2$$

which, for our purposes, can be rounded off conveniently to

$$k = 9 \times 10^9 \text{ n-m}^2/\text{coul}^2$$

when the medium in which the charges reside is a vacuum, and for most practical purposes also when the medium is air. In Chapter 3 we shall see how this constant is changed by the presence of other media.

For convenience (the reasons will become more apparent later), the value of k for a vacuum is written in theoretical equations as

$$k = \frac{1}{4\pi\epsilon_0}$$

where ϵ_0 is called the absolute dielectric constant of free space. Its value is readily found to be

$$\epsilon_0 = 8.85399 \times 10^{-12} \text{ coul}^2/\text{n-m}^2$$

which is easier to remember in the adequately accurate form,

$$\epsilon_0 = 10^{-9}/36\pi \text{ coul}^2/\text{n-m}^2$$

In Chapter 2, after we have defined capacitance, it will be found that the dimensions of ϵ_0 can also be expressed as farads/meter.

Because the factor 4π is included in the constant in Coulomb's law, later relations derived from it usually have simpler constants appearing in them. Such a choice of k is, of course, not a necessary one, and several other systems of units based on other choices have been and still are in use. The electrostatic system, for example, makes the simplest choice of $k = 1$. One virtue of the MKS system of units, perhaps most apparent in engineering applications, is that the electrical units for current and potential are the ampere and the volt, units in which the common laboratory instruments are calibrated. But in much of theoretical physics, other systems are also used, and the reader should ultimately become familiar with these too. They are described briefly in Appendix D, but will not be used in the body of this text. The author feels that the effort required in learning to switch from one system of units to another detracts from the understanding of the fundamental physics. When the student feels himself at home in the science of electromagnetism, it is time enough to practice the juggling of units. Such practice will, of course, ultimately be necessary for reading the literature of physics where several systems are in common use.

2. Vectors and Vector Notation

Electromagnetic theory is considerably simplified by the use of vector algebra and calculus. The elements of vector analysis will be developed as they are needed, but such topics will usually appear in a distinguishable format so that the reader who is already familiar with this mathematical tool may skip them (but some may nevertheless find these topics a profitable review). A summary of the more important vector relationships and identities appears in Appendix A for quick reference.

Vectors will be shown in bold-faced type, whereas scalars will appear in italic type (thus **R** is a vector whereas R is a scalar), but sometimes it will be necessary to show the scalar magnitude of a vector quantity by the use of vertical lines (thus $|\mathbf{R} + \mathbf{P}|$ is the scalar magnitude of the vector, $\mathbf{R} + \mathbf{P}$).

The addition of two vectors, **R** and **P**, is illustrated in Figure 2, and the subtraction in Figure 3. The magnitude of the resultant, Q, is obtained from

$$Q^2 = R^2 + P^2 \pm 2RP \cos \theta \tag{2-1}$$

where the positive sign goes with addition and the negative sign with subtraction. Evidently, when **R** and **P** are at right angles to each other, Equation 2-1 reduces to the familiar Pythagorean theorem.

Any vector can be resolved into the sum of two or more other vectors in non-parallel directions, and these are called components of the original vector. Such resolution into components is especially useful when the components are along mutually perpendicular directions, such as along the axes of a Cartesian coordinate system. It has become a common practice to define vectors of unit length along each of the three axes, x, y, and z, calling them **i**, **j**, and **k**, respectively (sometimes other symbols such as $\mathbf{1}_x$, $\mathbf{1}_y$, $\mathbf{1}_z$ are used), and to write the component of, say, **Q** along the x-axis as $\mathbf{i}Q_x$, that along the y-axis as $\mathbf{j}Q_y$, and that along the z-axis as $\mathbf{k}Q_z$. Then **Q** is written as the sum of its three components, thus:

$$\mathbf{Q} = \mathbf{i}Q_x + \mathbf{j}Q_y + \mathbf{k}Q_z \tag{2-2}$$

A similar convention of unit vectors is used with other systems of coordinates. In spherical polar coordinates, for example, a unit vector in the direction of increasing r is often written as $\mathbf{r}_u$ (sometimes $\mathbf{1}_r$), in the direction of increasing θ as $\boldsymbol{\theta}_u$ (sometimes $\mathbf{1}_\theta$), and in the direction of increasing ψ as $\boldsymbol{\psi}_u$ (sometimes $\mathbf{1}_\psi$).

The use of the component form simplifies the process of adding vectors. With the vectors of Figure 2 resolved along a set of Cartesian axes, the sum of **R** and **P** is written:

$$\mathbf{Q} = \mathbf{i}(R_x + P_x) + \mathbf{j}(R_y + P_y) + \mathbf{k}(R_z + P_z) \tag{2-3}$$

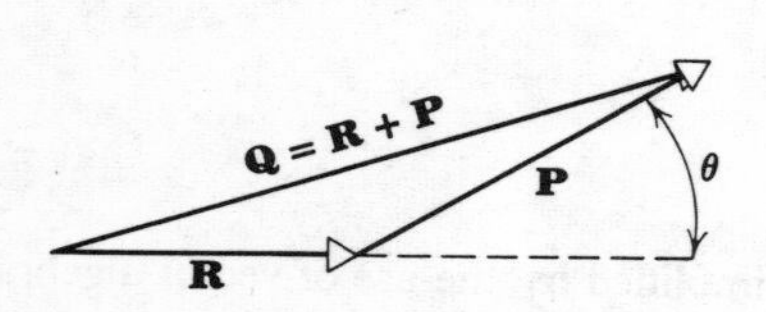

Figure 2. The addition of vectors.

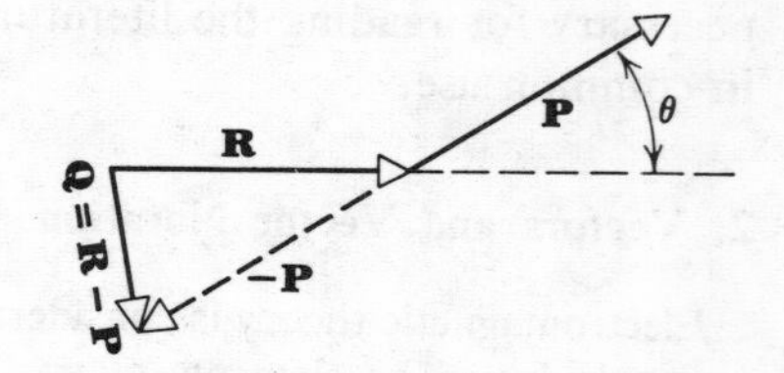

Figure 3. The subtraction of vectors.

from which it is seen that

$$Q_x = R_x + P_x \tag{2-4}$$
$$Q_y = R_y + P_y$$
$$Q_z = R_z + P_z$$

Evidently the order in which a vector sum is taken is immaterial; hence, vector addition is commutative.

3. Coulomb's Law in Vector Form—the Electric Field

Coulomb's law as it appears in Equation 1-1 contains an inherent ambiguity. There are two equal and opposite forces described by that equation, one acting on q_1 and the other on q_2. It is usually necessary to specify one of the two forces and also to indicate its direction. This can be done in a simple, concise fashion by using vector notation.

If we let $\mathbf{r}_{12}$ be a vector directed from q_1 to q_2 with a magnitude equal to the separation of q_1 and q_2, the force $\mathbf{F}_2$, on the charge q_2 is given in both magnitude and direction by

$$\mathbf{F}_2 = \frac{1}{4\pi\epsilon_0}\frac{q_1 q_2}{r_{12}^{\;3}}\mathbf{r}_{12} \tag{3-1}$$

where the denominator is a scalar magnitude. We could, as an alternative, define a vector of unit length, say, $\mathbf{r}_u$, in the direction from q_1 to q_2, so that $\mathbf{r}_{12} = \mathbf{r}_u r_{12}$, in which case the force on q_2 can be written:

$$\mathbf{F}_2 = \frac{1}{4\pi\epsilon_0}\frac{q_1 q_2}{r_{12}^{\;2}}\mathbf{r}_u \tag{3-2}$$

The reader should satisfy himself that the above two expressions for the force $\mathbf{F}_2$ are actually equivalent.

Obviously, the force on q_1 can be written in a similar fashion:

$$\mathbf{F}_1 = \frac{1}{4\pi\epsilon_0}\frac{q_1 q_2}{r_{21}^{\;3}}\mathbf{r}_{21} = -\frac{q_1 q_2}{4\pi\epsilon_0 r_{12}^{\;3}}\mathbf{r}_{12} = \frac{q_1 q_2}{4\pi\epsilon_0 r_{21}^{\;2}}\mathbf{r}_u{}' \tag{3-3}$$

where $\mathbf{r}_u{}'$ is a unit vector directed from q_2 to q_1, so that $\mathbf{r}_u{}' = -\mathbf{r}_u$.

For charges of like sign, either Equation 3-1 or Equation 3-2 show that the force is in the same direction as the vector $\mathbf{r}_{12}$, that is, it is a force of repulsion. The reverse is true for charges of opposite sign.

Experimentally, it is found that the force which one charge experiences by virtue of the presence of a second is unaffected by the presence of other charges in the neighborhood. Thus the net force on a charge q in the vicinity of $q_1, q_2, q_3, \ldots$, etc., is just the vector sum of the individual forces found by pairing each of the charges q_i with q.

We expect this to be true, no matter what the distribution of charges around q. Since each of the individual forces is proportional to the magnitude of q, so also will be the vector sum of these forces. It therefore follows that the ratio of the net force on q to the magnitude of q is a quantity independent of this magnitude, such a ratio being conveniently regarded as a property of the point where q resides and called the *electric field intensity* or *electric field strength*, $\mathbf{E}$, at this point.

This definition is expressed by the relation

$$\mathbf{F} = \mathbf{E}q \tag{3-3}$$

where $\mathbf{F}$ is the force which a point body carrying a charge q experiences solely by virtue of its charge. There may be other forces of a nonelectrical nature, such as gravitational forces, also acting on this body. These are not to be included in $\mathbf{F}$ in Equation 3-3.

The reader should clearly understand that, although we have approached the concept of an electric field through a discussion of Coulomb's law, the definition of electric field strength given in Equation 3-3 depends only on that part of the law which says that the force is proportional to the charge, not on the part which says that the force is proportional to the inverse square of the separation of the charges. Moreover, the definition does not require that we know the configuration of the charges which give rise to the field before we are able to measure the field strength experimentally. In principle, we can measure the electric field strength at a point in space by placing at this point an infinitesimal body carrying a test charge q, and measuring the total force on the body. We then remove the charge from the body and repeat the measurement. The difference between the two results is the force of electrical origin and the ratio of this electrical force to the charge q is the electric field strength. The direction of the electric field is the same as the direction of this force if q is positive, and opposite to it if q is negative.

It is sometimes stated in the literature that an infinitesimal charge must be used so that the test charge itself does not alter the field. This, however, is a warning to the experimenter, not to the theoretician. It merely emphasizes the fact that if the charges which give rise to the electric field happen to be free to move, then the force which they experience from the presence of the test charge will cause them to change their positions, which in turn will alter the field strength at the test charge. If the charges are fixed in space, then no such problem arises. This is no more than to say that in making any measurement, the experimenter must be careful to minimize the disturbance that his instrument (in this case the test charge) produces on what is being measured.

The defining Equation 3-3 determines, at least in principle, an experimental method for measuring electric field strength. A theoretical prediction of the field strength, on the other hand, can be made by combining Coulomb's law with 3-3, provided the values and locations of the charges giving rise to the field are known. Thus a group of charges, q_1, q_2, q_3, etc., which are at distances r_1, r_2, r_3, etc., from a point P, as in Figure 4, will give rise to an electric field at P that is the vector sum of the field strengths of the individual charges.

$$\mathbf{E}_P = \frac{1}{4\pi\epsilon_0}\left[\frac{q_1\mathbf{r}_1}{r_1^{\,3}} + \frac{q_2\mathbf{r}_2}{r_2^{\,3}} + \frac{q_3\mathbf{r}_3}{r_3^{\,3}} + \cdots\right] = \frac{1}{4\pi\epsilon_0}\sum_i \frac{q_i\mathbf{r}_i}{r_i^{\,3}} \tag{3-4}$$

where the summation includes all the charges which give rise to the field. Strictly, such a summation is over all the charges in the universe except any charge at the point P; in practice, Equation 3-4 is applied to cases where all charges, other than a particular group, are assumed to be far enough away so that their influence is negligible or to be paired positive to negative so that their forces cancel.

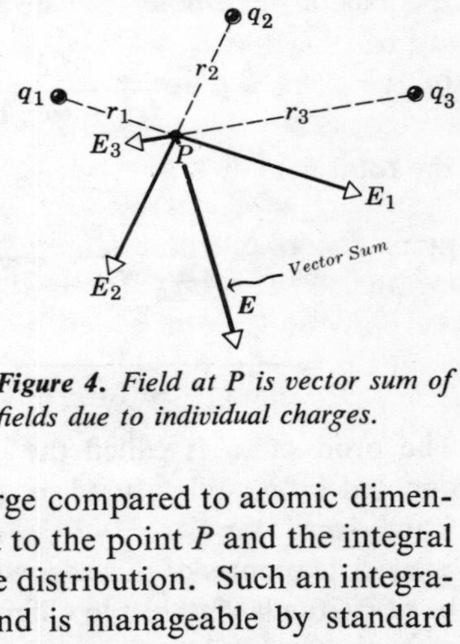

If the charges in the summation of Equation 3-4 form a sensibly continuous distribution (on an atomic scale this is, of course, never true), then Equation 3-4 may be written as an integral:

$$\mathbf{E} = \frac{1}{4\pi\epsilon_0}\int \frac{\rho\mathbf{r}\,d\mathscr{V}}{r^3} \tag{3-5}$$

Figure 4. Field at P is vector sum of fields due to individual charges.

where ρ is the charge per unit volume, $d\mathscr{V}$ is an infinitesimal volume (but still large compared to atomic dimensions) located at a distance $\mathbf{r}$ with respect to the point P and the integral is taken over the total volume of the charge distribution. Such an integration is, of course, a vector integration and is manageable by standard analytical methods, usually only for relatively simple distributions.

Several examples involving both discrete and continuous distributions of charge will now be illustrated.

4. Electric Fields Associated with Discrete Charges

An electric dipole is a configuration of two charges of equal magnitude but opposite sign, $+q$ and $-q$, separated by a distance d. We shall

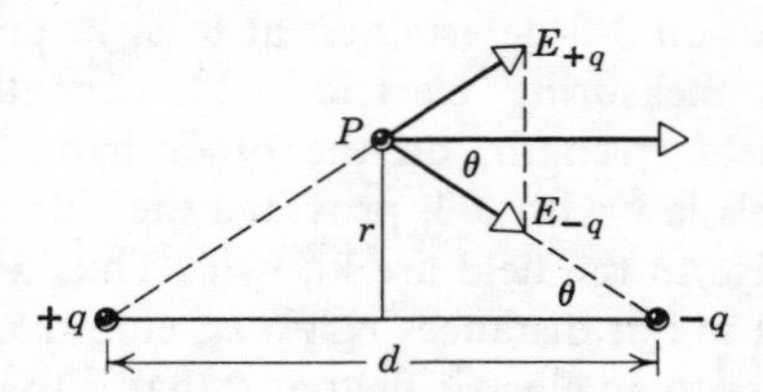

Figure 5. Fields on the perpendicular bisector of an electric dipole.

compute the electric field strength at the point P on the perpendicular bisector of the line joining $+q$ and $-q$, (Figure 5). That component of the total field due to $+q$ alone has a magnitude

$$E_{+q} = \frac{q}{4\pi\epsilon_0[r^2 + (d^2/4)]} \tag{4-1}$$

and is directed away from the positive charge. That due to $-q$ alone has the same magnitude and is directed toward the negative charge. From an inspection of the diagram it is evident that the resultant field is parallel to the axis of the dipole. Since

$$\cos\theta = \frac{d/2}{[r^2 + (d^2/4)]^{1/2}} \tag{4-2}$$

the total field at P is

$$E = \frac{2q}{4\pi\epsilon_0} \cdot \frac{1}{[r^2 + (d^2/4)]} \cdot \frac{d/2}{[r^2 + (d^2/4)]^{1/2}} \tag{4-3}$$

$$= \frac{1}{4\pi\epsilon_0} \frac{qd}{[r^2 + (d^2/4)]^{3/2}}$$

The product qd is called the electric dipole moment and is commonly denoted by p. When used as a vector, its sense is taken in the direction from $-q$ to $+q$.

Electric dipoles of this sort will be discussed again in Chapter 2 and will be of particular interest in Chapter 3, where we encounter them associated with the molecules of a material medium. Often with such molecular dipoles we are interested primarily in the fields which they produce at distances which are great compared to the charge separation itself. An examination of Equation 4-3 shows that when r is very much larger than d, the d in the denominator may be ignored as negligible and the field varies inversely as the third power of r. Although Equation 4-3 shows this property only for positions along the perpendicular bisector, we shall show in Chapter 2 that the inverse cube dependence is true for directions at any angle with the axis of the dipole so long as $r \gg d$. In fact, a field that varies with the inverse cube of the distance may be taken as identifying a dipole-like arrangement of charges.

Now, suppose we have a second such dipole, identical with the first except with the charges reversed, the two dipoles being placed end to end along the same line, as in Figure 6. Such a configuration is called a linear quadrupole.

An examination of the symmetry of Figure 6 shows that the net field at P, on the perpendicular bisector, will be a vector parallel to the r direction. The field component due to the center charge $2q$ is

$$E_{2q} = \frac{2q}{4\pi\epsilon_0 r^2} \tag{4-4}$$

and is directed away from $2q$. Each end charge $-q$ produces a field

$$E_{-q} = \frac{-q}{4\pi\epsilon_0[r^2 + d^2]} \tag{4-5}$$

directed toward the negative charge. Thus the total field is normal to the dipole with a magnitude

$$E_r = \frac{2q}{4\pi\epsilon_0}\left[\frac{1}{r^2} - \frac{r}{(r^2 + d^2)^{3/2}}\right] \tag{4-6}$$

When such quadrupoles occur in nature, they are usually associated with atoms or molecules and, as with dipoles, we are likely to be interested in the electric fields which they produce at distances large compared to the dimensions of the quadrupoles themselves. Equation 4-6 takes on an especially simple form under such circumstances. After $1/r^2$ is factored from the bracketed term, Equation 4-6 takes the form

$$E = \frac{2q}{4\pi\epsilon_0 r^2}\left[1 - \left(1 + \frac{d^2}{r^2}\right)^{-3/2}\right] \tag{4-7}$$

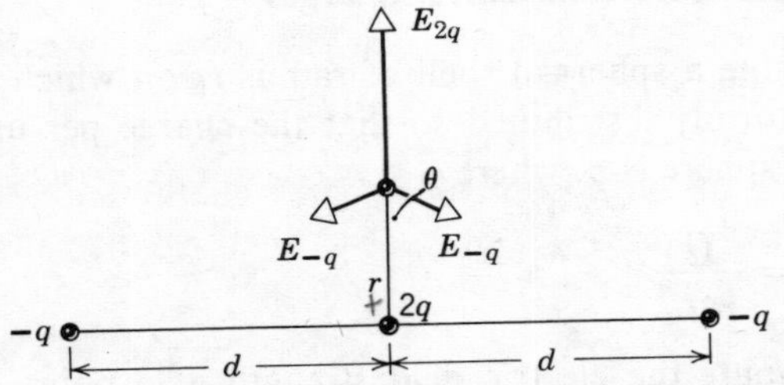

Figure 6. Electric field on bisector of linear quadrupole.

The binomial expansion formula

$$(1 + a)^n = 1 + na + \frac{n(n-1)}{2!}a^2 + \frac{n(n-1)(n-2)}{3!}a^3 + \cdots \tag{4-8}$$

is useful in this case, for if $a \ll 1$, the first two terms on the right constitute a good approximation to the value of the left side. Letting $(d/r)^2$ be a and with $n = -3/2$, then

$$\left(1 + \frac{d^2}{r^2}\right)^{-3/2} \approx 1 - \frac{3}{2}\frac{d^2}{r^2} \tag{4-9}$$

so that Equation 4-8 becomes very closely

$$E \approx \frac{3qd^2}{4\pi\epsilon_0 r^4} \tag{4-10}$$

Thus the remote electric field on the perpendicular bisector of a linear quadrupole drops off with the fourth power of the distance. By methods to be developed in Chapter 2 it will be shown that such is also the case for directions other than along the perpendicular.

It is helpful to imagine such a linear quadrupole as the result of starting with two oppositely directed but coincident dipoles, and then shifting one a distance d along the common line of the charges. Other types of quadrupoles may be formed by making such a shift in another direction, putting the four charges at the corners of a parallelogram. Although the computation of the field will differ in detail, it will be the same in principle as the method just outlined for the linear quadrupole. It turns out that at great distances this field will also drop off as the inverse fourth power of the distance and that it will also be proportional to the product of the magnitude of one of the charges and the square of a dimension of the quadrupole. By analogy, this process may be extended to form an octopole from two quadrupoles, and so on to still higher order poles.

5. Electric Field from Distributed Charges

Let us imagine a spherical shell of radius r_0 on which there is a total charge Q uniformly distributed so that the charge per unit area on the surface of the sphere is σ, where

$$\sigma = \frac{Q}{4\pi r_0^{\,2}} \tag{5-1}$$

We shall compute the electric field strength at a point P, a distance r from the center of the sphere (Figure 7). The point P is drawn in the figure

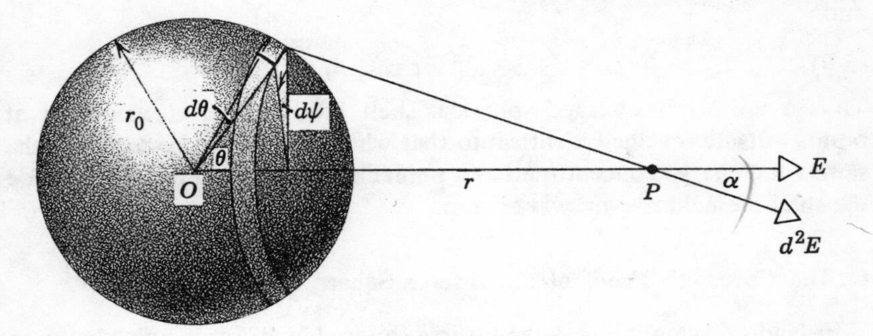

Figure 7. Field due to a charged spherical shell.

outside the sphere but may as well be inside. Spherical polar coordinates are clearly the most convenient to use to locate elements of charge on the sphere. The direction OP will be the direction of the reference axis for these coordinates, with the angle θ and ψ taken relative to this direction.

An infinitesimal region on the sphere of area $r_0^2 \sin\theta\, d\theta\, d\psi$ is shown in the figure, the charge within the area being $\sigma r_0^2 \sin\theta\, d\theta\, d\psi$. The field at P due to the charge on this element of area is

$$(5\text{-}2)\qquad d^2E = \frac{\sigma r_0^2 \sin\theta\, d\theta\, d\psi}{4\pi\epsilon_0[r^2 + r_0^2 - 2rr_0\cos\theta]}$$

We find the total electric field strength at P by taking the vector sum of all such elements of field from all points over the spherical shell. From the symmetry of the figure we see that the sum of the components of the field elements normal to the line OP is zero, and we need to concern ourselves only with the components parallel to OP, the parallel component of Equation 5-2 being $d^2E\,(\cos\alpha)$, where

$$(5\text{-}3)\qquad \cos\alpha = \frac{r - r_0\cos\theta}{(r^2 + r_0^2 - 2rr_0\cos\theta)^{1/2}}$$

The net field at P is thus

$$(5\text{-}4)\qquad E_P = \int_0^{2\pi}\int_0^{\pi} \frac{\sigma r_0^2(r - r_0\cos\theta)\sin\theta\, d\theta\, d\psi}{4\pi\epsilon_0(r^2 + r_0^2 - 2rr_0\cos\theta)^{3/2}}$$

In the result of this integration the expression $(r^2 + r_0^2 - 2rr_0)^{1/2}$ occurs. This expression must be written as $(r - r_0)$ or as $(r_0 - r)$, depending on whether P is outside or inside the sphere, respectively, in order to avoid imaginary quantities. The result is

$$(5\text{-}5)\qquad E_P = \frac{\sigma r_0^2}{\epsilon_0 r^2} = \frac{Q}{4\pi\epsilon_0 r^2} \qquad \text{for } r > r_0$$

and

$$E_P = 0 \qquad \text{for } r < r_0 \tag{5-6}$$

Thus a uniformly charged spherical shell produces an electric field at points outside the shell identical to that which would be produced by the same total charge concentrated as a point charge at the center, while inside the shell the field is everywhere zero.

6. The Cavendish Proof of the Inverse Square Law

Instead of assuming that the inverse square law of force holds between charges, let us assume the force law involves the inverse nth power and compute the electric field to be expected within the sphere of Figure 7. The field due to a point charge q is then

$$E = \frac{q}{4\pi\epsilon_0\, r^n} \tag{6-1}$$

where n is not necessarily an integer. By applying Equation 6-1 to the uniformly charged spherical shell of Figure 7, we get the equivalent of Equation 5-2:

$$d^2E_P = \frac{\sigma r_0^{\,2} \sin\theta\, d\theta\, d\psi}{4\pi\epsilon_0(r^2 + r_0^{\,2} - 2rr_0\cos\theta)^{n/2}} \tag{6-2}$$

and the equivalent of 5-4:

$$E_P = \int_0^{2\pi}\int_0^{\pi} \frac{\sigma r_0^{\,2}(r - r_0\cos\theta)\sin\theta\, d\theta\, d\psi}{4\pi\epsilon_0(r^2 + r_0^{\,2} - 2rr_0\cos\theta)^{(n+1)/2}} \tag{6-3}$$

which, when the integration is performed becomes:

$$E_P = \frac{\sigma r_0}{2\epsilon_0 r^2(3-n)(1-n)}\{(r + r_0)^{2-n}[r(2-n) - r_0] - (r - r_0)^{2-n}[r(2-n) + r_0]\} \tag{6-4}$$

for the case where $r > r_0$, and

$$E_P = \frac{\sigma r_0}{2\epsilon_0 r^2(3-n)(1-n)}\{(r_0 + r)^{2-n}[r(2-n) - r_0] + (r_0 - r)^{2-n}[r(2-n) + r_0]\} \tag{6-5}$$

for $r < r_0$.

It is easily checked that Equation 6-4 reduces to Equation 5-5 for $n = 2$. Similarly Equation 6-5 is zero for $n = 2$, but not for other values of n. In particular, Equation 6-5 is indeterminate for $n = 1$ and $n = 3$.

The above is a rather free translation into modern language and notation of the reasoning on which Cavendish based his experiment to demonstrate the inverse square law. He constructed two concentric, conducting spherical shells, the outer shell consisting of two well-fitting hemispheres which could be separated so that he could get to the inner sphere. A conducting wire connected the inner with the outer sphere while a static charge was being placed on the outer sphere from an electrostatic machine. At the completion of the charging process the wire was removed, the outer hemispheres were separated, and the amount of charge on the inner sphere was measured. Within the sensitivity of the instruments that Cavendish had available, he found no charge present on the inner sphere.

Equation 6-5 shows that the electric field within the charged sphere is positive for values of n between 1 and 2 and is negative for values of n between 2 and 3. Hence, if n were less than 2, the inner sphere should have shown evidence of a negative charge; and conversely, the inner sphere should have shown a positive charge if n were greater than 2. On the basis of the lack of measurable charge on the inner sphere, Cavendish concluded that the law of force involved the inverse square of the distance.

Tests of the force law using this reasoning have been reperformed several times with more refined instrumentation. Maxwell* found that n differed from 2 by no more than 5×10^{-5}; and in 1936 Plimpton and Lawson,† with highly refined techniques, showed that $n = 2 \pm 10^{-9}$. Their reasoning is expressed in terms of the potential concept, to be discussed in the next chapter. Following that discussion, the reader will find it interesting and profitable to read their paper critically.

7. Lines and Tubes of Force

Scientists of the Nineteenth Century were uneasy with the concept of "action at a distance." Although there was no logical necessity for their objection, "action at a distance" was probably associated in their minds with the magician and with the more primitive religions, associations that they were especially anxious to avoid. Faraday particularly disliked the idea and preferred to think of the electric field in terms of a structure or medium that "carried" the force between charges.

He imagined lines whose tangents at every point were parallel to the direction of the electric field at that point, the lines having arrows to show this direction. As an example, the field of a single positive point charge, far away from any other charges, has lines which come out radially from

---

* Maxwell, *Treatise on Elec. and Mag.*, Vol. 1, Art. 74*a*–74*e*.

† Plimpton and Lawson, *Phys. Rev.* **50**, 1066 (1936).

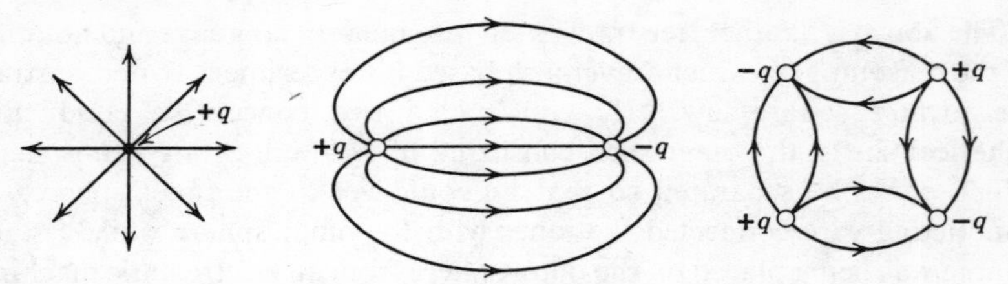

Figure 8. Examples of electric lines of force.

the charge; in the case of a negative point charge, the arrows point inward (see Figure 8).

Two properties follow from the way in which these lines are defined. First, under static conditions any one line must start on a positive charge and end on a negative charge (with time-varying conditions this statement is not necessarily true); and second, lines of force cannot cross. (Why?)

Lines of force provide a useful mental picture of the electrostatic field, a usefulness that is enhanced by introducing a further quantitative restriction. We imagine tubes of varying cross sections drawn in the field in such a way that the sides of the tubes are everywhere parallel to lines of force. No line of force ever crosses a tube wall. We imagine the whole space filled with such tubes, with their cross sections chosen so that the number of tubes per unit area normal to the field at a point in space is equal to the magnitude of the field. It can be shown that if the number of tubes per unit area normal to the field at any point is equal to the field strength at that point, the same will be true everywhere along a tube length. The tubes will have a small cross section and be densely packed in regions where the field strength is high, and a large cross section where the field strength is low. It should be clearly understood that there is nothing in electromagnetic theory that requires the use of lines and tubes, but the picture can be a help in thinking about fields.

Tubes, however, are difficult to draw in a diagram, and it has become the practice, instead, to draw lines which are considered to pass along the "center" of a tube, so that the field is *pictured* as a discrete number of lines, beginning on positive charges and ending on negative charges. Some examples of such field diagrams are shown in Figure 8, where lines which pass out of the picture are imagined as ending on charges not shown. Because of the way these lines have been defined, it is common practice to speak of the strength of an electric field at a point in space in terms of a number of lines per unit area, in spite of the logical difficulty that arises when "lines per unit area" is not a whole number. This difficulty does not occur when we speak of tubes per unit area, and "lines per unit area" should always be mentally translated to tubes.

8. Gauss's Law

A relation between charges and field strength, called Gauss's law, is of fundamental importance in the later development of electromagnetic theory. In addition, this relation makes possible some convenient short cuts in computing electric fields in special cases.

Let us imagine a point charge q within a closed surface of arbitrary shape, as in Figure 9. We fix our attention on an infinitesimal area of this surface at a point which is a distance r from the charge. The electric field strength at this point is

$$E = \frac{q}{4\pi\epsilon_0 r^2} \tag{8-1}$$

which we can think of as a density of field tubes at this point. The number of such tubes over this infinitesimal area, $d\mathscr{A}$, is called the flux of the electric field, dN, through the area $d\mathscr{A}$. It is the same as the flux over the projection of $d\mathscr{A}$ on the plane perpendicular to the direction of the field at this point on the surface. With θ as the angle between the normal to $d\mathscr{A}$ and the direction of the field, this projection is $d\mathscr{A}\cos\theta$, and the flux through $d\mathscr{A}$ becomes

$$dN = E\,d\mathscr{A}\cos\theta = \frac{q}{4\pi\epsilon_0}\frac{d\mathscr{A}\cos\theta}{r^2} \tag{8-2}$$

Let us draw a line from q to some point on the perimeter of $d\mathscr{A}$ and then move this line around the perimeter so that it describes a conical surface with apex at q. By analogy with plane angles, which are a measure of the "openness" between two intersecting lines on a plane, we define a solid angle as a measure of the "openness" of a cone in space. A plane angle is measured by the ratio of the length of an intercepted arc to the radius of this arc; so, similarly, a solid angle is measured by the ratio of an

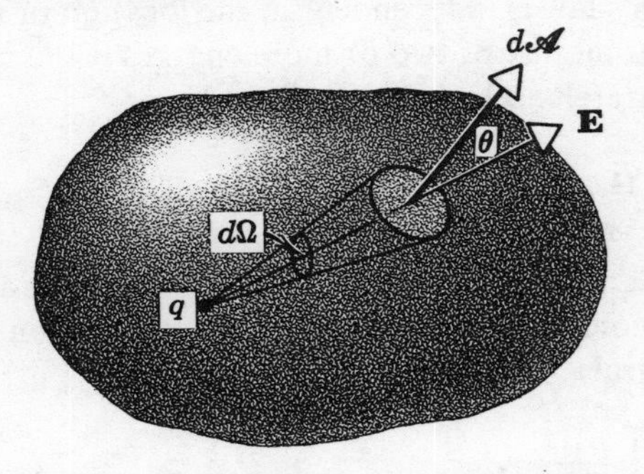

Figure 9. Gaussian surface enclosing charge q.

intercepted spherical area to the square of the radius of the sphere of which the intercepted area is a part. Both measures are dimensionless quantities, in spite of our practice of referring to plane angles in radians and to solid angles in *steradians*. For an *infinitesimal* solid angle, the capping area need not be spherical, but it must be normal to a radius drawn from the apex.

Thus the solid angle $d\Omega$ of the cone associated with the area $d\mathscr{A}$ in Figure 9 is given by

$$d\Omega = \frac{d\mathscr{A} \cos \theta}{r^2} \tag{8-3}$$

in terms of which Equation 8-2 can be written:

$$dN = \frac{q}{4\pi\epsilon_0} \, d\Omega \tag{8-4}$$

The total electric field flux through the entire surface area of the closed volume is the integral of Equation 8-4 over the entire solid angle surrounding the charge q. The total solid angle around a point is the area of a sphere around that point divided by the square of the radius of the sphere; that is, 4π. Consequently, the total flux over the closed area, which is the total number of field tubes originating at the charge q, is given by

$$N = \frac{q}{4\pi\epsilon_0} \int d\Omega = \frac{q}{\epsilon_0} \tag{8-5}$$

This total flux is proportional to the charge and independent of the size or shape of the volume used in computing it. It is also independent of the location of q within the volume.

As a simple example, suppose the surface in Figure 9 is a sphere of radius r, with the charge q at the center. The field strength, $q/4\pi\epsilon_0 r^2$, is uniform over the surface of the sphere. Since the area of the sphere is $4\pi r^2$, the total flux over the sphere is q/ϵ_0, consistent with Equation 8-5.

Because Gauss's law is independent of the location of the charge within the volume, the total flux of two or more charges is the sum of the flux of each charge separately. Or, in general, we have

$$N = \frac{q_1}{\epsilon_0} + \frac{q_2}{\epsilon_0} + \cdots = \frac{1}{\epsilon_0} \sum_i q_i \tag{8-6}$$

If, moreover, the charges are continuously distributed with a volume density of ρ coulombs per cubic meter, the summation in Equation 8-6 becomes an integral:

$$N = \int \frac{\rho \, d\mathscr{V}}{\epsilon_0} \tag{8-7}$$

where $d\mathscr{V}$ is an element of volume and the integration is over the closed volume.

The significance of Gauss's law is not immediately apparent from the above expressions, but the reader will appreciate it shortly when we use it to compute electric fields. Later it will be seen how it forms one of the fundamental building blocks of electromagnetic theory.

9. The Scalar Product of Vectors

The notation in Gauss's law can be simplified by introducing one of several kinds of products between vectors, the scalar product. This product is defined as the product of the magnitudes of the two vectors times the cosine of the angle between them, and is indicated by a dot between the vector quantities:

$$\mathbf{R} \cdot \mathbf{P} = RP \cos \theta \tag{9-1}$$

where θ is the angle between R and P as shown in Figure 2. The product is a scalar quantity and is not affected by the sense of θ, since $\cos \theta = \cos(-\theta)$. Consequently, a scalar product, like algebraic multiplication, is commutative:

$$\mathbf{R} \cdot \mathbf{P} = \mathbf{P} \cdot \mathbf{R} \tag{9-2}$$

Such a scalar product can be zero either because one of the two vectors is zero or because the two vectors are at right angles to one another.

Scalar multiplication is also distributive:

$$\mathbf{Q} \cdot (\mathbf{R} + \mathbf{P}) = \mathbf{Q} \cdot \mathbf{R} + \mathbf{Q} \cdot \mathbf{P} \tag{9-3}$$

When the vectors in a scalar product are written out in component form, the product has nine terms:

$$\begin{aligned} \mathbf{R} \cdot \mathbf{P} &= (\mathbf{i}R_x + \mathbf{j}R_y + \mathbf{k}R_z) \cdot (\mathbf{i}P_x + \mathbf{j}P_y + \mathbf{k}P_z) \\ &= \mathbf{i} \cdot \mathbf{i}R_xP_x + \mathbf{i} \cdot \mathbf{j}R_xP_y + \mathbf{i} \cdot \mathbf{k}R_xP_z + \mathbf{j} \cdot \mathbf{i}R_yP_x + \mathbf{j} \cdot \mathbf{j}R_yP_y \\ &\quad + \mathbf{j} \cdot \mathbf{k}R_yP_z + \mathbf{k} \cdot \mathbf{i}R_zP_x + \mathbf{k} \cdot \mathbf{j}R_zP_y + \mathbf{k} \cdot \mathbf{k}R_zP_z \end{aligned} \tag{9-4}$$

But since **i**, **j**, and **k** are mutually perpendicular vectors, it follows that:

$$\mathbf{i} \cdot \mathbf{j} = \mathbf{i} \cdot \mathbf{k} = \mathbf{j} \cdot \mathbf{k} = \mathbf{j} \cdot \mathbf{i} = \mathbf{k} \cdot \mathbf{i} = \mathbf{k} \cdot \mathbf{j} = 0 \tag{9-5}$$

and

$$\mathbf{i} \cdot \mathbf{i} = \mathbf{j} \cdot \mathbf{j} = \mathbf{k} \cdot \mathbf{k} = 1$$

so that Equation 9-4 becomes simply

$$\mathbf{R} \cdot \mathbf{P} = R_xP_x + R_yP_y + R_zP_z \tag{9-6}$$

from which it also follows that the scalar product of a vector with itself, $\mathbf{R} \cdot \mathbf{R}$, is just the square of the magnitude of $\mathbf{R}$:

$$\mathbf{R} \cdot \mathbf{R} = R_xR_x + R_yR_y + R_zR_z = R_x^2 + R_y^2 + R_z^2 \tag{9-7}$$

and may also be simply written as R^2.

10. The Vector Form of Gauss's Law

A vector may be associated with an area, since an area, like a line, has both a magnitude and an orientation in space. The length of such a vector is proportional to the magnitude of the area, and the direction is normal to the plane of the area. Which direction is taken as the positive sense of this vector depends on the nature of the area. If the area is part of a closed surface, then the outward drawn normal is taken as the positive sense. If the area is not part of a closed surface, it is always possible to define a positive sense in which to draw the perimeter; then the positive sense of the vector is given by the thumb of the right hand when the fingers are curved in the positive sense of the perimeter.

Since the infinitesimal area $d\mathscr{A}$ of Figure 9 is part of a closed surface, its associated vector is along the outward drawn normal to this surface. This fact, together with the definition 9-1 permits Equation 8-2 to be written:

$$dN = E\,d\mathscr{A}\cos\theta = \mathbf{E}\cdot d\mathscr{A} \tag{10-1}$$

and Equation 8-7 takes the form

$$N = \int \mathbf{E}\cdot d\mathscr{A} = \int \frac{\rho\,d\mathscr{V}}{\epsilon_0} \tag{10-2}$$

11. Computation of Fields with Gauss's Law

One of the simplest computations using Gauss's law is to find the electric field around a uniformly charged spherical shell, a case already worked out in Section 5 by a different method. A comparison of the two methods shows the power of Gauss's law.

The spherical shell of Section 5 had a radius r_0 and carried a uniformly distributed charge, Q. We shall compute the electric field strength at a point r meters from the center of the sphere. We draw an imaginary sphere, a so-called Gaussian surface, of radius r, concentric with the charged shell. From symmetry we expect the electric field to have the same strength at all points on this Gaussian sphere and to be everywhere normal to the sphere. Therefore, from Equation 10-2, the total flux over the Gaussian sphere is

$$N = \int \mathbf{E}\cdot d\mathscr{A} = \mathbf{E}\cdot\int d\mathscr{A} = E4\pi r^2 \tag{11-1}$$

This total flux is related to the total charge enclosed within the Gaussian sphere. If $r > r_0$, the enclosed charge is the entire charge Q, so that

$$N = E4\pi r^2 = \frac{Q}{\epsilon_0} \tag{11-2}$$

from which

$$E = Q/4\pi\epsilon_0 r^2 \tag{11-3}$$

whereas if $r < r_0$, there is no enclosed charge, and hence the field strength is zero.

These results have already been found in Section 5 by a more laborious process, and the simplification introduced by Gauss's law is evident.

As a second example, consider the case of an infinite plane on which charge is uniformly distributed with a density σ coulombs per square meter. From the uniformity of charge distribution we expect the field tubes to extend normally away from the plane on either side. Let us erect a Gaussian surface in the form of a cylinder whose generators are normal to the charged plane and extend on either side of the plane. The cross section of this cylinder is an arbitrary area, $\mathscr{A}$, and the end planes of the cylinder are parallel to the charged plane, the point at which we wish to compute the field strength lying in one of the end planes (Figure 10).

Since the electric field is normal to the charged plane, the integral of the flux density over the sides of the Gaussian cylinder is zero. But over each end the flux integral is $E\mathscr{A}$. Since the total charge within the cylinder is $\sigma\mathscr{A}$, the application of Gauss's law to this surface gives:

$$N = 2E\mathscr{A} = \frac{\sigma\mathscr{A}}{\epsilon_0} \tag{11-4}$$

from which we find the electric field strength:

$$E = \frac{\sigma}{2\epsilon_0} \tag{11-5}$$

showing that the electric field strength of an infinite, uniformly charged

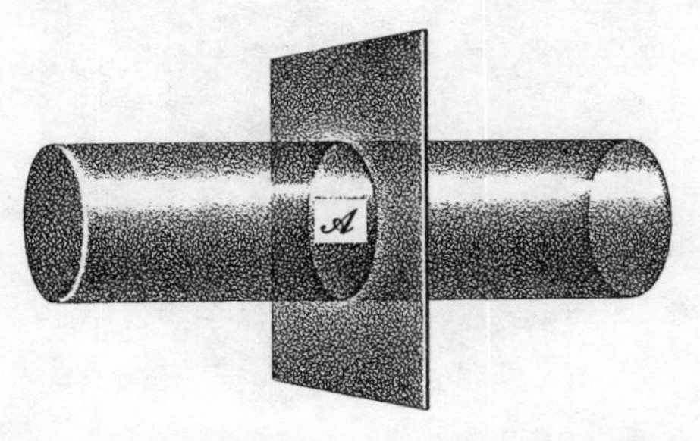

Figure 10. Gaussian surface for finding field of a charged infinite plane.

plane is independent of the distance from the plane. No such infinite plane exists, but the results can be applied to a finite, uniformly charged plane for a point whose distance from the plane is small compared to its distance from the edges.

These two cases illustrate how Gauss's law provides a basis for a simple way of computing some types of electric fields. Other cases appear as problems at the end of this chapter. What makes possible the use of Gauss's law in such cases is the existence of some element of symmetry from which a predictable uniformity of the field can be inferred. Where such symmetry does not exist, Gauss's law is of little value in field computation, even though the relation it expresses is a general one.

12. The Point Form of Gauss's Law

Gauss's law as it appears in Equation 10-2 associates physical quantities measured over a finite region of space. In theoretical physics, however, equations which show a relationship among quantities measured at the same point are often needed, and Gauss's law can also be formulated into an equation of this type.

Figure 11 shows an infinitesimal volume, $dx\,dy\,dz$, related to a Cartesian set of axes. Without being concerned with what produces it, let us assume an electric field over this region, the components of which are E_x, E_y, and E_z at the center of $dx\,dy\,dz$. Suppose also that there is a distributed

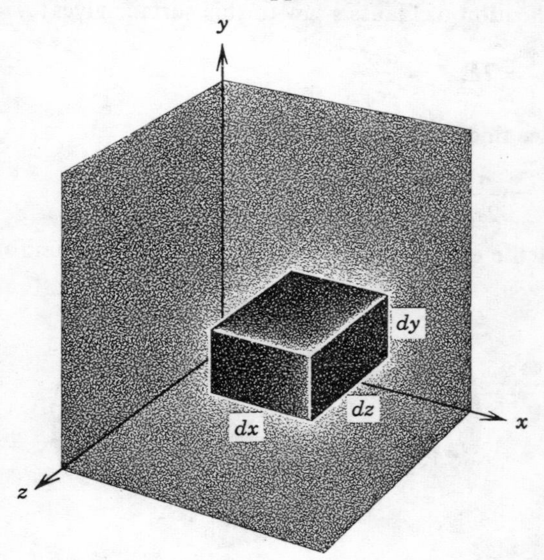

Figure 11. Infinitesimal Gaussian volume.

charge of volume density ρ, where ρ is a scalar function of position. We apply Gauss's law to this infinitesimal volume by summing the flux over each face and relating this sum to the total charge within the volume.

First consider the left-hand face whose edges are dy and dz. The area vector of this face points in the negative x-direction, and the x component of the electric field, which is parallel to this area vector, has at this face the value $E_x - (\partial E_x/\partial x)(dx/2)$. Hence, the total flux over this face is

$$-\left(E_x - \frac{\partial E_x}{\partial x}\frac{dx}{2}\right) dy\, dz \tag{12-1}$$

At the opposite face the x-component of the field is given by

$$E_x + \frac{\partial E_x}{\partial x}\frac{dx}{2}$$

so that the total flux over this face is:

$$\left(E_x + \frac{\partial E_x}{\partial x}\frac{dx}{2}\right) dy\, dz \tag{12-2}$$

Over only these two faces, the net outward flux is thus:

$$\frac{\partial E_x}{\partial x}\, dx\, dy\, dz \tag{12-3}$$

There are similar expressions for the front-back and the top-bottom pairs of faces, the sum of the three expressions being the total flux over the surface of $dx\, dy\, dz$. Thus Gauss's law for this infinitesimal volume becomes:

$$\left(\frac{\partial E_x}{\partial x} + \frac{\partial E_y}{\partial y} + \frac{\partial E_z}{\partial z}\right) dx\, dy\, dz = \frac{\rho}{\epsilon_0}\, dx\, dy\, dz \tag{12-4}$$

from which we get:

$$\frac{\partial E_x}{\partial x} + \frac{\partial E_y}{\partial y} + \frac{\partial E_z}{\partial z} = \frac{\rho}{\epsilon_0} \tag{12-5}$$

This is the differential or point form of Gauss's law. It relates the rates of change of the electric field components at a point in space to the density of charge at the same point. It was not, of course, necessary to use Cartesian coordinates in this derivation; the same kind of reasoning can be used with any other system; or, more simply, Equation 12-5 can be transformed to a new coordinate system by standard procedures which are discussed in any work on the calculus. This differential form does not, fundamentally, present new information over the integral form; rather, added significance can be inferred from the new way of showing the relationship between charge and electric field.

13. The Operator Del and the Divergence

We are now led to some further vector notation—notation that, at the very least, makes possible a happy reduction in symbols that need to be written, and sometimes suggests mathematical possibilities that are not always so readily evident in the conventional notation.

Symbols that indicate mathematical operations are called operators. In particular the symbols $\partial/\partial x$, $\partial/\partial y$, $\partial/\partial z$ are called differential operators, and it has become the practice to combine them into a kind of synthetic vector:

$$\mathbf{i}\frac{\partial}{\partial x} + \mathbf{j}\frac{\partial}{\partial y} + \mathbf{k}\frac{\partial}{\partial z} = \nabla \tag{13-1}$$

where the symbol ∇ is read "del". For example, the scalar product of ∇ with a vector such as $\mathbf{E}$ becomes

$$\begin{aligned} \nabla \cdot \mathbf{E} &= \left(\mathbf{i}\frac{\partial}{\partial x} + \mathbf{j}\frac{\partial}{\partial y} + \mathbf{k}\frac{\partial}{\partial z}\right) \cdot (\mathbf{i}E_x + \mathbf{j}E_y + \mathbf{k}E_z) \\ &= \frac{\partial E_x}{\partial x} + \frac{\partial E_y}{\partial y} + \frac{\partial E_z}{\partial z} \end{aligned} \tag{13-2}$$

The right side of Equation 13-2 has already been encountered in the point form of Gauss's law. Equation 13-2 is called the "divergence of **E**" and is also frequently read as "del dot **E**". In this notation the point form of Gauss's law becomes simply

$$\nabla \cdot \mathbf{E} = \frac{\rho}{\epsilon_0} \tag{13-3}$$

We can see a physical interpretation to the divergence, if we remember that an electric charge is a source of field tubes (a negative charge is a negative source). Equation 13-3 shows that the divergence is a measure of the strength of the field source at a point. If the divergence is zero, there are no sources; hence, no field lines start or end at that point. This interpretation is mathematically general and is not confined just to electric fields. In problems on fluid flow, a nonzero value of divergence of a function of the fluid velocity indicates a source of fluid. Later we shall encounter the magnetic vector **B**, whose divergence is everywhere zero, showing that the **B** field has no sources of this type and that its field lines always form closed curves.

14. Gauss's Theorem

Let us integrate Equation 13-3 over an arbitrary volume,

$$\int \nabla \cdot \mathbf{E} d\mathscr{V} = \int \frac{\rho \, d\mathscr{V}}{\epsilon_0} \tag{14-1}$$

But a comparison of Equation 14-1 with 10-2 shows that

$$\int \nabla \cdot \mathbf{E} d\mathscr{V} = \int \mathbf{E} \cdot d\mathscr{A} \tag{14-2}$$

a relation that is quite generally true for any vector, although we have developed it by using the electric field. It says that we may always convert an integral of the divergence of a vector over a closed volume into an integral of the vector itself over the area surrounding this closed volume. This relation is known as Gauss's *theorem* (or the divergence theorem) in mathematics and is not to be confused with Equations 10-2, 12-5, or 13-3, which are different versions of Gauss's *law* in physics. Gauss's theorem is an important mathematical tool that is often used to bring out significant features of field equations.

15. Some Comments in Summary

We have seen how the law of force between charges developed first (as far as published results are concerned) out of direct measurements with charged spheres. In these experiments the forces followed approximately the inverse square law, and it was postulated that the law holds exactly for *point* charges. On the other hand, the Cavendish type of experiment assumed that an inverse nth power law holds for *point* charges, and it was then shown that the experimentally observed field within a charged spherical shell was only consistent with a value $n = 2$.

Although the inverse square law has been very successful in predicting electrostatic behavior of charges, we should recognize the limitations of its experimental foundations. One such limitation lies in the assumption of point charges, since there are no point charges in nature. In the Cavendish type of test there is yet another limitation. The theory of this experiment depended on an integration whose limit process required a smoothly continuous charge density function; but at the molecular level there are violent discontinuities in the charge density. An attempt was made to get around this difficulty by saying that the infinitesimal volumes used in the integration are to be small, but not so small as to be comparable to molecular dimensions. This is, of course, a subterfuge. However, it is a subterfuge that works within the accuracy with which it has been tested.

Another approximation was implicit in the theory behind the tests of Coulomb's law. Mason and Weaver have pointed out that the charges we have consistently assumed to be at rest are, in actual fact, in motion, often probably in rather violent motion. These motions are at least of thermal origin, even if we do not raise the question of whether electrons are or are not in motion within the atom and whether protons move within the nucleus. In later chapters we shall see that fields produced by charges in motion are not the same as for charges at rest. The best we can say about the tests of Coulomb's law is that they are statistically valid; that is, the measurements must cover a long enough period of time to average out the small fluctuations in charge density that are produced by the random motions of the charges. When the instruments are sufficiently sensitive and have a short enough reaction time, these statistical fluctuations can be observed. For example, in high-gain amplifiers, an irreducible minimum of noise found in the output can be attributed to the random motions of charges in certain resistors, particularly resistors at the input stage.

Another and more philosophic question is also worth mentioning. The definition of the electric field strength at a point requires the measurement of a force on a charge placed at that point. In the strictest sense are we, therefore, justified in speaking of the electric field strength in empty space? In physics it is an accepted principle that only such quantities as can be measured have physical reality; and in order to measure the electric field, we need a charge present; then the space is not empty.

The language we shall use in discussing electric fields (and the same comments apply equally well for magnetic fields) will seem to impute to these fields a physical reality of their own, apart from any charges. One pragmatic justification for this approach is quite simply brevity. Moreover, we do get verifiable predictions about experimental measurements, but it should be kept in mind that these predictions are fundamentally predictions of the behavior of charges influenced by the fields. Our experimental data are measurements on charges; we do not have any way of observing the fields without the use of charges. It will be a healthy mental discipline for the reader always to relate the predicted behavior of electric and magnetic fields to the possible physical measurements that can be made to observe this behavior.

Discussion Questions

1. Would you say, on the basis of any information you have from previous work in physics, that there has been a definitive answer to the question of whether an electric charge is a manifestation of a single "fluid" or of two "fluids"?

2. Obtain from the library a copy of the papers of Henry Cavendish, as edited by Maxwell, and read the original arguments Cavendish uses in demonstrating the inverse square law. The mathematical vocabulary is of particular interest.

3. Examine critically the justifications for the inverse square law. In what kinds of experiments would it be wise to emphasize that the law is being *assumed* to hold rather than that the law is *known* to hold?

4. In your previous work in physics you have studied the Rutherford scattering experiment, in which Rutherford observed the deflection of high-speed alpha particles as they passed through thin films and deduced that the atom consisted of a small, heavy nucleus of positive charge surrounded by a cloud of negative charges. Would you judge that this experiment provides a test of Coulomb's law within atomic distances?

5. A conducting sphere carrying no net charge is in an electric field. The electric field occupies a volume which is large compared to the volume of the sphere and is essentially uniform except in the vicinity of the sphere. (It would be uniform everywhere except for the presence of the sphere.) Use what knowledge and intuition you can muster to sketch out the pattern of the field in the vicinity of the sphere. What do you expect the distribution of charges on the surface of the sphere to be? How might such a uniform field be produced?

6. Equation 12-5 showed that the electric field in front of a uniformly charged plane was $\sigma/2\epsilon_0$ where σ is the charge per unit area on the plane, and that this field is in a direction normal to the plane. Suppose that there is another plane, parallel to the first and charged with $-\sigma$ coulombs/square meter. These latter charges are in the field of the charges on the first plane and, hence, experience a force. What is this force per unit area on the second plane?

So long as the two planes are separated by a distance that is small compared to their smallest linear dimension, this argument can be applied to a pair of finite planes, with but slight error from the nonuniformity of the field at the edges. Neglecting this error, what will be the force on one plate of a parallel-plate capacitor with each plate of area A, separation of plates d, and air or vacuum in the space between the plates?

7. Find references in your library to the Guard-Ring Capacitor.* Then discuss how this device is used to reduce the nonuniformity of the field at the edge of a pair of capacitor plates and thus minimize the error in the computation discussed in 6.

Problems

1. Show that if a charge is moved in a straight line from the origin of coordinates to the point (3, −2, 1) and if an electric field in this region has the x, y, and z components of 100, 50, and −200 Newtons/coulomb, respectively, the path of the charge is at all times normal to the electric field.

2. Find the electric field strength at a point which is a distance r from the center of the dipole illustrated in Figure 5, where r is on the extension of the axis

* Scott, *Physics of Electricity and Magnetism*, Wiley, p. 88. Harrington, *Introduction to Electromagnetic Engineering*, McGraw-Hill, p. 68.

of the dipole. Show that when r is much larger than d the field drops off with the inverse third power of r. How far from the center of the dipole is the point where the exact and the approximate values of the field differ by, say, 1%?

3. Determine the electric field strength at a point a distance r from the center of the linear quadrupole of Figure 6 when r is on the extension of the line of charges. Show that for large distances, this field drops off with the inverse fourth power of r. How far from the center charge is the point where the exact and the approximate values of field differ by 1%?

4. Determine by integration the field strength at a distance r from a uniformly charged wire of infinite length. Make the same determination by using Gauss's law.

5. What is the electric field strength at a point a distance r from the axis of an infinitely long uniformly charged cylinder of radius R, the charge being on the surface of the cylinder? Use Gauss's law. Consider the cases where $r > R$ and $r < R$.

6. What is the field strength at a point a distance r along the perpendicular bisector of a wire of *finite* length L, the wire being uniformly charged? Can Gauss's law be used in this solution? Give reasons.

7. Find the field strength at a point which lies a distance y beyond the end of a uniformly charged wire of length L, on the line of the wire extended.

8. Determine the field strength at a point P which is located a perpendicular distance d from one end of a uniformly charged wire of length L (see Figure P8).

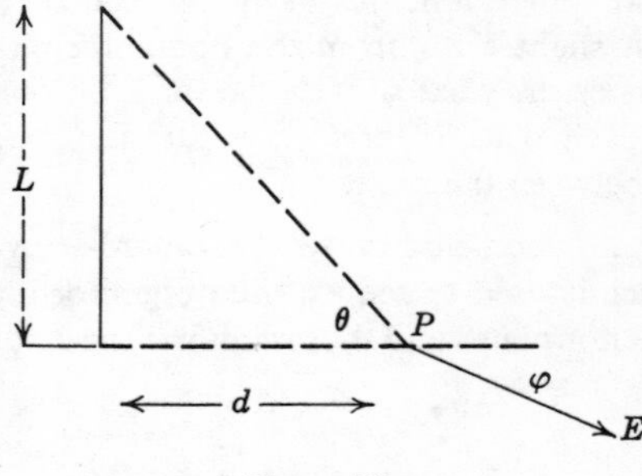

Figure P8

Show that this field makes an angle φ with the perpendicular d, where $\varphi = \theta/2$, θ being the angle subtended by the wire at the point where the field is being determined.

9. A uniformly charged spherical shell has a radius a. Concentric with it and carrying an equal charge of opposite sign is a conducting shell of radius b, where b is greater than a. Determine the electric field at a distance r from the common center, where r lies between a and b. How does this field compare with the field that would exist if the outer sphere were not present (or of infinite radius)?

10. An infinite plane is uniformly charged with a surface density of charge σ. We can divide the plane into infinitesimal elements of area $d\mathscr{A}$, on each of which there is a charge $\sigma d\mathscr{A}$, which approximates a point charge. Determine by integration the field strength at a point P a distance r from the plane, and show that the result agrees with that obtained from reasoning based on Gauss's law.

11. Two infinite planes are parallel to each other. One is uniformly charged with a charge of density $+\sigma$ and the other with a charge of density $-\sigma$. Show that the field strength between the two planes has a value σ/ϵ_0, and that the field strength outside the two planes is zero.

12. Two equal charges of opposite sign have a fixed separation d to form a dipole of moment p. This dipole is in a uniform electric field of strength E, the direction of vector dipole moment making an angle θ with the direction of E. Show that the torque on this dipole is given by $pE \sin \theta$.

13. An electron of charge $-e$ and mass m moves in a uniform electric field E. At time $t = 0$ the electron velocity is v_0 and this velocity makes an angle θ_0 with the direction of the field. Set up appropriate axes and determine the equation of the trajectory of this electron.

14. Show that the surface density of charge on a conductor of arbitrary shape is related to the electric field strength at the surface by $\sigma = \epsilon_0 E$. Use Gauss's law.

REFERENCES

1. Max von Laue, *History of Physics* (translated by Ralph Oesper), Academic Press.
2. Max Mason and Warren Weaver, *The Electromagnetic Field*, University of Chicago Press. (Paperbound edition, Dover Publications.)
3. William Francis Magie, *A Source Book of Physics*. McGraw-Hill Book Company.
4. Morris H. Shamos, *Great Experiments in Physics*, Henry Holt.
5. *The Scientific Papers of the Hon. Henry Cavendish*, Vol. I, Electrical Researches, edited by James Clerk Maxwell. Revised by Joseph Larmor. Cambridge University Press. First edition, 1879; revised edition, 1921.
6. Maxwell, *Treatise on Electricity and Magnetism*, Art. 36–37, on the one- and two-fluid theories, Dover Publications.

2

Potential energy, potential, and capacitance

As mass has potential energy in a gravitational field, so has a charged body potential energy in an electric field. The study of this potential energy develops a deeper insight into the physics of electric charges; and in addition much information about the behavior of charges can be obtained rather simply from energy equations. These, because they are scalar relations, are usually easier to manipulate than equations relating vector forces.

1. Electrical Potential Energy and Potential

Our analysis of electrostatic forces started with the simplest case, the force between point charges. In like manner, the energy of such point charges initiates our analysis of electrical potential energy.

A charge q resides on a point mass m in an electric field $\mathbf{E}$. At the same time it is acted on by some outside agent exerting a force $\mathbf{F}$. Such an agent might be the belt in a Van de Graaff generator transporting the charge toward the high-voltage sphere, a photon causing an electron to be emitted from a metal surface, or, perhaps only symbolically, the experimenter's finger pushing the charged body around. The net force on the mass m, $\mathbf{F} + q\mathbf{E}$, is related to the velocity $\mathbf{v}$ by Newton's second law

$$\text{(1-1)} \qquad \mathbf{F} + q\mathbf{E} = \frac{d}{dt}(m\mathbf{v})$$

When the mass moves from point A to point B (Figure 1), the work done

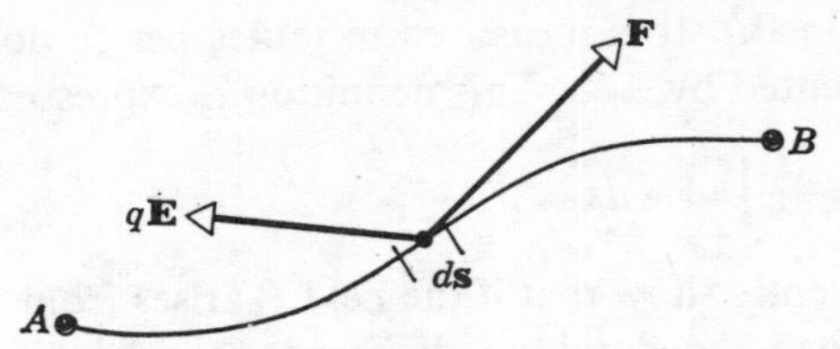

Figure 1. Forces in computing line integral from A to B.

by the outside agent force **F** is, by definition

$$W = \int_A^B \mathbf{F} \cdot d\mathbf{s} = \int_A^B \frac{d}{dt}(m\mathbf{v}) \cdot d\mathbf{s} - \int_A^B q\mathbf{E} \cdot d\mathbf{s} \tag{1-2}$$

where $d\mathbf{s}$ is an element of the path A-B. The integral,

$$\int_A^B \frac{d}{dt}(m\mathbf{v}) \cdot d\mathbf{s}$$

is the difference between the kinetic energy of m at A and the kinetic energy at B, as is seen from

$$\int_A^B \frac{d}{dt}(m\mathbf{v}) \cdot d\mathbf{s} = \int_A^B m\mathbf{v} \cdot d\mathbf{v} = \tfrac{1}{2} m(v_B^2 - v_A^2) \tag{1-3}$$

where v_B and v_A are the velocities of m at B and at A, respectively. If the mass m starts from rest at A and stops at B, then this term is zero and the work done by the force **F** is equal to $-\int_A^B q\mathbf{E} \cdot d\mathbf{s}$ (we are assuming that there are no frictional forces). In this case, the force **F** only does work against the electric field and this work appears as potential energy stored in the system consisting of the charge q and whatever gives rise to the field **E**. Such potential energy, measured in joules or newton-meters, is always proportional to q. So long as attention is directed on the charge q, the rest of the system remaining fixed, it is customary to assign this energy to q itself, rather than to the system as a whole.

The potential energy per unit positive charge,

$$-\int_A^B \mathbf{E} \cdot d\mathbf{s},$$

called the change in *potential* or the *difference in potential* between the points A and B, is a convenient function. It is the work *per unit positive charge* which an outside agent does in moving (without change of velocity) a charge from A to B. (Compare the analogous way in which electric field strength was defined.) It is measured in joules per coulomb or volts and is usually represented by ΔV. This definition is expressed by

$$\Delta V = -\int_A^B \mathbf{E} \cdot d\mathbf{s} \tag{1-4}$$

and we shall presently show that if the field **E** arises from a distribution of static charges, then the potential difference is solely a function of the positions of the end points of the path, A and B, and not of the shape of the path chosen for the integration.

Inherent in the definition of potential difference is an *experimental* method for measuring this quantity. The work required to move a test charge q from A to B without change of velocity is measured and this work is divided by the charge q. It is not necessary to know anything about the charges that give rise to the field acting on the test charge, except that they remain fixed as q moves. If, however, the charge distribution is known, it becomes possible by Coulomb's law to compute the field **E** and from that to make a theoretical prediction of the potential difference between A and B.

2. Potential Difference in the Field of a Single Point Charge

Where the field E (Equation 1-4) arises from an isolated point charge, Q, the computation of potential difference is particularly simple. The field strength at a distance r from such a charge was found in Chapter 1 to be $Q/4\pi\epsilon_0 r^2$ and to be directed along radial lines from Q. For two points A and B which lie on the same radial line from Q and are at distances r_A and r_B, respectively, from Q, the work per unit positive charge in moving a test charge from A to B (see Figure 2) is:

$$\Delta V = -\int_A^B \frac{Q}{4\pi\epsilon_0 r^2}\, dr = \frac{Q}{4\pi\epsilon_0}\left(\frac{1}{r_B} - \frac{1}{r_A}\right) \tag{2-1}$$

As expected, it requires a positive amount of work on the part of the outside agent to move a test charge closer to a positive point charge. Clearly the work to move the test charge from B to A differs only in a reversal of sign.

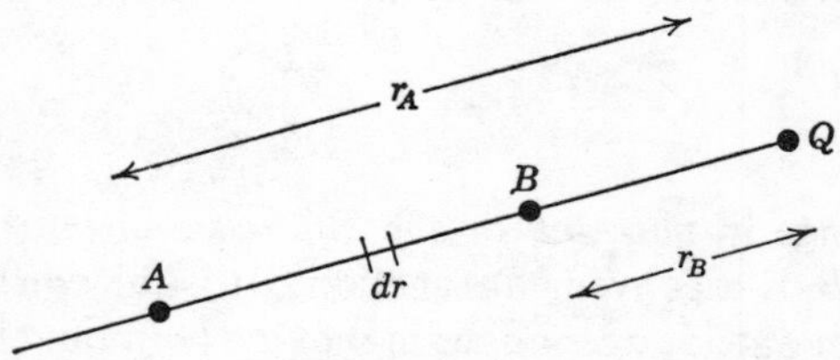

Figure 2. Computing potential difference in field of a point charge.

3. Potential Difference Independent of Path of Integration

Suppose, now, that point A were located on another radial line, but still at a distance r_A from Q, and suppose we chose as a path of integration a path of arbitrary shape, starting at A and ending at B, such as path I in Figure 3. We can approximate this path as closely as we please by a succession of alternate arc and radial increments. But the line integral over the increments of arc is zero, because the field is normal to these displacements, whereas the line integral over the radial increments gives the same result as in Equation 2-1. Thus, at least for the field of a point charge, the potential difference between two points depends only on the location of these points and not on the path of integration used in computing the potential difference. We are free to choose the path that provides the simplest computation.

The fact that the potential difference is independent of the path of integration has a fundamental significance. Suppose the contrary were true. In Figure 3, for example, suppose it required less work to move a test charge from A to B along path I than along path II. Since a reversal of direction on a path results only in a change of sign of the work that is required of the outside agent, the charge could be moved from A to B along path I and returned from B to A along path II at a net profit in work, even though the system at the end of the operation is unchanged from what it was at the beginning. We should then be gaining energy from the system without any change in the system itself. Clearly, this cyclic process could be repeated as often as we pleased, and we should have a most attractive violation of the law of conservation of energy. (Should this be possible, a large amount of the energy so obtained would be required to rewrite most of the scientific literature of the past century!)

Equation 3-1 succinctly expresses the fact that in an electrostatic field the difference of potential between two points is not dependent on the path used in computing it:

$$\oint \mathbf{E} \cdot d\mathbf{s} = 0 \tag{3-1}$$

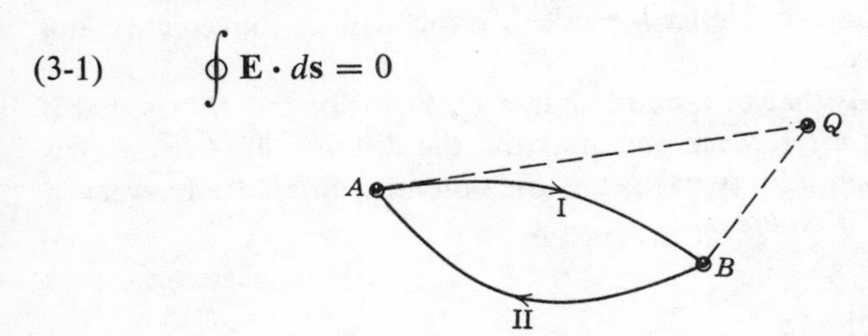

Figure 3. Closed path in field of a point charge.

where the circle on the integral sign is used to indicate that the path of integration is over a closed circuit. It will be evident shortly that the validity of Equation 3-1 is not confined to the field of a single point charge, but applies generally to any configuration of static charges.

4. Absolute Potential

The relation

$$\Delta V = -\int_A^B \mathbf{E} \cdot d\mathbf{s} \tag{4-1}$$

is called the potential difference between the points A and B, suggesting that there is a *potential* at A and a *potential* at B, whose difference is being measured. As with other physical quantities, the *potential at a point* depends on the arbitrary fixing of a reference from which it is to be measured. This is the same as saying that the *difference* of potential is the only physically measurable quantity, but if we choose to measure all differences from a common point, then we can assign "absolute" values of potential to each point in space.

There is nothing special about this. We can only measure differences in position in space, but if we measure all positions relative to a common set of reference axes, we speak of *the* position of a point. Similarly, we can measure physically only a change in velocity (and hence only a change in kinetic energy), but we can speak of *the* velocity of a body relative to a reference which is taken to be standing still, although the reference itself may be moving with respect to some other reference system. The reader should think of other examples.

The choice of a reference for the measurement of potential is a matter of convenience. Often in electrostatics, where the electric field is associated with a finite distribution of charges, the most convenient choice is to assign a potential of zero to a point infinitely distant from the charge distribution. (In some idealized cases, examples of which appear later, where the charge distribution itself extends to infinity, such a reference is not useful.) In other cases, it is convenient to take the potential of the earth as zero. Whatever choice is made, it is dictated by convenience, not by logical necessity.

Where the field is that of a point charge Q, Equation 2-1 shows that if point A is used as a reference potential and the distance of A from Q is allowed to approach infinity, we get as the potential difference between A and B, that is, *the absolute potential* of B:

$$V_B = \frac{Q}{4\pi\epsilon_0 r_B} \tag{4-2}$$

a function solely of the distance of the point B from the charge Q.

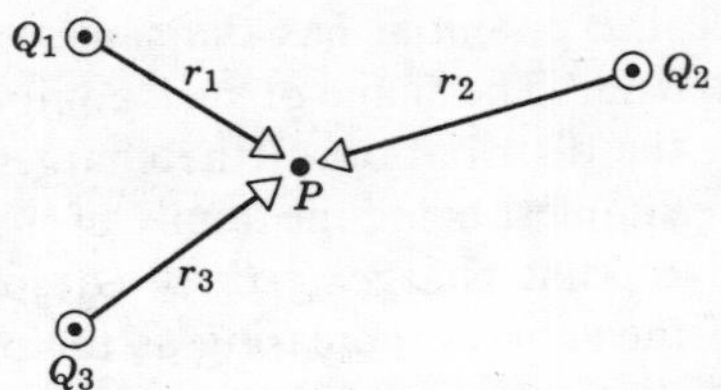

Figure 4. Potential at P is scalar sum of fields of individual charges.

For a group of point charges, as in Figure 4, the scalar nature of potential requires that the total potential at a point such as P is the scalar sum of the separate absolute potentials. In general, then:

$$V_P = \frac{1}{4\pi\epsilon_0} \sum_i \frac{Q_i}{r_i} \tag{4-3}$$

With a sensibly continuous distribution of charge, as in Figure 5, the density of charge at any point being ρ, the potential at the point P is given by:

$$V_P = \frac{1}{4\pi\epsilon_0} \int \frac{\rho \, d\mathcal{V}}{r} \tag{4-4}$$

where r is the distance of P from the volume element $d\mathcal{V}$, and the integration is over the volume occupied by the charge distribution.

It is left as a problem for the reader to show that if Equation 3-1 holds in the field around a single point charge, it must also hold for a distribution of discrete charges and for a continuous charge distribution.

5. Equipotential Surfaces

A pictorial representation of potential functions is just as useful as a pictorial representation of field patterns. Equipotential surfaces, surfaces

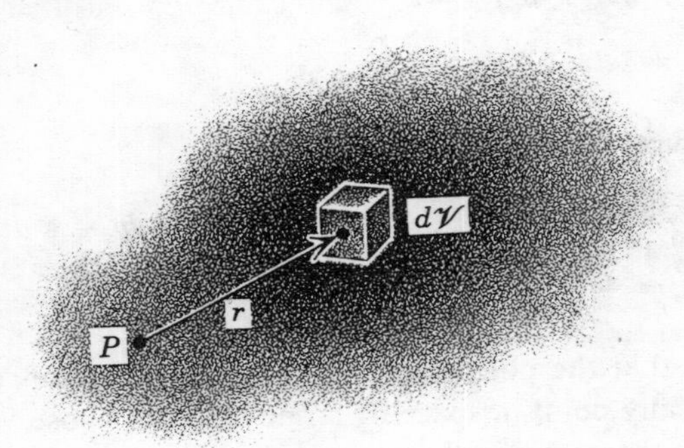

Figure 5. Potential at P due to charge in volume element $d\mathcal{V}$.

at all points of which the potential has the same value, can be used to visualize the potential field. The shape of such equipotential surfaces will, of course, depend on the distribution of the charges associated with the potential function, the simplest being the family of surfaces around a point charge. For a positive point charge Q these equipotential surfaces are spheres, the radius of the spheres increasing as the potential decreases.

6. The Potential Gradient and the Electric Field Strength

Before proceeding to show some examples of the computation of potential for various charge distributions, it is in order to develop further the implications of the definition of potential difference expressed in Equation 4-1,

$$\Delta V = -\int_A^B \mathbf{E} \cdot d\mathbf{s}$$

This is an integral relation. By a process analogous to that in which we developed a point relation out of the integral form of Gauss's law, we can develop a point relation from Equation 4-1 between the electric field strength at a point and the rate of change of potential at that point.

To obtain this differential relation, we use Equation 4-1 to find the potential difference dV between two points separated by an infinitesimal distance $d\mathbf{s}$:

$$dV = -\mathbf{E} \cdot d\mathbf{s} \tag{6-1}$$

With $\mathbf{E}$ and $d\mathbf{s}$ expressed in terms of their components in a Cartesian system of axes,

$$\mathbf{E} = \mathbf{i}E_x + \mathbf{j}E_y + \mathbf{k}E_z \tag{6-2a}$$

$$d\mathbf{s} = \mathbf{i}\,dx + \mathbf{j}\,dy + \mathbf{k}\,dz \tag{6-2b}$$

Equation 6-1 becomes

$$\begin{aligned} dV &= -(\mathbf{i}E_x + \mathbf{j}E_y + \mathbf{k}E_z) \cdot (\mathbf{i}\,dx + \mathbf{j}\,dy + \mathbf{k}\,dz) \\ &= -(E_x\,dx + E_y\,dy + E_z\,dz) \end{aligned} \tag{6-3}$$

We have seen that the potential function V is a function whose value is determined at any point in space by a line integral whose value in turn is independent of the path of integration and depends only on the end points of the path. It is demonstrated in texts on advanced calculus that the differential of a function with these properties can be expressed in Cartesian coordinates

(6-4) $$dV = \frac{\partial V}{\partial x}\,dx + \frac{\partial V}{\partial y}\,dy + \frac{\partial V}{\partial z}\,dz$$

A comparison of Equations 6-3 and 6-4 results in the following relations for the components of the electric field strength:

(6-5) $$E_x = -\frac{\partial V}{\partial x} \qquad E_y = -\frac{\partial V}{\partial y} \qquad E_z = -\frac{\partial V}{\partial z}$$

so that the electric field vector can be written:

(6-6) $$\mathbf{E} = -\left(\mathbf{i}\frac{\partial V}{\partial x} + \mathbf{j}\frac{\partial V}{\partial y} + \mathbf{k}\frac{\partial V}{\partial z}\right)$$

or $$\mathbf{E} = -\left(\mathbf{i}\frac{\partial}{\partial x} + \mathbf{j}\frac{\partial}{\partial y} + \mathbf{k}\frac{\partial}{\partial z}\right)V$$

The term in parentheses will be recognized as the operator ∇, introduced in Section 14 of Chapter 1, making possible a concisely written point relation:

(6-7) $$\mathbf{E} = -\nabla V$$

in which the right-hand side is commonly read as "the negative gradient of V, and often more briefly as "minus del V." The gradient of V is a vector function of position.

The geometrical significance of the gradient is readily appreciated if we first combine Equations 6-7 and 6-1:

(6-8) $$dV = \nabla V \cdot d\mathbf{s}$$

showing that the change in V over a displacement $d\mathbf{s}$ is the scalar product of the gradient of V with $d\mathbf{s}$. This is illustrated in Figure 6 where ∇V is shown at a point A where the potential is V. At a neighboring point B, displaced $d\mathbf{s}$ from A, the potential is $V + dV$. Writing Equation 6-8 as:

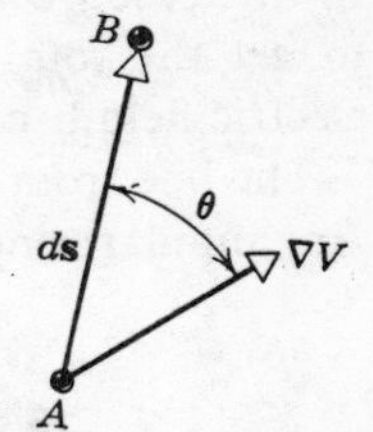

Figure 6. Relation of infinitesimal potential difference to the potential gradient.

(6-9) $$dV = |d\mathbf{s}|\,|\nabla V|\cos\theta$$

where θ is the angle between $d\mathbf{s}$ and ∇V, shows that dV has a maximum value when $d\mathbf{s}$ is parallel to ∇V. Identifying this particular value of $d\mathbf{s}$ by $d\mathbf{s}'$, shows that

(6-10) $$dV/ds' = |\nabla V|$$

from which we conclude that ∇V is a vector whose direction is the direction of the maximum rate of change of the potential V, and whose magnitude is equal to this rate of change. Then Equation 6-7 shows that the electric field is a vector whose direction is opposite to the direction of maximum rate of change of V and whose magnitude is this maximum rate of change.

It is a corollary to the foregoing argument that the component of the electric field strength in an arbitrary direction is the negative of the rate of change of the potential in that direction.

There are thus two ways of obtaining both the electric field and the potential function: (1) We may use Equation 4-3 or 4-4 to determine the potential function and then Equation 6-7 to obtain the electrostatic field strength; or (2) we may use Coulomb's law or Gauss's law to compute the electrostatic field strength, and then Equation 4-1 to obtain the potential field. Usually (1) is somewhat easier because of the scalar nature of the potential function, but in such cases where the charge configuration has sufficient symmetry to warrant the use of Gauss's law in computing the electric field, (2) may be the simpler method. In the examples to follow each method will be illustrated.

7. Potential Functions for Continuous Charge Distributions

A SPHERICAL SHELL. A spherical shell has a radius r_0 and carries a uniform charge of density σ. Finding the potential at a point P, a distance r from the center of the shell (Figure 7), has certain similarities with the procedure followed in Chapter 1 in finding the electric field at the point P. In this case we sum up the effects of infinitesimal elements of the sphere to get the total potential, a somewhat simpler operation than for the electric field because of the scalar nature of the potential function.

The line from the center of the shell to the point P is the reference axis. An annular ring of width $r_0\,d\theta$, radius $r_0 \sin\theta$, and centered on the

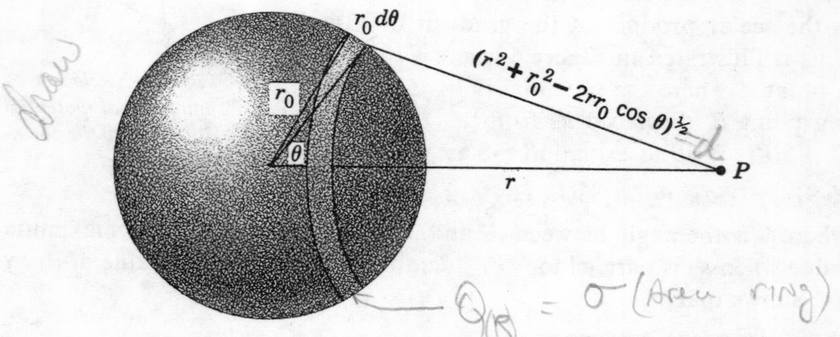

Figure 7. Computing electric field of a charged spherical shell.

reference axis is drawn on the shell. All points on this ring are at a distance $(r^2 + r_0^2 - 2rr_0 \cos \theta)^{1/2}$ from P, and the total charge on this ring is $2\pi\sigma r_0^2 \sin \theta \, d\theta$. From Equation 4-4, the absolute potential at P due to this ring of charge is

$$dV = \frac{2\pi\sigma r_0^2 \sin \theta \, d\theta}{4\pi\epsilon_0(r^2 + r_0^2 - 2rr_0 \cos \theta)^{1/2}} \tag{7-1}$$

and the total potential, on integrating Equation 7-1 from $\theta = 0$ to $\theta = \pi$, is

$$V = \frac{2\pi\sigma r_0}{4\pi\epsilon_0 r}[(r^2 + r_0^2 + 2rr_0)^{1/2} - (r^2 + r_0^2 - 2rr_0)^{1/2}] \tag{7-2}$$

Each radical within the brackets represents a positive distance—the distances from the point P to the far and near points of the sphere, respectively—and the quantity under the root sign is a perfect square in each case. The first radical is thus $r + r_0$; but the second is either $r - r_0$ or $r_0 - r$, depending on whether $r > r_0$ or $r < r_0$. Since the total charge Q is $4\pi r_0^2\sigma$, the potential function outside the sphere becomes

$$V_o = \frac{4\pi\sigma r_0^2}{4\pi\epsilon_0 r} = \frac{Q}{4\pi\epsilon_0 r} \qquad r > r_0 \tag{7-3}$$

and inside the sphere:

$$V_i = \frac{4\pi\sigma r_0}{4\pi\epsilon_0} = \frac{Q}{4\pi\epsilon_0 r_0} \qquad r < r_0 \tag{7-4}$$

Since we found in Chapter 1 that the electric field outside a uniformly charged sphere behaves as if the total charge were concentrated at the center of the sphere, it is not surprising to find that the potential function outside also has the form to be expected from a point charge Q at the center. We also found in Chapter 1 that the electric field inside the sphere is zero; it is therefore plausible that the potential inside is a constant, because it requires no additional work against electrical forces to move a test charge from infinity to a point inside the sphere than to move a test charge from infinity to the surface of the sphere.

The above method of finding the potential function for a charged shell illustrates the use of Equation 4-4. We might also have started from the known electric field function for this shell and applied Equation 4-1, integrating from infinity to an arbitrary distance r from the center. It is left to the reader to show that the results are the same.

A UNIFORMLY CHARGED, INFINITELY LONG CYLINDER. In Problem 4 of Chapter 1 the electric field strength at a distance r from the axis of an infinite cylinder of radius R was found. For a charge per unit length of the cylinder, λ, this field strength is

$$E = \frac{\lambda}{2\pi\epsilon_0 r} \tag{7-5}$$

outside the cylinder and zero inside, provided the charge resides entirely on the surface of the cylinder. For reasons which will shortly become apparent we shall determine the *difference* of potential between two points, r_1 and r_2, rather than the absolute potential at one point. Let r_2 be greater than r_1. Using the value of E from Equation 7-5 in the defining Equation 4-1, the potential difference is

$$\Delta V_{r_2 \to r_1} = -\int_{r_2}^{r_1} \frac{\lambda\, dr}{2\pi\epsilon_0 r} = \frac{\lambda}{2\pi\epsilon_0} \ln \frac{r_2}{r_1} \tag{7-6}$$

In this case, if we were to let r_2 approach infinity in the hope of using a point at infinity as a zero reference for the absolute potential, all finite points around the cylinder would have infinite absolute potential, an illustration of the remark that the usefulness of putting the zero reference at infinity disappears when the charge distribution giving rise to the potential field itself extends to infinity, as does this infinite cylinder. It will, of course, be possible in particular cases to choose some finite distance from the cylinder as a zero reference for the potential. This is done in effect in some computations with cylindrical capacitors to be discussed later in this chapter.

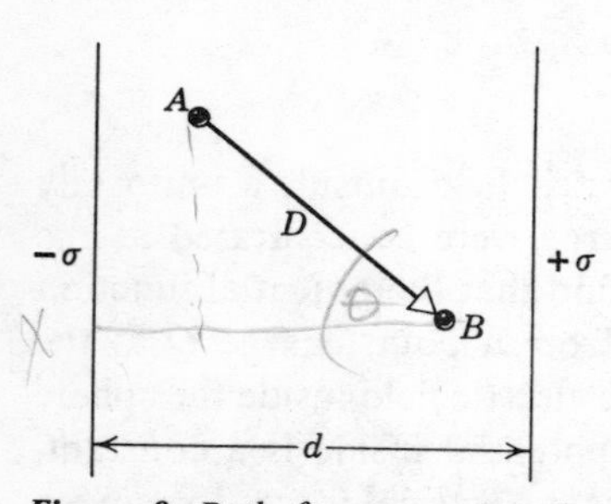

Figure 8. Path for computing the potential difference between A and B within a parallel plate capacitor.

INFINITE PARALLEL UNIFORMLY CHARGED PLATES. In Problem 11 at the end of Chapter 1, the electric field in the region between two infinite planes, uniformly and oppositely charged with σ coulombs/meter^2, was found to be σ/ϵ_0. Let the two parallel lines in Figure 8 represent traces of two such planes separated by a distance d. Between the plates, let A and B be two points, D meters apart, with the direction of D making an angle θ with the normal to the plates. From Equation 4-1 the difference of potential between A and B is

$$\Delta V_{AB} = \frac{\sigma D \cos\theta}{\epsilon_0} \tag{7-7}$$

Thus when point A lies in the negative plane and B in the positive plane, the potential difference between the planes is

$$\Delta V = \frac{\sigma d}{\epsilon_0} \tag{7-8}$$

It will be left for the reader to verify that for an infinite, isolated, charged plane, the absolute potential at a finite point, using infinity as a reference, is infinite.

8. Potential of a Discrete Charge Distribution

We now discuss another example in which the potential function is found by means of Equation 4-3 after which the electric field components are obtained by differentiation. Although the electric field on the axis and on the bisector of a dipole can easily be obtained directly, as was done in Chapter 1, and although the same method can be used to obtain the field at arbitrary points, the resulting expressions are awkward to manipulate, and the following procedure, using spherical coordinates is simpler and reduces readily to simple expressions for points far from the dipole.

The dipole, with charges $+q$ and $-q$ separated by a distance d, is shown in Figure 9, where the point P is located by the polar coordinates r and θ (because of symmetry we need not concern ourselves with the longitude angle). The potential at P is

$$V = \frac{q}{4\pi\epsilon_0}\left[\frac{1}{(r^2 + (d^2/4) - rd\cos\theta)^{1/2}} - \frac{1}{(r^2 + (d^2/4) + rd\cos\theta)^{1/2}}\right] \tag{8-1}$$

Using the fact that the component of the electric field strength in a given direction is the negative of the rate of change of the potential function in that direction, the electric field strength in the increasing r and increasing

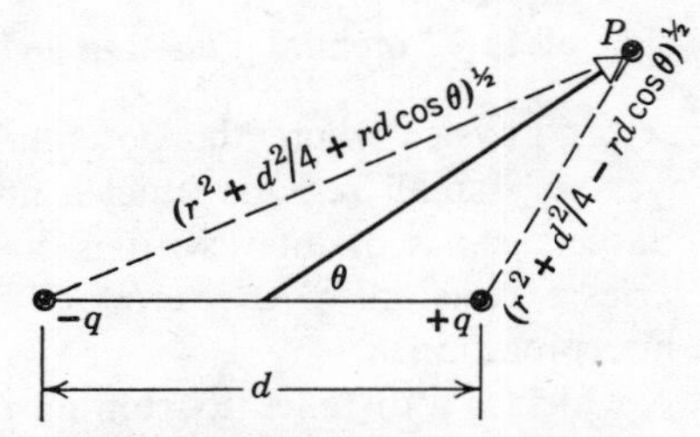

Figure 9. Potential at P due to an electric dipole.

θ directions may be obtained from Equation 8-1 by differentiation. There is, however, greater interest and usefulness in the simpler expressions for the potential and the electric field at large distances where $r \gg d$, and these expressions will now be developed.

By factoring an r out of each denominator of Equation 8-1, we get

$$V = \frac{q}{4\pi\epsilon_0 r}\left[\left(1 + \frac{d^2}{4r^2} - \frac{d}{r}\cos\theta\right)^{-1/2} - \left(1 + \frac{d^2}{4r^2} + \frac{d}{r}\cos\theta\right)^{-1/2}\right] \tag{8-2}$$

and an application of the binomial theorem to each term in the brackets, neglecting terms in $(d/r)^2$ and higher powers, gives very closely

$$V = \frac{q}{4\pi\epsilon_0 r}\left[1 - \frac{1}{2}\frac{d^2}{4r^2} + \frac{1}{2}\frac{d}{r}\cos\theta - 1 + \frac{1}{2}\frac{d^2}{4r^2} + \frac{1}{2}\frac{d}{r}\cos\theta\right] \tag{8-3}$$

$$= \frac{qd\cos\theta}{4\pi\epsilon_0 r^2} = \frac{p\cos\theta}{4\pi\epsilon_0 r^2}$$

where $p = qd$.

In vector notation, if $\mathbf{r}$ is directed toward P and $\mathbf{d}$ from $-q$ toward $+q$, Equation 8-3 becomes:

$$V = \frac{\mathbf{p}\cdot\mathbf{r}}{4\pi\epsilon_0 r^3} \tag{8-4}$$

The components of the electric field strength are then:

$$E_r = -\frac{\partial V}{\partial r} = \frac{2qd\cos\theta}{4\pi\epsilon_0 r^3} = \frac{2p\cos\theta}{4\pi\epsilon_0 r^3} \tag{8-5}$$

$$E_\theta = -\frac{\partial V}{r\,\partial\theta} = \frac{qd\sin\theta}{4\pi\epsilon_0 r^3} = \frac{p\sin\theta}{4\pi\epsilon_0 r^3}$$

which vary inversely as the cube of r. Note that Equation 8-5 reduces to the results of Section 4 and Problem 1 of Chapter 1 when $\theta = \pi/2$ and when $\theta = 0$, respectively.

9. Multipole Expansion of the Potential Function

In atomic and nuclear physics where the potential due to groups of charges extending over very small regions (nuclei, atoms, molecules, or small aggregates of molecules) is of interest, it is convenient to express Equation 4-4 as a series expansion, to the terms of which can be given interesting physical interpretations.

Consider a point S which is a distance $\mathbf{R}$ from an origin located in or near to a small aggregate of charges, as in Figure 10. $d\mathcal{V}$ is an infinitesimal

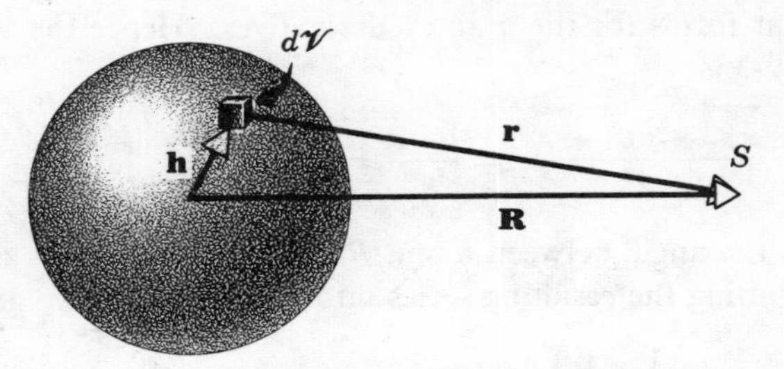

Figure 10. Computing potential at S due to charge in volume element d𝒱 of an aggregate of charge.

volume element whose position is given by the vector **h**. In the approximation of a continuous distribution of charges within the group, with the density of the charge at a point being ρ, the charge in the volume $d\mathcal{V}$ is $\rho\, d\mathcal{V}$, and the potential at S is given by

$$\frac{1}{4\pi\epsilon_0}\int \frac{\rho\, d\mathcal{V}}{r}$$

So long as S is far from the charge group, **r** will differ only by a small amount from **R**, and the quantity $1/r$ can be expanded in a Taylor series in terms of $1/R$.

We use Cartesian coordinates and identify the position of S with the coordinates x', y', and z', and the position of the volume element $d\mathcal{V}$ by x, y, and z. Then the vectors **R** and **h** are written:

$$\mathbf{R} = \mathbf{i}x' + \mathbf{j}y' + \mathbf{k}z' \qquad \mathbf{h} = \mathbf{i}x + \mathbf{j}y + \mathbf{k}z \tag{9-1}$$

and the magnitude $1/r$ is

$$\frac{1}{r} = [(x' - x)^2 + (y' - y)^2 + (z' - z)^2]^{-1/2} \tag{9-2}$$

The three-dimensional Taylor series for the function $1/r$ is

$$\frac{1}{r} = \sum_{n=0}^{\infty} \frac{1}{n!}\left(x\frac{\partial}{\partial x} + y\frac{\partial}{\partial y} + z\frac{\partial}{\partial z}\right)^n \frac{1}{r} \tag{9-3}$$

where each derivative is to be evaluated at $x = y = z = 0$. The operation is a bit tedious but, nevertheless, straightforward. The term for $n = 1$, for example, is obtained as follows. A typical derivative is

$$\frac{\partial}{\partial x}\left[\frac{1}{[(x' - x)^2 + (y' - y)^2 + (z' - z)^2]^{1/2}}\right]_{x=y=z=0} \tag{9-4}$$

$$= \frac{(x' - x)}{[(x' - x)^2 + (y' - y)^2 + (z' - z)^2]^{3/2}}\Bigg|_{x=y=z=0} = \frac{x'}{R^3}$$

with similar forms for the y and z derivatives. Hence the $n = 1$ term in Equation 9-3 is

$$\frac{xx' + yy' + zz'}{R^3} = \frac{\mathbf{h} \cdot \mathbf{R}}{R^3} = \frac{h \cos \theta}{R^2} \tag{9-5}$$

where θ is the angle between h and R. The procedure for other terms is similar. Putting the resulting series into Equation 4-4, we get

$$V = \frac{1}{4\pi\epsilon_0}\left\{\frac{1}{R}\int \rho P_0 \, d\mathscr{V} + \frac{1}{R^2}\int \rho h P_1 \, d\mathscr{V} + \frac{1}{R^3}\int \rho h^2 P_2 \, d\mathscr{V} + \frac{1}{R^4}\int \rho h^3 P_3 \, d\mathscr{V} + \cdots\right\} \tag{9-6}$$

where the P_i's, called Legendre polynomials, are functions of $\cos \theta$, the first four of which are:

$$\begin{aligned} P_0 &= 1 \\ P_1 &= \cos \theta \\ P_2 &= \tfrac{1}{2}(3 \cos^2 \theta - 1) \\ P_3 &= \tfrac{1}{2}(5 \cos^3 \theta - 3 \cos \theta) \end{aligned} \tag{9-7}$$

Those terms of the series that depend on h also depend in general on the location of the origin of coordinates, and the series usually converges more rapidly the closer the origin is to the "center of gravity" of the charge aggregate.

A physical interpretation of the terms in Equation 9-6 can be made by direct inspection. The integral in the first term, ($n = 0$), is $\int \rho \, d\mathscr{V}$, the total charge in the aggregate, and the whole term is the potential at S if the total charge were concentrated at the origin of coordinates.

In the second term ($n = 1$), define $\rho \mathbf{h} \, d\mathscr{V}$ as the vector moment about the origin of the charge in the volume element $d\mathscr{V}$. Then the scalar quantity, $\rho h \cos \theta \, d\mathscr{V}$, is the component of this moment along the **R**-direction, and the integral is the net R component of the moments of the charges about the origin. If the aggregate consists of equal amounts of positive and negative charge, the charges may be paired into dipoles, with the result that the integral represents the sum of the components of the dipole moments along R. For an unbalanced charge distribution, the integral is still called the dipole moment component along **R**, and the second term as a whole is the potential at S of a dipole of this strength located at the origin. In this latter case, however, the integral depends on the location of the origin, whereas for equal positive and negative charges, it does not.

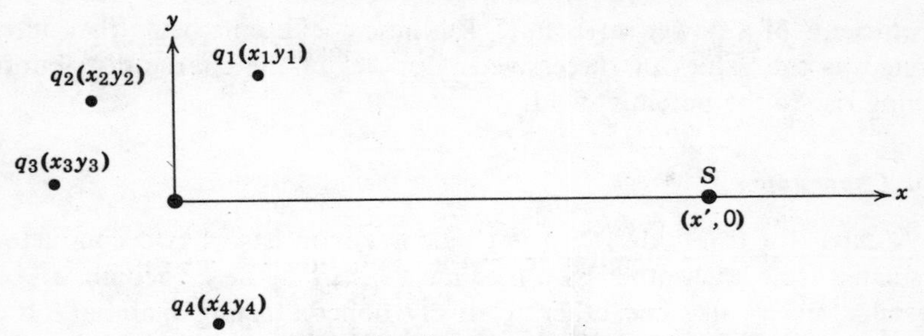

Figure 11. A charge aggregate.

As an example, consider four point charges whose values and locations are q_1 at (x_1y_1), q_2 at (x_2y_2), q_3 at (x_3y_3), and q_4 at (x_4y_4) (Figure 11). For simplicity, let the coordinate of S be $(x', 0)$. The second term of Equation 9-6 is then a summation, and the integrand may conveniently be written by using the first expression of Equation 9-5. Thus,

$$\text{(9-8)} \qquad \frac{1}{R^2}\int \rho h \cos\theta \, d\mathscr{V} = \frac{1}{x'^2}(q_1x_1 + q_2x_2 + q_3x_3 + q_4x_4)$$

$$= \frac{1}{x'^2}(p_{1x} + p_{2x} + p_{3x} + p_{4x})$$

where $\mathbf{p}_i$ is the moment of the charge q_i about the origin. The sum in Equation 9-8 is not in general independent of the location of the origin; but suppose that $q_1 = Q = -q_2$ and $q_3 = q = -q_4$. Then 9-8 becomes

$$\text{(9-9)} \qquad \frac{1}{R^2}\int \rho h \cos\theta \, d\mathscr{V} = \frac{1}{x'^2}[Q(x_1 - x_2) + q(x_3 - x_4)]$$

Here $\sum_i q_i = 0$, and it is seen not only that the numerator is independent of the location of the origin, but also that the numerator is the sum of the x components of the dipoles formed by the pairs of charges.

By analogous reasoning, the third term is the quadrupole potential, the fourth term the octopole potential, and so on, each being the potential that would exist at S if all the poles of its particular type were concentrated at the origin. We are thus saying that an arbitrary charge group may be treated for purposes of computing the potential at some distant point as if it consisted of multipoles concentrated at the origin of coordinates. The reasoning may also be reversed: measurements, direct or indirect, on a potential field of a charge aggregate can be used to determine the

coefficients of a power series in $(1/R)$, these coefficients being then interpreted as the values of successive multipoles of the charge distribution giving rise to the potential field.

10. Capacitance

A capacitor (the older term is condenser) consists of two conductors insulated from each other by a medium which may be a vacuum, an oil, paraffin, mica, or some other relatively nonconducting material. It is found experimentally that, whatever the shape or the relative orientation of the conductors, if they carry equal charges of opposite sign, the potential difference between them is proportional to the magnitude of the charge on either plate. The ratio of the magnitude of the charge to the potential difference is called the capacitance (older term, capacity), C. Thus

$$C = Q/\Delta V \qquad (10\text{-}1)$$

where Q is the magnitude of the charge on one conductor and ΔV is the potential difference between the conductors. Where no ambiguity is introduced, Equation 10-1 will be written simply as $C = Q/V$, where V is the potential difference. Very often the distinction between absolute potential and potential difference is obvious from the nature of the problem and need not be made explicitly.

An especially simple example of a capacitor is the two parallel planes considered in Section 7, page 40.

The two parallel planes are separated by a distance d and charged oppositely with σ coulombs per square meter. The difference of potential between the planes was found to be $\sigma d/\epsilon_0$ when each was infinite in extent. The field and the potential difference are negligibly changed if there are two plates of finite area, provided that the separation d is very small compared to the smallest linear dimension of either plate. (It is true that the charge distribution is not quite uniform near the edges, but to a first approximation this can be neglected.) If the area of the finite plate is A and the total charge is Q, then $\sigma = Q/A$, and the potential difference becomes

$$\Delta V = \frac{\sigma d}{\epsilon_0} = \frac{Qd}{A\epsilon_0} \qquad (10\text{-}2)$$

A comparison of Equation 10-2 with 10-1 shows that the capacitance of a parallel plate capacitor is

$$C = \frac{\epsilon_0 A}{d} \qquad (10\text{-}3)$$

with a slight error because of edge effects.

For two concentric spheres, the inner with a charge Q and a radius r_1, and the outer with a charge $-Q$ and radius r_2, the difference of potential from Equation 7-3 is

$$\Delta V = \frac{Q}{4\pi\epsilon_0}\left(\frac{1}{r_1} - \frac{1}{r_2}\right) \tag{10-4}$$

from which the capacitance is

$$C = 4\pi\epsilon_0 \frac{r_1 r_2}{r_2 - r_1} \tag{10-5}$$

P 107
KIP

When the outer sphere is enlarged indefinitely so that its radius approaches infinity, Equation 10-5 shows that the capacitance of the combination, which is now the capacitance of an isolated sphere, becomes

$$C = 4\pi\epsilon_0 r_1 \tag{10-6}$$

and thus is directly proportional to the radius. In early work in electrostatics the capacitance of an isolated sphere of unit radius was used as a unit of capacitance. This was particularly convenient with the electrostatic system of units, in which the constant $k = 1/4\pi\epsilon_0$ appearing in Coulomb's law was unity, making the capacitance of an isolated sphere numerically equal to its radius. Thus the unit of capacitance in the electrostatic system of units is the centimeter (although the name statfarad is also used).

In the MKS system of units capacitance is expressed in farads, in honor of Michael Faraday, and is defined by Equation 10-1. A capacitor has a capacitance of one farad if the potential difference is one volt when there is one coulomb of charge on each conductor. Unfortunately, this is a very large unit. In common use are the submultiples microfarad (10^{-6} farad), abbreviated μfd or μF, and picofarad(10^{-12} farad), abbreviated as pfd or pF (older term is micromicrofarad).

The general case of several charged conductors will be mentioned briefly. When three or more conducting bodies have charges Q_1, Q_2, Q_3, $\cdots$, their absolute potentials V_1, V_2, V_3, $\cdots$ are linearly related to the charges by

$$\begin{aligned} V_1 &= a_{11}Q_1 + a_{12}Q_2 + \cdots \\ V_2 &= a_{21}Q_1 + a_{22}Q_2 + \cdots \\ &\cdots\cdots\cdots\cdots\cdots\cdots \end{aligned} \tag{10-7}$$

and these equations may be solved for the Q's in terms of the V's:

$$\begin{aligned} Q_1 &= c_{11}V_1 + c_{12}V_2 + \cdots \\ Q_2 &= c_{21}V_1 + c_{22}V_2 + \cdots \\ &\cdots\cdots\cdots\cdots\cdots\cdots \end{aligned} \tag{10-8}$$

Here the coefficients, c_{ij}, have the dimensions of farads and are called coefficients of capacitance.

When we speak of "a capacitor," we are assuming that the two conducting bodies are so close compared to their distances from other bodies that the influence of such other bodies is negligible.

11. Capacitance and Energy Storage

Charging a capacitor requires work, work done in separating the opposite charges. This work is stored in the capacitor as potential energy in a way that is formally analogous to the way that potential energy is stored in a compressed spring.

In whatever way the charging process is accomplished, it is always equivalent to transporting, seriatim, infinitesimal charges, dq, from one plate to the other, leaving the first plate negatively charged and the second positively. At an arbitrary time during this charging process a charge q has been carried from one plate to the other, with a resulting difference of potential v between the plates. To transfer the next element of charge, dq, will require $v\,dq$ joules of work. The sum of such elements of work, starting with a state of zero charge and ending with a magnitude of charge Q on each plate, will be the total energy stored in the capacitor. Thus,

$$W = \int_0^Q v\,dq \tag{11-1}$$

Since $v = q/C$, where C is the capacitance, Equation 11-1 becomes

$$W = \int_0^Q \frac{q\,dq}{C} = \frac{1}{2}\frac{Q^2}{C} \tag{11-2}$$

which may be written in three alternative ways:

$$W = \tfrac{1}{2}Q^2/C = \tfrac{1}{2}CV^2 = \tfrac{1}{2}QV \tag{11-3}$$

using $Q = CV$, where V is the final potential difference. The form used in a particular problem is a matter of convenience.

As an example of the use of Equations 11-3, let us examine the situation in Figure 12, where a capacitor whose capacitance is C_1 is first to be connected by means of the double-throw switch to the battery whose terminal potential difference is V. As a result, the total charge on C_1 is

$$Q_T = C_1V \tag{11-4}$$

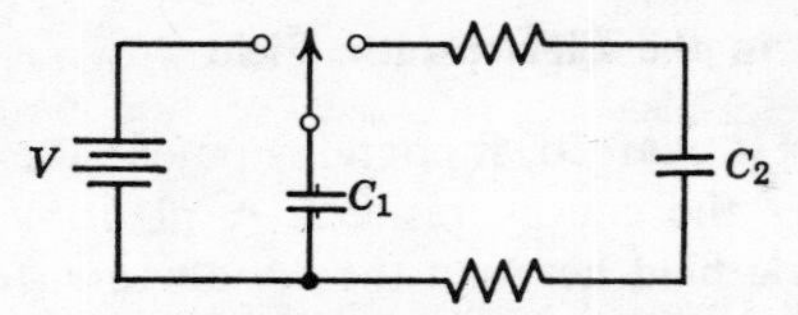

Figure 12. Capacitor C_1 is charged to potential V and then shares charge with C_2.

The switch is then thrown to the other side, connecting C_1 to another—initially uncharged—capacitor of capacitance C_2. (The symbols for resistance in the figure merely indicate the resistance of the connecting wires. The resistance does not enter into the computation.) Since there is no initial difference of potential across C_2, charge will flow off C_1 to C_2 until both capacitors reach the same potential difference, say, V'. Because no charge has been lost,

$$Q_T = Q_1 + Q_2 \tag{11-5}$$

or

$$C_1V = C_1V' + C_2V' \tag{11-6}$$

Thus

$$V' = V\frac{C_1}{C_1 + C_2} \tag{11-7}$$

Initially the energy stored in capacitor C_1 was, from Equation 11-3:

$$W = \tfrac{1}{2}C_1V^2 \tag{11-8}$$

After the new equilibrium is established, the sum of the energies stored in the two capacitors is

$$W' = \tfrac{1}{2}C_1V'^2 + \tfrac{1}{2}C_2V'^2 = \tfrac{1}{2}V'^2(C_1 + C_2) \tag{11-9}$$

which, in view of Equations 11-7 and 11-8, is

$$W' = \frac{1}{2}\frac{C_1^2V^2}{C_1 + C_2} = \frac{WC_1}{C_1 + C_2} \tag{11-10}$$

which is less than the original energy, W, on C_1. It seems that an amount of energy

$$W - \frac{WC_1}{C_1 + C_2} = \frac{WC_2}{C_1 + C_2} \tag{11-11}$$

has disappeared. What happened to this energy is a question that is left to the reader to reflect upon.

12. Energy Density in the Electrostatic Field

When a capacitor is charged, it contains potential energy, and we may ask just where does the energy reside? A plausible answer is that it resides in the electric field between the conductors, for it is the electric field that is present after charging and was not present before.

Consider again the parallel-plate capacitor, a simple example because, except for the region near the edges, the electric field was found to be uniform within the volume between the plates. In the expression $W = QV/2$, let us replace Q with $\epsilon_0 EA$ and V with Ed (see Section 1), where A is the area of a plate and d is the separation of the plates. Then

$$W = QV/2 = \tfrac{1}{2}(\epsilon_0 EA)(Ed) = \tfrac{1}{2}\epsilon_0 E^2 Ad \tag{12-1}$$

Here the product Ad is the volume between the plates, and evidently the potential energy is proportional to this volume (at least to the approximation, usually a very good one, that the edge effects are negligible). It is plausible to regard the energy as being stored throughout the volume, with an energy per unit volume, the energy density, w, of

$$w = \tfrac{1}{2}\epsilon_0 E^2 \tag{12-2}$$

Although this expression for the energy density has been obtained from the special case of a parallel-plate capacitor, it is quite generally applicable to electric fields in a vacuum. In later chapters we shall see that it is a special form of a more general expression that applies for any medium.

Because the total energy of the capacitor is a function of the electric field strength, it was said "it is plausible to regard the energy as being stored throughout the volume." It is not, however, possible to measure the energy in a part of a total volume. Only the total energy is physically measurable. Nevertheless, the concept of energy density will be found to be a real convenience in our thinking about energy relationships and will be used frequently in electromagnetic theory.

13. Poisson's and Laplace's Equations

We now recall Gauss's law,

$$\nabla \cdot \mathbf{E} = \left(\mathbf{i}\frac{\partial}{\partial x} + \mathbf{j}\frac{\partial}{\partial y} + \mathbf{k}\frac{\partial}{\partial z}\right) \cdot (\mathbf{i}E_x + \mathbf{j}E_y + \mathbf{k}E_z) = \frac{\rho}{\epsilon_0} \tag{13-1}$$

and the point relation between electric field strength and potential,

$$\mathbf{E} = -\nabla V = -\left(\mathbf{i}\frac{\partial}{\partial x} + \mathbf{j}\frac{\partial}{\partial y} + \mathbf{k}\frac{\partial}{\partial z}\right) V \tag{13-2}$$

Each of these expressions is of general validity. By combining them, we can eliminate **E** and get a relation between the potential and the charge density:

$$\nabla \cdot (-\nabla V) = -\left(\mathbf{i}\frac{\partial}{\partial x} + \mathbf{j}\frac{\partial}{\partial y} + \mathbf{k}\frac{\partial}{\partial z}\right) \cdot \left(\mathbf{i}\frac{\partial}{\partial x} + \mathbf{j}\frac{\partial}{\partial y} + \mathbf{k}\frac{\partial}{\partial z}\right) V \tag{13-3}$$

$$= -\left(\frac{\partial^2}{\partial x^2} + \frac{\partial^2}{\partial y^2} + \frac{\partial^2}{\partial z^2}\right) V = -\left(\frac{\partial^2 V}{\partial x^2} + \frac{\partial^2 V}{\partial y^2} + \frac{\partial^2 V}{\partial z^2}\right)$$

$$= \frac{\rho}{\epsilon_0}$$

The dot product of ∇ with itself represents the sum of the second derivatives in the x, y, and z directions and is commonly written as ∇^2. The differential equation

$$\nabla^2 V = -\frac{\rho}{\epsilon_0} \tag{13-4}$$

is known as Poisson's equation, named for Simeon Denis Poisson, who developed it in 1812 for use with a gravitational field rather than with an electric field. In the above form, it relates the second space rates of change of the potential at a point to the density of charge at that point. In uncharged regions the right side of Equation 13-4 is zero; this form of the equation was obtained earlier by Pierre Simon Marquis de Laplace, and it is customary to call the equation as it applies to uncharged regions

$$\nabla^2 V = 0 \tag{13-5}$$

Laplace's equation. Separate names for Equations 13-4 and 13-5 are purely a matter of historical custom. It is also customary to refer to the operator ∇^2 as the Laplacian, with the left side of Equations 13-4 or 13-5 usually read as "the Laplacian of V."

Just what has been gained by introducing Poisson's (or Laplace's) equation? Here is a differential equation from which, given the distribution of charge density at every point, we can solve for the potential function V. But we have already developed methods of finding the potential function from the charge distribution in earlier sections of this chapter. In fact the expression

$$V = \frac{1}{4\pi\epsilon_0} \int \frac{\rho \, d\mathscr{V}}{r}$$

is a particular solution to Poisson's equation.

The methods for finding the potential function that were discussed earlier in this chapter require a complete knowledge of the charge configuration. There are, however, situations where our knowledge of these charges is incomplete but where, instead, we may know the charge distribution within a limited region and either the potential or the electric field at points on the boundary of this region. By the use of Poisson's (or Laplace's) equation the potential function within such a region can be found, without requiring complete knowledge of the location of charges that contribute to the boundary conditions.

Let us call the solution of Poisson's equation that expresses the potential as a function of the charge distribution V_P, and the solution to Laplace's equation V_L. Then, clearly, $V_P + V_L$ is also a solution to Poisson's equation. V_P is called the particular solution and V_L the complementary function, the latter containing the two arbitrary constants of integration that any complete solution to a second-order differential equation must have. There are a large number of types of functions which will satisfy Laplace's equation, and the choice among them must be made by whether the boundary conditions of the problem can be satisfied.

There are many problems for which it is not possible to find an analytic function as a solution to Poisson's or Laplace's equations. For such cases there have been developed various numerical techniques for finding approximate values of the potential point by point over the region of interest. Usually points arranged in a particular pattern are assigned values of potential by guess or intuition, and then these values are tested for their consistency with Poisson's equation. The testing process suggests an approximate correction at each point, and the testing process is repeated on the new set of values, resulting in a better approximation, etc. How good the end result is depends on the quality of the original guess and the number of times the testing process is repeated.

Whether the problem is solved numerically or analytically, the solution sought is the one which simultaneously satisfies Poisson's equation and the given boundary conditions. An example of an analytic solution is given in the next section.

The reader who has had limited experience in finding an analytic solution for a differential equation will find comfort in the fact that the finding of such a solution is basically a process of educated guessing. The better educated the guess, the more quickly the solution is likely to be found. There is an extensive literature on techniques and tricks for finding solutions to the Poisson and Laplace equations under a large variety of circumstances, the amount of the literature attesting to the importance of these equations in physics. But in all cases the validity of the solution does not depend on how it was found. Whether it has been obtained by classical methods or by gazing into a crystal ball, the solution is valid if, and only if, it satisfies the differential equation and fits the known boundary conditions of the problem.

14. An Example of the Use of Laplace's Equation

Imagine a uniform electric field in the x-direction and place into this field an uncharged conducting sphere of radius r_0 (Figure 13), with the center of the sphere at the origin of coordinates. Since the sphere is a conductor, its surface (and its entire volume) must be an equipotential surface and this potential will be used as our reference potential. We expect the presence of the sphere to alter the nature of the electric field in its vicinity from what the field would be without the sphere, but this should be a local effect, with the field being negligibly affected at very large distances. Thus the boundary conditions for this problem are the known potential at the surface of the sphere (zero) and the known field (known rate of change of potential) at sufficiently large distances from the sphere. The problem is best treated in terms of spherical coordinates with the polar axis in the x-direction, as indicated in Figure 13. Clearly, there is symmetry about this axis, so that we need only the radius, r, and the co-latitude, θ, as coordinates.

The boundary conditions can be expressed analytically as

$$V = 0 \qquad \text{at } r = r_0 \tag{14-1a}$$

$$\left.\begin{aligned} E_r &= -\frac{\partial V}{\partial r} \to E\cos\theta \\ E_\theta &= -\frac{\partial V}{r\,\partial\theta} \to -E\sin\theta \end{aligned}\right\} \text{as } r \to \infty \tag{14-1b}$$

Condition 14-1b is equivalent to

$$V \to -Er\cos\theta \qquad \text{as } r \to \infty \tag{14-1c}$$

We seek a potential function for the charge-free region outside the sphere; hence, we must look for a solution to Laplace's equation, 13-5. This equation, however, must be expressed in terms of spherical coordinates obtained by a standard transformation from Equation 13-5:

$$\frac{1}{r^2}\frac{\partial}{\partial r}\left(r^2\frac{\partial V}{\partial r}\right) + \frac{1}{r^2\sin\theta}\frac{\partial}{\partial\theta}\left(\sin\theta\frac{\partial V}{\partial\theta}\right) = 0 \tag{14-2}$$

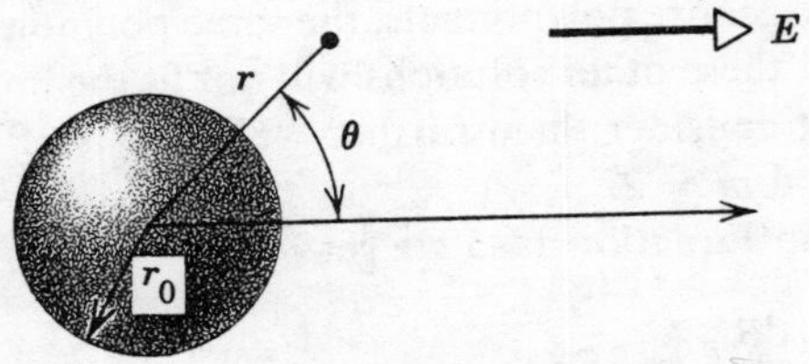

E

r

θ

r_0

Figure 13. A spherical conductor in an electric field.

where we have omitted the term involving the longitude angle since symmetry shows that derivatives with respect to this angle are zero. We look for some function $V(r, \theta)$ that will satisfy Equation 14-2, will go to zero at $r = r_0$, and will approach $-Er \cos \theta$ as r becomes very large. It should not surprise us, in view of the latter condition, if the function $V(r, \theta)$ involves the angle θ in the form of $\cos \theta$.

In cases where some of the boundary conditions can be expressed in terms of a single variable, it is often found that a solution can be expressed in the form of a product of two functions, each involving only one of the variables. We therefore look for a solution which is a product of a function R, involving only r, and a function T, involving only θ:

$$V = R(r)T(\theta) \tag{14-3}$$

Substituting Equation 14-3 into 14-2, we get after some rearrangement:

$$\frac{1}{R}\frac{\mathrm{d}}{\mathrm{d}r}\left(r^2 \frac{\mathrm{d}R}{\mathrm{d}r}\right) + \frac{1}{T \sin\theta}\frac{\mathrm{d}}{\mathrm{d}\theta}\left(\sin\theta \frac{\mathrm{d}T}{\mathrm{d}\theta}\right) = 0 \tag{14-4}$$

in which the first term on the left involves only the variable r and the second term involves only the variable θ, and for any combination of r and θ the sum of the two terms is zero. This can only be true if each term separately is a constant. Let the first term be equal to m and the second to $-m$, where m can have any value so far as Equation 14-4 is concerned (but we shall be free to choose that value of m which will make possible the satisfying of the boundary conditions). We thus have two differential equations, each in one variable:

$$\frac{\mathrm{d}}{\mathrm{d}r}\left(r^2 \frac{\mathrm{d}R}{\mathrm{d}r}\right) = mR \tag{14-5}$$

$$\frac{\mathrm{d}}{\mathrm{d}\theta}\left(\sin\theta \frac{\mathrm{d}T}{\mathrm{d}\theta}\right) = -mT\sin\theta \tag{14-6}$$

Inspection and trial show that $\cos\theta$ is a solution of Equation 14-6, provided $m = 2$. For other values of m, there are other solutions that involve higher powers of $\cos\theta$ (as a group the solutions for integer values of m are called Legendre polynomials, the same polynomials that appeared in Section 8), but these other solutions will not fit the boundary conditions and we shall not consider them further. Thus, out of Equation 14-6 we get $T = \cos\theta$ and $m = 2$.

Using $m = 2$ in Equation 14-5 we get

$$\frac{\mathrm{d}}{\mathrm{d}r}\left(r^2 \frac{\mathrm{d}R}{\mathrm{d}r}\right) = 2R \tag{14-7}$$

for which trial shows that $R = r^n$ is a solution, provided that $n = 1$ or $n = -2$. Any solution of Laplace's equation can be multiplied by an arbitrary constant and remain a solution; thus there are two solutions in the form of Equation 14-3:

$$V = Ar \cos \theta \quad \text{and} \quad V = \frac{B \cos \theta}{r^2} \tag{14-8}$$

where A and B are arbitrary constants. Since Laplace's equation is a linear differential equation, the sum of these two solutions is itself a solution:

$$V = Ar \cos \theta + \frac{B \cos \theta}{r^2} \tag{14-9}$$

and we need now only find out if the constants A and B can be so chosen as to fit the boundary conditions.

At very large r the second term in Equation 14-9 approaches zero, so that condition 14-1c requires that $A = -E$; whereupon condition 14-1a requires that $B = Er_0^3$. Hence Equation 14-9 becomes

$$V = Er \cos \theta \left(\frac{r_0^3}{r^3} - 1 \right) \tag{14-10}$$

which is the desired potential function. By differentiation the r and θ components of the electric field are readily found.

15. The Uniqueness of the Solution

Now that a potential function has been found, a good question to ask is: Have we found *the* potential function for this problem, or might there not be another function or functions that will also satisfy the boundary conditions and also Laplace's equation? Since the finding of a solution to a differential equation is a matter of educated guessing, and since we know there are many solutions to Laplace's equation, what confidence have we that another guess might not serve as well? The following reasoning provides the answer.

We shall assume that two solutions *do* exist and then demonstrate that they must in fact be identical. Let us call these two functions V_1 and V_2. If they are solutions to the problem, then necessarily each must satisfy Laplace's equation, so that:

$$\nabla^2 V_1 = 0 \quad \text{and} \quad \nabla^2 V_2 = 0 \tag{15-1}$$

or

$$\nabla^2 (V_1 - V_2) = 0 \tag{15-2}$$

and each solution must also fit the boundary conditions of the problem, so that *at the boundary*

$$(V_1 - V_2) = 0 \tag{15-3}$$

For example, in the problem of the previous section, both V_1 and V_2 must be zero at the surface of the sphere and both approach $-Er \cos \theta$ for large r, but we postulate that at least at some places within these boundaries V_1 and V_2 do not have the same value.

We form the vector

$$\mathbf{S} = (V_1 - V_2)\nabla(V_1 - V_2) \tag{15-4}$$

and then recall Gauss's theorem, developed at the end of Chapter 1, which equates the integral of the divergence of any vector over a volume to the integral of the vector itself over the bounding surface of this volume. Applying this theorem to the vector **S**, we get:

$$\int \nabla \cdot \mathbf{S}\, d\mathscr{V} = \int \mathbf{S} \cdot d\mathscr{A} \tag{15-5}$$

or

$$\int \nabla \cdot [(V_1 - V_2)\nabla(V_1 - V_2)]\, d\mathscr{V} = \int (V_1 - V_2)\nabla(V_1 - V_2) \cdot d\mathscr{A}$$

where the integral on the left is over the entire volume in which the functions are valid and the integral on the right is over the surface of this volume, that is, the surface over which the boundary conditions apply. Again referring to the previous example, this boundary is the surface of the sphere and the surface at infinity. Clearly, by virtue of Equation 15-3, the surface integral on the right of Equation 15-5 is zero.

We now need the following vector identity:

$$\nabla \cdot (\psi \mathbf{A}) = (\nabla \psi) \cdot \mathbf{A} + \psi(\nabla \cdot \mathbf{A}) \tag{15-6}$$

(Equation A-1, Appendix A.)

which the reader can readily justify by performing the indicated operations (ψ is any scalar function and **A** is any vector function). For present purposes, let $\psi = (V_1 - V_2)$ and $\mathbf{A} = \nabla(V_1 - V_2)$, so that Equation 15-5 becomes

$$\int [(V_1 - V_2)\nabla^2(V_1 - V_2) + \nabla(V_1 - V_2) \cdot \nabla(V_1 - V_2)]\, d\mathscr{V} = 0 \tag{15-7}$$

Because of Equation 15-2, the first term in the integrand is zero, and we are left with

$$\int [\nabla(V_1 - V_2)]^2\, d\mathscr{V} = 0 \tag{15-8}$$

whose integrand, being a square, can only be either positive or zero. But if its value is greater than zero anywhere, it is not possible for the integral over the whole volume to be zero. Hence, we conclude that $\nabla(V_1 - V_2)$—the maximum rate of change of $(V_1 - V_2)$—must be zero everywhere and, consequently, that:

$$(V_1 - V_2) = \text{constant} \tag{15-9}$$

Since $(V_1 - V_2)$ was postulated to be zero on the boundary, it is zero everywhere and hence the two solutions are identical. We can thus be confident that, if we have found a solution to Laplace's equation that fits the boundary conditions of a problem, this solution is unique.

Discussion Questions

1. Two infinite planes are parallel and are held at a potential difference of, say, V. In one there is cut a half-cylindrical groove, as in Fig. Q1. From your

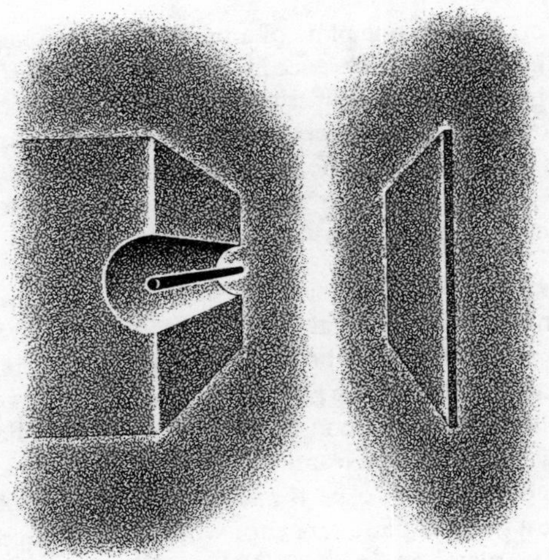

Figure Q1

best knowledge and intuition, sketch the qualitative pattern of the equipotential surfaces between these planes. Also sketch the electric field lines. (If a heated wire, placed within and parallel to the groove, is a source of electrons, and if the grooved plate has a negative potential with respect to the flat plate, the electrons will form a stream moving toward the flat plate, the stream being partially focused. Such a system has been used in X-ray tubes to concentrate the electron beam on the target.)

2. Similarly sketch the equipotentials and the field lines between the ends of two cylinders arranged on a common axis, as in the following diagram, when there

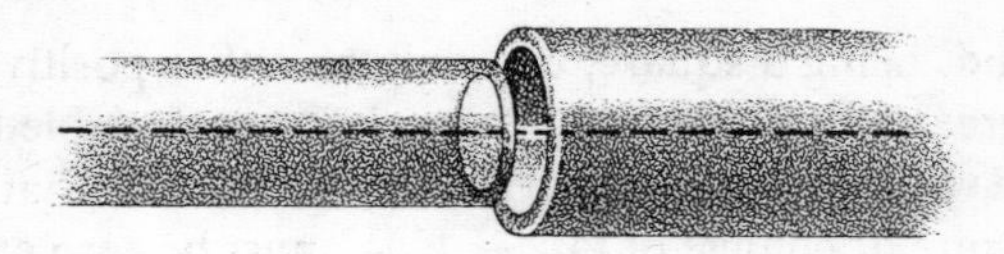

Figure Q2

is a potential difference between them. (This arrangement will tend to focus an electron beam traveling parallel to the common axis. Discuss possible applications.)

3. In Discussion Question 6 of Chapter 1, the force on a plate of a parallel-plate capacitor was found in terms of the dimensions of the capacitor and the charge. This same result may be obtained from energy considerations, by equating the work required to move one plate a distance dx further away from the other plate and equating this work to the added stored energy in the additional volume between the plates. Do you find one method preferable to the other or not, for reasons of logic, universality, or neatness?

4. Express the force which one plate of a parallel-plate capacitor exerts on the other in terms of the potential difference between the plates. If one plate of such a capacitor is hung from, say, the arm of an analytical balance while the other is held fixed, this force of attraction between the plates can be measured. Design an instrument in which potential difference can be measured directly in terms of mass and length. (Include in your design the advantages of the guard-ring capacitor studied in Discussion Question 7 of Chapter 1.)

5. In the next chapter it will be shown that when the medium in which charges are immersed is an isotropic material, the expressions for the force, the electric field, the potential, the energy, and the energy density retain the same form as for a vacuum, but the absolute dielectric constant, ϵ_0, is multiplied by a dimensionless constant, κ (the relative dielectric constant), whose value is characteristic of the material and is always greater than unity. For most purposes it is sufficient to consider the value of κ for air to be the same as for a vacuum, that is, unity.

Suppose we have a parallel-plate capacitor whose capacitance with air between the plates is C_0. The plates are in vertical planes. A block of ceramic material of relative dielectric constant κ is of such a size that it fits between the plates and fills the space of the capacitor. It is free to move with negligible friction into the capacitor volume. The capacitor plates carry a charge Q and are insulated so that the charge Q cannot change. From energy considerations, decide whether the ceramic block will be pulled into or repelled from the capacitor, and discuss its motion under friction-free conditions.

6. Now consider the same system as in 5, except that the capacitor is connected to a battery of fixed potential V. Thus any change of capacitance that results from the presence of the ceramic block produces a change in charge on the plates instead of a change in potential difference. Discuss the forces in this case.

Problems

1. An electric field in a region is given by:

$$\mathbf{E} = \mathbf{i}(6x^2 - 6xy - 6y^2) + \mathbf{j}(-3x^2 + 3y^2 - 12xy)$$

Form the line integral of this field over the perimeter of the rectangle formed by the points $(0, 0)$, $(x_0, 0)$, (x_0, y_0), $(0, y_0)$, and show that this line integral is zero. Find the potential function associated with this field.

2. Show that the electric field vector at a point must be normal to the equipotential surface that passes through that point.

3. From Equation 8-1, plot carefully on graph paper the trace of an equipotential surface for a dipole where the approximation for large distances is not valid. Also, on a different scale, plot two equipotential surfaces from Equation 8-3 at large distances from the dipole.

4. Differentiate Equation 8-1 to get the components of the electric field of a dipole. Then apply the approximation $r \gg d$ and the binomial theorem to find the components of the electric field at large distances. Show that these results agree with Equation 8-5.

5. Below what value of the ratio d/r is the expression 8-3 accurate to better than 1% for points on the axis of a dipole? Compare with results of Problem 2, Chapter 1.

6. A circular disk of radius a is uniformly charged with σ coulombs per square meter. Find the potential at a point on the perpendicular through the center of the disk, a distance d meters from the center. What happens if the radius a is allowed to become infinite, the charge density remaining the same.

7. A system of n point charges, $q_1, q_2, \cdots, q_n$, form an assemblage in which the ith charge is a distance r_{i1} from q_1, r_{i2} from q_2, etc. Find the work done against electrostatic forces when these charges are assembled into this configuration from an initial state in which all the charges are at infinity. Demonstrate that this work is independent of the order in which the charges are brought into position.

8. Obtain the third and fourth terms of Equation 9-6, using Equation 9-3.

9. Compute the potential at an arbitrary point in the plane of the quadrupole of Problem 3, Chapter 1, for the case where $r \gg d$, by adding the potentials produced by each constituent dipole separately. From this potential function determine the components of the electric field and show that the result reduces to the answer found in Problem 3, Chapter 1, for points on the axis.

10. Using Coulomb's law, show that the absolute dielectric constant, ϵ_0, has the units of farads/meter or (coul)(volt)$^{-1}$(meter)$^{-1}$.

11. A metal cylinder is 50 cm long. A metal sphere of radius 2 cm, suspended by a long silk thread, is charged to an absolute potential of 100 volts and then placed in contact with the inner surface of the cylinder at a point 20 to 25 cm from an end. If, after ten such operations, the absolute potential of the cylinder is 80 volts, find the electrical capacitance of the cylinder and the energy of its stored charge.

12. Compute the capacitance per unit length of two infinite coaxial cylinders of radii a and b, with $a < b$.

13. The axes of two parallel cylinders are separated by a distance d. Each cylinder is of radius a; one carries a charge λ per unit length and the other $-\lambda$ per unit length. Show that the capacitance per unit length is given by:

$$C = \frac{\pi\epsilon_0}{\ln\left(\frac{d-a}{a}\right)}$$

Assume either that the cylinders are perfect insulators and that the charge has been applied symmetrically about the axis, or that the cylinders are metal and that d is very large compared to a, so that the distribution on each cylinder is negligibly affected by the presence of the other.

14. A spherical capacitor consists of two concentric spheres of radii a and b. Let the inner sphere carry a charge Q and the outer a charge $-Q$. Show that the electrostatic energy stored in this capacitor is given by the integral over the volume between the spheres of the energy density in the electric field.

15. In order for a given function to be a valid potential function, that is, it could be associated with an electrostatic field, what condition must it satisfy? In a three-dimensional, charge-free region (sources of field outside the region) determine which of the following functions could be valid potential functions. For those that are valid, determine the components of the electric field.

(*a*) $V = 5x^2 - 6y^2 + z^2$

(*b*) $V = \dfrac{x^2 - y^2}{x^2 + y^2}$

(*c*) $V = x^2 + 5y - 3xz^2$

(*d*) $V = \dfrac{1}{(x^2 + y^2 + z^2)^{1/2}}$

(*e*) $V = \arctan\dfrac{2ay}{a^2 - x^2 - y^2} + \arctan\dfrac{2ax}{a^2 - x^2 - y^2}$

(*f*) $V = \dfrac{\cos^2\theta}{r^2}$

(*g*) $V = \dfrac{\cos\theta}{r^4}$

(*h*) $V = \dfrac{\cos^3\theta}{r^4}$

16. What linear combination of (*g*) and (*h*) in Problem 15 is a valid potential function?

17. For what values of A and B is the following a valid potential function in a charge-free region?

$$V = \frac{A \cos^2 \theta - B}{r^3}$$

18. The electrostatic potential in a certain region is given by:

$$V = \frac{ke^{-ar}}{r}$$

where k and a are constants. How is the charge distributed in this region? Test your answer with Gauss's Law. What physical interpretation can be given to the constant k?

19. A point charge q is a distance d in front of an infinite conducting plane whose potential is zero. A second charge q' is placed in back of the plane. Where must q' be placed and what must be the value of q' so that, if the conducting plane is removed, the electric field and the potential distribution in the region which was originally in front of the plane will be unchanged?

20. Using the results of Problem 14, Chapter 1, find the surface charge density on the plane of Problem 19 as a function of the distance from the foot of the perpendicular drawn from q.

21. A point charge q is a distance d from the center of a conducting sphere of radius a, where $d > a$. The sphere is at zero potential. It is possible to locate a charge q' at a point within the sphere, such that with the sphere removed the electric field and the potential distribution in the region outside the sphere will be unchanged. What is the value of q' and where must it be placed?

22. Suppose $V = V_0$ is the potential at a point 0 within a charge-free region where the potential function is $V(r, \theta, \varphi)$. Imagine a small sphere of arbitrary radius r drawn with 0 as center. The average value of the potential over the surface of this sphere is given by:

$$\overline{V} = \frac{1}{4\pi r^2} \int_0^{\pi} \int_0^{2\pi} V r^2 \sin \theta \, d\theta \, d\varphi$$

and since r is a constant, the right side of this expression is independent of r, except as r appears in the function V. Take the derivative of both sides of this equation with respect to r and show that the right side of the resulting equation is zero by virtue of Gauss's law. From this, show that the average value of V over this sphere is a constant independent of r and, hence, must be equal to V_0.

23. From the results of Problem 22, show that a potential function cannot have a maximum or a minimum at a point not occupied by a charge.

REFERENCES

1. Scott, *The Physics of Electricity and Magnetism*, Wiley, New York, 1959. Chapters 1 and 4. Section 3 of Chapter 3 gives a summary of the multipole expansion.
2. Slater and Frank, *Electromagnetism*, McGraw-Hill, New York, 1947. Appendix VI on multipoles.

3. Jackson, *Classical Electrodynamics*, Wiley, New York, 1962. An advanced book with very clear treatment.
4. Smythe, *Static and Dynamic Electricity*, McGraw-Hill, New York, 1950. Potential Problems treated extensively in Chapters IV and V.
5. Maxwell, *Treatise on Electricity and Magnetism*, Dover. First published in 1873. Article 74*c* for the theory of the Cavendish experiment in terms of potential concepts.
6. Moon and Spencer, *Foundations of Electrodynamics*, Van Nostrand, Princeton, 1960. Chapter 4 contains some good examples of field-potential mappings.
7. Abraham and Becker, *The Classical Theory of Electricity and Magnetism*, Hafner, New York, 1932. An old but still excellent standard text.

3

Dielectric media

Heretofore the space between electric charges has been assumed to be empty, the field and potential functions being determined by the locations of the charges. In this chapter we find that another independent variable is introduced when space is partially or wholly occupied by neutral matter. The field and the potential at points both inside and outside such matter are affected by its presence.

In one sense the methods of Chapters 1 and 2 for empty space are still applicable, because in introducing matter into the field we have, from the electrical point of view, introduced additional charges—the negative charges of the electrons and the positive charges of the atomic nuclei. To compute the electric field at a point in space, we should sum up the contributions to the field of all the charges present. But, clearly, our information is insufficient to do this. Moreover, even had we information in sufficient fineness of detail about the location of charges we are not at this level of detail dealing with a static condition. In any material body the atoms are in thermal motion and the electrons in orbital motion. The simple laws of *electrostatics* should not be expected to hold. How, then, can we avoid the necessity of acquiring such detailed data about the distribution and motions of the charges of a neutral body?

When neutral matter is in mechanical as well as thermal equilibrium and when the gross measurements made on such matter require a time that is long compared to the periods of motion of the charges within it, it is not difficult to justify the use of a simplified picture in which the atoms and electrons are in fixed positions determined by time averages over their motions. With a few exceptions—such as the observation of background noise in some electronic equipment, which is associated with the thermal motion of electrons in conductors—electrical measurements normally are slow enough to justify such a time-averaged interpretation. But even in this static picture our information is still inadequate about the locations of individual charges. Fortunately, however, on a macroscopic scale these charges are so close together, and there are so many of them, that we can make the further approximation that the material medium may be treated as a continuum.

In effect, when we wish to obtain the field and potential functions in the presence of neutral matter, we first compute these functions as if the matter were not present, and then introduce a correction factor. Ultimately the value of this correction factor, which is characteristic of the nature of the medium, must be obtained experimentally, but it is the principal task of this chapter to develop a theoretical justification for it by analyzing the relation between the atomic and the simplified continuum pictures of neutral matter.

We shall begin by imagining neutral matter to consist of atoms or molecules whose positive nuclei are surrounded by negatively charged clouds. No assumptions will be made about the forces holding the negatively charged cloud to the nucleus except for one discussion in Section 8, where these forces are assumed to be purely electrical in nature. The structure of the atoms or molecules is assumed not to be perfectly rigid; rather, in the presence of an external electric field, there is a slight shift of the negative charge cloud against the field and a slight shift of the positive nucleus with the field, thereby producing an electric dipole moment in the atom or molecule. In addition, such dipoles as may already exist prior to the application of the electric field experience a torque tending to align them with the field.

A complete picture of neutral matter must also include some electrons that are free of attachment to any one atom, but still are confined within the boundary of the material as a sort of electron gas. These electrons give rise to electrical conduction (see Chapter 4). For the present we confine our attention to materials having so few free electrons as to approximate perfect insulators. Such materials are called dielectrics.

1. The Polarization of Matter

It is convenient to distinguish two general classes of dielectrics—those with polar and those with nonpolar molecules. Polar molecules are those in which the center of gravity of the electron cloud does not coincide with the center of gravity of the nuclei, so that even without an external electric field these molecules have an electric dipole moment. Nonpolar molecules are those with such symmetry that they have no permanent dipole moment. Included in this class are all monatomic molecules, as well as many that are not monatomic.

In both kinds of molecules, the presence of an external electric field will induce electric dipoles because of the shifting in opposite directions of the negative and positive charges. But on the polar molecules the electric field also exerts a torque which tends to align their permanent dipoles with the field. Thermal agitation as well as other interatomic forces will oppose the complete alignment of such dipoles so that the degree of such alignment may be expected to be temperature-dependent. The production of induced dipoles, on the other hand, is independent of temperature. With a few exceptions, the fact that a substance may consist of molecules with permanent dipole moments is not detectable in the absence of an electric field, because these dipoles will be oriented at random. There are, however, some substances, called electrets, in which the dipoles, once aligned, will remain so for an appreciable time after the aligning influence has been removed.

Dipole moment is a vector quantity which may be summed over the molecules in a region to obtain a gross dipole moment. A region in which this sum is not zero is said to be "polarized." By summing the dipole moments over a small volume $d\mathscr{V}$ (which nevertheless contains a large number of molecules) and then dividing by $d\mathscr{V}$, we obtain the dipole moment per unit volume, $\mathbf{P}$, called the *polarization*, with units of coulombs/meter^2. It is an important quantity in describing the electrical behavior of a dielectric.

2. An Isotropic Dielectric in a Parallel Plate Capacitor

Before we develop a general theory of the polarization property of a dielectric, let us examine qualitatively a simple example. Imagine a dielectric between two parallel charged plates (Figure 1). It is easier, but not essential, to take it to be of the nonpolar type. In Figure 1 the external field is directed from left to right, and the direction of the induced molecular dipoles is thus also from left to right. Associated with each dipole is a local electric field that varies from point to point but tends to be directed toward the left. Consequently, the electric field in the space between the molecules, while fluctuating greatly in detail, will on the average be less than the field that would exist if the dielectric were not present.

An average electric field in the dielectric is more precisely defined in terms of two points, say, A and B, separated by a distance large compared

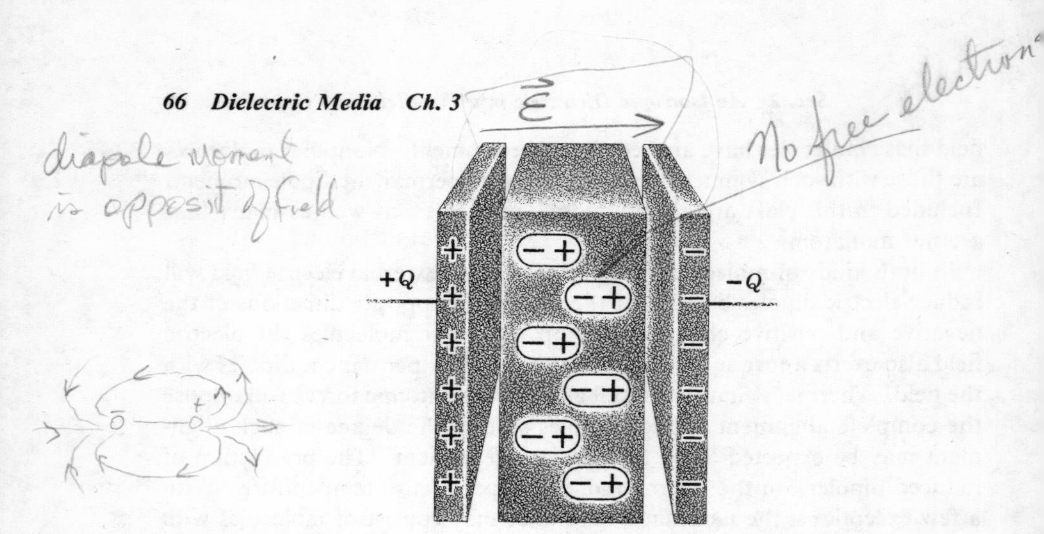

Figure 1. A polarized dielectric slab between parallel charged plates.

to intermolecular distances. The difference of potential between A and B is

$$\Delta V = -\int_A^B \mathbf{E}_d \cdot d\mathbf{l} \tag{2-1}$$

where $\mathbf{E}_d$ is the microscopic field in the dielectric. The average field, $\mathbf{E}$, is the uniform field which will give the same potential difference between A and B:

$$\Delta V = -\int_A^B \mathbf{E}_d \cdot d\mathbf{l} = -\mathbf{E}\cdot\int_A^B d\mathbf{l} \tag{2-2}$$

When for differing separations of A and B along a given line the field $\mathbf{E}$ is the same, we have a uniform average field in the dielectric. When such is not the case, then successive averages must be taken over small separations of A and B to get an $\mathbf{E}$ which is a smoothly varying function of position, but the steps must necessarily be large compared to intermolecular distances.

As already suggested above, this average field is less than the field which would exist if the dielectric were not present. In the simple case of Figure 1 the difference of potential between the opposite surfaces of the dielectric is thus less than it would be in the absence of the dielectric, and from the definition of capacitance, $C = Q/V$, the introduction of the dielectric results in an increase of capacitance between the two plates.

The ratio of the capacitance with the dielectric to the capacitance if the space is empty is a property characteristic of the material and is called the relative dielectric constant, to which we shall give the symbol κ. It is a dimensionless quantity whose value is unity for free space, and whose value is always greater than unity for a material substance. For constant charges on the plates the average electric field within the dielectric is less by the factor κ over what it would be without the dielectric present.

Because Coulomb's law is modified by the presence of a dielectric, the electric field of a charge, q, imbedded in a dielectric, becomes

$$E = \frac{q\mathbf{r}}{4\pi\kappa\epsilon_0 r^3} \tag{2-3}$$

so long as the position vector r does not extend beyond the limits of the dielectric. Since we have called ϵ_0 the absolute dielectric constant of free space, it is consistent to call the product $\kappa\epsilon_0$ the absolute dielectric constant of the dielectric substance.

3. Polarization Charges

We now examine in more detail the polarized dielectric illustrated in Figure 1. In spite of there being a shift of charges in each molecule, the dielectric as a whole is still electrically neutral. We may expect that in any small volume (which nevertheless contains many molecules) there will be on the average as much charge moving into the volume as there is moving out, provided the electric field and the dielectric properties are constant in space. At a surface, however, the properties of the medium change sharply. In a small volume at, say, the left-hand surface, the negative charges that shift to the left are not compensated for by positive charges moving to the right, because there are no charges in the empty space outside the dielectric. As a consequence, a layer of negative charge develops on the surface adjacent to the positively charged plate, and similarly a layer of positive charge develops on the surface adjacent to the negatively charged plate. The reduced electric field within the dielectric can be regarded as a consequence of the reduction of effective charge on the plates by these induced charges on the surface of the dielectric.

Should either the average electric field or the properties of the dielectric be nonuniform, it is possible for more charges to move into a given region within the dielectric than moves out, resulting in a local charge density different from zero. Such a charge density is called a polarization charge density, and is related to the dipole moment per unit volume, **P**. (The integral of such a polarization charge density over the whole volume of the dielectric must, of course, be zero for an uncharged block of material.)

Figure 2 represents such an element of volume of dimensions dx, dy, dz, within a dielectric. Although small compared to the total volume of the dielectric, it still contains many molecules, so that any conclusions we draw will be averages which do not reflect microscopic variations. We

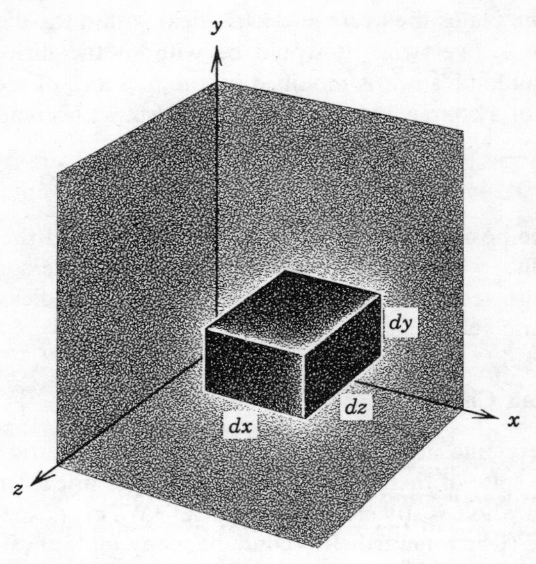

Figure 2. A volume element within a dielectric medium.

assume neither a uniform external electric field nor that the properties of the dielectric are constant from point to point.

Under the influence of the applied field, positive charges will, on the average, be displaced an amount $\mathbf{s}_p$ and negative charges an amount $\mathbf{s}_n$ from their equilibrium positions, where $\mathbf{s}_p$ and $\mathbf{s}_n$ are given by:

$$\begin{aligned} \mathbf{s}_p &= \mathbf{i}s_{px} + \mathbf{j}s_{py} + \mathbf{k}s_{pz} \\ \mathbf{s}_n &= \mathbf{i}s_{nx} + \mathbf{j}s_{ny} + \mathbf{k}s_{nz} \end{aligned} \tag{3-1}$$

We represent the density of *positive* charges within the medium as ρ_p and the density of negative charges as ρ_n, where, in an electrically neutral medium,

$$\rho_p = -\rho_n = \rho \tag{3-2}$$

ρ being the magnitude of charge density of either sign. $\mathbf{s}_p$, $\mathbf{s}_n$, ρ_p, ρ_n are, in general, functions of position.

We shall first treat only the x component of the charge displacement. At the left-hand face of the volume element of Figure 2, those positive charges which were, before the application of the field, within a parallelepiped of height s_{px} and base $dy\,dz$ at the left face will move into the volume upon application of the field.. Thus, for the x component of displacement, the amount of positive charge entering the volume at the left face is

$$\rho_p s_{px}\,dy\,dz \tag{3-3}$$

On the opposite face the amount of positive charge leaving the volume is:

$$\rho_p s_{px}\, dy\, dz + \frac{\partial}{\partial x}(\rho_p s_{px}\, dy\, dz)\, dx \tag{3-4}$$

where the second term is the change of the first term over the distance dx. It is immaterial whether this change arises from a nonuniform applied field or from nonuniformity in the dielectric. The difference between Equations 3-3 and 3-4,

$$\rho_p s_{px}\, dy\, dz - \left[\rho_p s_{px}\, dy\, dz + \frac{\partial}{\partial x}(\rho_p s_{px}\, dy\, dz)\, dx\right] = -\frac{\partial}{\partial x}(\rho_p s_{px})\, dx\, dy\, dz \tag{3-5}$$

is the net positive charge that has entered the volume $dx\, dy\, dz$ due to the x components of positive charge displacement.

In a similar way we find the net increase in the negative charge within the volume because of the x component of the displacement of negative charges to be

$$-\frac{\partial}{\partial x}(\rho_n s_{nx})\, dx\, dy\, dz \tag{3-6}$$

The total increase in charge resulting from the x displacements of both negative and positive charges is

$$-\frac{\partial}{\partial x}(\rho_p s_{px} + \rho_n s_{nx})\, dx\, dy\, dz \tag{3-7}$$

With Equation 3-2, 3-7 becomes

$$-\frac{\partial}{\partial x}[\rho(s_{px} - s_{nx})]\, dx\, dy\, dz \tag{3-8}$$

The components s_{px} and s_{nx} have opposite sense, and their difference is the x component of the total charge separation produced, on the average, in a molecule. If the charge of one sign in a molecule is q, then $q(s_{px} - s_{nx})$ is the average x component of the dipole moment induced in a molecule and $\rho(s_{px} - s_{nx})$ is the x component of the dipole moment per unit volume, P_x. Thus Equation 3-8 becomes

$$-\frac{\partial}{\partial x} P_x\, dx\, dy\, dz \tag{3-9}$$

The extension of this argument to the y and z components of charge displacement is evident. Combining the increases of charge for all three components of displacement gives us the total charge increase in the volume $dx\,dy\,dz$

$$-\left(\frac{\partial P_x}{\partial x} + \frac{\partial P_y}{\partial y} + \frac{\partial P_z}{\partial z}\right) dx\,dy\,dz \tag{3-10}$$

so that the charge density resulting from polarization of the medium is

$$\rho_P = -\left(\frac{\partial P_x}{\partial x} + \frac{\partial P_y}{\partial y} + \frac{\partial P_z}{\partial z}\right) \tag{3-11}$$

which is more concisely written (Section 13, Chapter 1):

$$\rho_P = -\nabla \cdot \mathbf{P} \tag{3-12}$$

Equation 3-12 shows that there is a polarization charge density wherever the divergence of P is not zero. Within a uniform medium in a uniform electric field there are no rates of change of the components of **P**, and hence although the medium is polarized, no volume charge density is developed. However, at the surface of such a medium there is a change in the polarization because there is a sharp change in the properties of the medium. Here $\nabla \cdot \mathbf{P}$ is not zero. These conclusions are consistent with the qualitative ideas developed at the beginning of this section.

4. Induced Surface Charge

Sometimes it is more convenient to think of the induced charge at the surface of a dielectric medium in terms of a charge per unit area of surface, rather than in terms of a volume charge density in the very thin boundary layer. Figure 3 shows a trace of the surface of a dielectric on which is

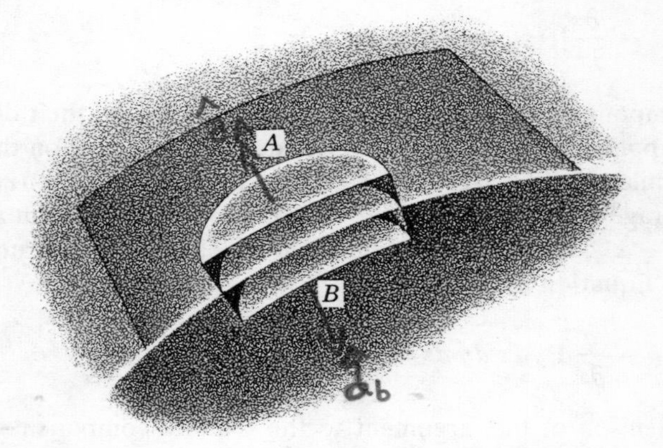

Figure 3. Section of a Gaussian volume including part of the surface of a dielectric.

indicated a Gaussian volume. The two surfaces, A and B, of this volume are parallel to the dielectric surface with A being just outside and B just inside the material surface, with the sides of the Gaussian volume connecting the surfaces A and B being of infinitesimal height.

From Equation 3-12 the total polarization charge within this Gaussian volume is $-\int \nabla \cdot \mathbf{P}\, d\mathscr{V}$, which by Gauss's theorem becomes $-\int \mathbf{P} \cdot d\mathscr{A}$, where the integration is to be carried out over the area enclosing the Gaussian volume. Inasmuch as the sides contribute negligibly to the integral, it consists of two parts, one over the surface A and one over the surface B. The integral over A is zero if the medium outside (a vacuum, say) is unpolarized, and we are left with $-\int_B \mathbf{P} \cdot d\mathscr{A}_B$ over the surface B for the total charge within the Gaussian surface. Write this integral as $-\int_B P_n'\, d\mathscr{A}_B$, where P_n' is the component of P along the normal to the surface B, which normal is directed outward from the Gaussian volume and hence into the dielectric. Evidently, then, $-P_n'$ can be interpreted as the polarization charge per unit area on the surface of the dielectric.

Custom requires, however, that, when we speak of the component of a vector normal to the surface of a medium, we shall refer the vector to the outward drawn normal. Let the component of $\mathbf{P}$ along the outward normal be P_n. Evidently $P_n = -P_n'$, and we have,

$$\sigma_P = P_n \tag{4-1}$$

where σ_P is the surface charge density induced on the dielectric.

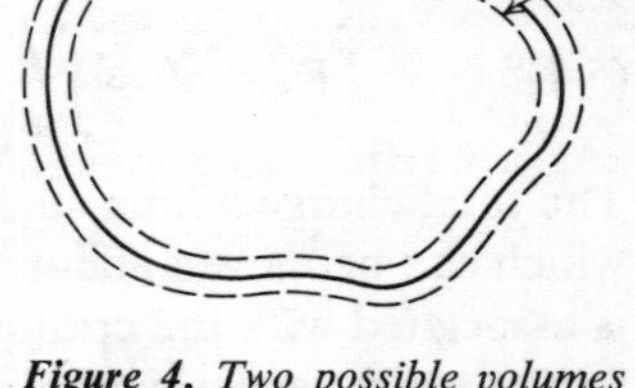

Figure 4. Two possible volumes for integrating the polarization charge density in a dielectric body.

Such a surface polarization charge is not to be regarded as different in any fundamental respect from the volume charge density of Equation 3-12. It is only a more convenient way of expressing the induced charge at the surface, where the rate of change of $\mathbf{P}$ with distance is large. There are available to us two ways of summing up the charge over the entire block of dielectric material. In Figure 4, for example, where the heavy line is intended to suggest a block of dielectric, we may integrate the divergence of $\mathbf{P}$ over a volume suggested by surface A, which is everywhere just outside the block of material. Then the total induced charge is

$$-\int_A \nabla \cdot \mathbf{P}\, d\mathscr{V} \tag{4-2}$$

Or we may integrate the volume charge density over the volume suggested by surface B, just inside the surface of the material, and add to it the integral of the surface charge density over the surface of the block. Thus,

$$-\int_B \nabla \cdot \mathbf{P}\, d\mathcal{V} + \int_B P_n\, d\mathcal{A} \tag{4-3}$$

Either method gives zero total charge for a neutral dielectric. However, for the reasons suggested, the second method may be easier to handle. The application of Gauss's theorem to either term of 4-3 clearly shows that the sum is zero, but the application of this theorem to Equation 4-2 shows the same result, inasmuch as the value of the polarization on the surface A is everywhere zero.

The extension of this reasoning to the determination of the surface charge at the interface between two polarized media is left to the reader.

5. The Displacement Vector

In its integral form Gauss's law (Section 10, Chapter 1) relates the surface integral of the electric field over a closed surface to the total charge within the surface:

$$\int \mathbf{E} \cdot d\mathcal{A} = \int \frac{\rho}{\epsilon_0}\, d\mathcal{V} \tag{5-1}$$

The total charge within such a surface can be of two types—free charge, which can be moved about independently, and polarization charge, which is associated with the creation or alignment of molecular dipoles. Henceforth we reserve the symbol ρ, without a subscript, for free-charge densities only, and use the symbol ρ_P for polarization charge densities. With this distinction, Equation 5-1 takes the form

$$\int \mathbf{E} \cdot d\mathcal{A} = \int \frac{\rho + \rho_P}{\epsilon_0}\, d\mathcal{V} = \int \frac{\rho}{\epsilon_0}\, d\mathcal{V} - \int \frac{\nabla \cdot \mathbf{P}}{\epsilon_0}\, d\mathcal{V} \tag{5-2}$$

By virtue of Gauss's theorem, the left side can be written as the integral of the divergence of $\mathbf{E}$ over the volume enclosed within the Gaussian surface. After multiplying by ϵ_0 and transferring the term involving $\nabla \cdot \mathbf{P}$ to the left, Equation 5-2 becomes

$$\int \nabla \cdot (\epsilon_0 \mathbf{E} + \mathbf{P})\, d\mathcal{V} = \int \rho\, d\mathcal{V} \tag{5-3}$$

Again, because this integral relation is true for any volume of integration, it follows that

$$\nabla \cdot (\epsilon_0 \mathbf{E} + \mathbf{P}) = \rho \tag{5-4}$$

The divergence of the vector sum $(\epsilon_0\mathbf{E} + \mathbf{P})$ is thus a function only of the free charge density and not of the polarization charge density, a conclusion that at first glance may seem strange. This vector sum, $(\epsilon_0\mathbf{E} + \mathbf{P})$ is called the displacement vector and is identified by the symbol **D**. Equation 5-4 is more concisely written as

$$\nabla \cdot \mathbf{D} = \rho \tag{5-5}$$

In certain cases Equation 5-5 has simple and rather interesting implications. For example, in the capacitor illustrated in Figure 1, the vector **D** is the same in the space between the plates whether the dielectric is present or not (for a given free charge on the plates). What is changed by the introduction of the dielectric is the electric field strength and the polarization. Similarly, **D** in the region around an isolated point charge is not affected by the nature of the medium surrounding the charge. Even if the point charge is located at the center of a dielectric sphere of radius R, the value of **D** inside or outside R is independent of the presence of the dielectric. Both of these examples are, however, somewhat special cases in which the vector **D** is normal to the boundary between the dielectric and free space. There will later be occasion to show that in less symmetrical situations where **D** is not normal to a boundary, the direction of **D** undergoes a kind of refraction as it passes from one medium to another. It is the divergence of **D**, not always **D** itself, that is independent of the medium.

The significance of the displacement vector will be better understood when we discuss situations in which electric and magnetic fields change with time. From Equation 5-5 we see that the dimensions of **D** are charge per unit area, from which it follows that the time rate of change of **D** will have the dimensions of current per unit area. Such a current is called a displacement current to distinguish it from the more familiar conduction current in which actual charges are moving. It is possible to have a current over an area where there is no motion of free charges through that area. Maxwell pointed out that any current, whether displacement or conduction, gives rise to a magnetic field, and this fact, together with the fact that a changing magnetic field gives rise to an electric field, made possible the prediction of electromagnetic waves consisting of mutually supporting electric and magnetic fields, even in regions where there are no free charges.

6. The Dielectric Constant

The relation of the displacement to the electric field depends on the relation of the electric field to the polarization. For this latter relation no general statement can be made, although with many substances a simple and very good approximation is available.

The polarization depends on the magnitude and direction of the charge separations in the individual molecules. As long as the charge separation is small, the forces opposing separation very closely obey Hooke's law—the restoring force is proportional to the separation—and the induced dipole moment in the molecule is proportional to the average electric field within the dielectric. Consequently the polarization **P** from induced dipoles is also proportional to the electric field. For many materials the polarization from the alignment of permanent molecular dipoles is similarly proportional to the electric field. We write this relation

(6-1) $$\mathbf{P} = \chi\epsilon_0\mathbf{E}$$

where χ is a dimensionless constant, the electric susceptibility. The displacement **D** then becomes

(6-2) $$\mathbf{D} = \epsilon_0\mathbf{E} + \mathbf{P} = \epsilon_0\mathbf{E} + \chi\epsilon_0\mathbf{E} = \epsilon_0\mathbf{E}(1 + \chi) = \kappa\epsilon_0\mathbf{E}$$

where

(6-3) $$\kappa = 1 + \chi$$

is called the relative dielectric constant. It must be emphasized that Equation 6-2 is a very good relation for many materials but does not hold for all. It assumes that the polarization vector is directed parallel to the electric field, and this is not the case for some materials, particularly certain crystals. In such materials it is found that each rectangular component of **D** becomes a function of all three components of **E**:

(6-4)
$$D_x = \epsilon_0(\kappa_{xx}E_x + \kappa_{xy}E_y + \kappa_{xz}E_z)$$
$$D_y = \epsilon_0(\kappa_{yx}E_x + \kappa_{yy}E_y + \kappa_{yz}E_z)$$
$$D_z = \epsilon_0(\kappa_{zx}E_x + \kappa_{zy}E_y + \kappa_{zz}E_z)$$

where the κ_{ij} are dimensionless numbers that are components of a tensor relating **D** and **E**. In Chapter 11 such materials are shown to have the property of birefringence or double refraction.

There may also be departures from the simple linearity of Equation 6-2 for materials whose molecules show a permanent dipole moment when the temperature is very low and the forcing field is very high, so that the alignment of the molecules may approach saturation. But Equation 6-2 holds very closely for many common materials, and its simplicity makes it attractive in theoretical analysis. It is not, however, a fundamental law of electromagnetism.

7. E and D at a Boundary

Since no material medium is of infinite extent, it is desirable to formulate the conditions governing the changes in **E** and the changes in **D** at a

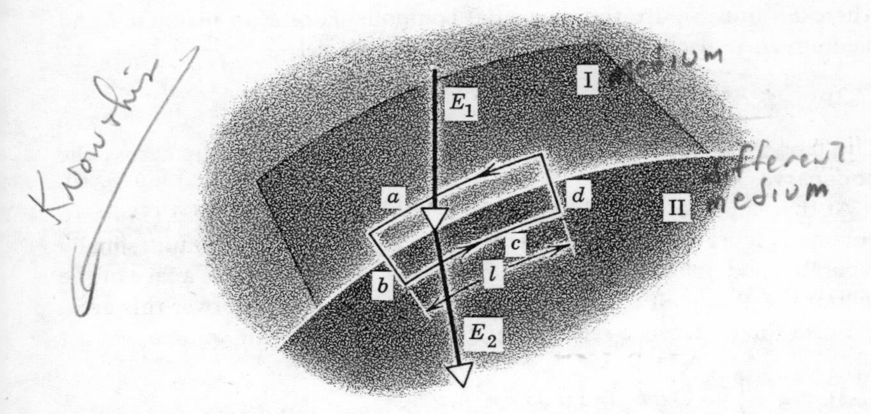

Figure 5. A closed path drawn at an interface between two dielectric media.

boundary between two media. We shall treat only a sharply defined boundary, but the basic laws apply to any type of transition from one medium to another. We found that the line integral of the electric field around a closed path of integration always vanishes for electrostatic fields, and that the divergence of the displacement is equal to the free charge density. These are general relations which must hold in a region of transition between media as well as within a medium. We shall show that the vanishing of the line integral of the electric field requires that there cannot be an abrupt change in that component of the electric field tangential to a boundary; and the divergence relation for the displacement requires that there cannot be an abrupt change in the normal component of the displacement at a boundary unless the boundary carries free charge.

In Figure 5 is shown such a boundary between medium *I* and medium *II*, on which has been described a closed path around which we shall integrate the electric field. This path consists of four segments, (*a*) a segment of length l parallel to the surface and short enough so that the electric field is effectively constant along its length; (*b*) a segment of negligible length, penetrating the surface; (*c*) a segment of length l parallel to the first segment, but in medium *II*; and (*d*) a segment of negligible length, again penetrating the boundary and closing the curve. The line integral of the electric field around this closed path is zero for static fields (we shall find later that for time-varying fields it is also zero). Using only the contributions of paths (*a*) and (*c*) in the line integral, since those of (*b*) and (*d*) are negligible, we get

$$E_{t1}l - E_{t2}l = 0 \tag{7-1}$$

where E_{t1} and E_{t2} are the tangential components of **E** in medium *I* and medium *II*, respectively. From Equation 7-1 we get

$$E_{t1} = E_{t2} \tag{7-2}$$

The tangential component of the electric field is continuous across the boundary.

At this same point on the boundary, let there be described a Gaussian volume (Figure 6) of pillbox shape, with its two surfaces infinitesimally separated and on either side of the boundary. Let the area of a face of the pillbox, $\mathscr{A}$, be small enough that **D** is effectively constant over this area.

From the divergence relation applied to this Gaussian volume, we get

$$\int \nabla \cdot \mathbf{D}\, d\mathscr{V} = \int \rho\, d\mathscr{V} = \sigma\mathscr{A} \tag{7-3}$$

where σ is the surface density of free charge on the boundary. By Gauss's theorem the left side of Equation 7-3 becomes an integral over the area bounding the Gaussian volume. The contributions to the surface integral of the edges of the pillbox volume are negligible, so that Equation 7-3 becomes

$$-D_{n1}\mathscr{A} + D_{n2}\mathscr{A} = \sigma\mathscr{A} \quad \text{or} \quad D_{n2} - D_{n1} = \sigma \tag{7-4}$$

where D_{n1} and D_{n2} are the normal components of the displacement in medium *I* and in medium *II*, respectively. Thus, only when free charge exists on the boundary does the normal component of the displacement undergo an abrupt change. At an uncharged surface the normal component is continuous.

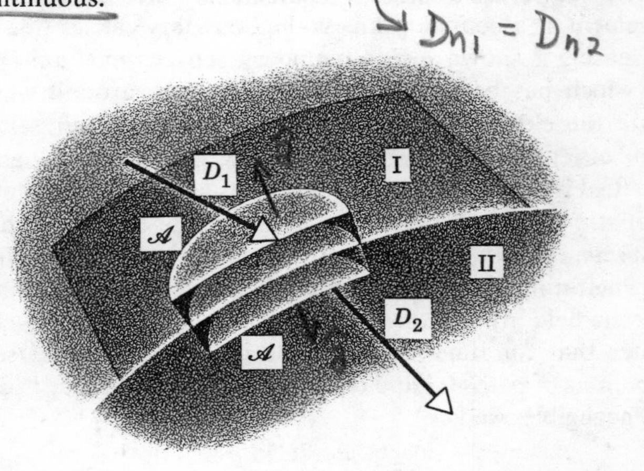

Figure 6. *A Gaussian volume at the interface between two dielectric media.*

For two media in each of which **D** and **E** are parallel to each other and for an uncharged interface, the directions of these vectors relative to the boundary are readily obtained. In medium *I*, let $D_1 = \kappa_1\epsilon_0 E_1$ and in medium *II* let $D_2 = \kappa_2\epsilon_0 E_2$. From Equation 7-4

$$\kappa_2\epsilon_0 E_{n2} - \kappa_1\epsilon_0 E_{n1} = 0 \tag{7-5}$$

But if $\mathbf{E}_1$ makes an angle θ_1 with the normal to the surface and $\mathbf{E}_2$ an angle θ_2 with the normal, then

$$\begin{aligned} E_{t1} &= E_1 \sin\theta_1 \qquad & E_{n1} &= E_1 \cos\theta_1 \\ E_{t2} &= E_2 \sin\theta_2 \qquad & E_{n2} &= E_2 \cos\theta_2 \end{aligned} \tag{7-6}$$

Combining Equations 7-2, 7-5, 7-6, we get

$$\frac{\tan\theta_2}{\tan\theta_1} = \frac{\kappa_2}{\kappa_1} \tag{7-7}$$

which is a refraction law for electric fields. An electric field directed from a medium of lower toward a medium of higher dielectric constant has its direction changed away from the normal to the interface.

8. Actual Electric Field at a Molecule

The field **E** that appears in the defining relation for the displacement **D** is a smoothed out or average field within the medium and is not at a particular molecule the same as the actual field that induces the dipole moment in that molecule. In certain simple configurations it is possible and instructive to compute the actual field and, by doing so, relate certain molecular properties to gross measurements made on a dielectric.

We use a dielectric in a parallel plate capacitor, such as in Figure 1. By assuming a uniform, isotropic dielectric with sides parallel to the capacitor plates, we are assured of uniform polarization. In this dielectric we imagine a sphere drawn with center at the molecule where we wish to find the electric field strength. The net electric field at this molecule consists of four parts:

(*a*) The field due to the free charges on the capacitor plates, as if the dielectric were not present.

(*b*) The field due to the surface charges developed on the outer surfaces of the dielectric.

(*c*) The field associated with the surface charges that would exist, if the medium within the sphere were removed without affecting the polarization of the remaining material.

(*d*) The field due to the molecular dipoles of the medium within this sphere.

Before performing these computations in detail in the next section, let us discuss them qualitatively, with a special comment on the assumption involved.

The computations of the fields in (*a*) and (*b*) are straight forward, following the procedure used for obtaining the parallel plate capacitor field in Chapter 2.

In (*c*) we imagine the dielectric within the sphere about the molecule removed (there are no special restrictions on the size of this sphere, other than that it must not extend beyond the boundary of the dielectric) without changing the polarization of the remaining medium. At the surface of the resulting spherical cavity there will be uncompensated ends of molecular dipoles and, consequently, a surface charge density, positive over one hemisphere and negative over the other. We find the integrated electric field at the center of the cavity associated with these polarization charges.

In (*d*) we sum the fields produced at the center of the sphere by all the molecular dipoles contained in that part of the medium within the sphere except for the dipole of the molecule at the center. The expressions for the electric field of one dipole was obtained in Chapter 2, the form to be used here being that for distances from the dipole which are large compared to the dimension of the dipole itself.

The use of the approximate equations for the dipole field may at first glance seem to have doubtful justification, inasmuch as the distances of at least some of the dipoles contributing to the field at the center will be of the order of molecular dimensions. It is therefore worth interrupting our discussion to examine this assumption in some detail.

By making some simple assumptions about a molecule it is possible to

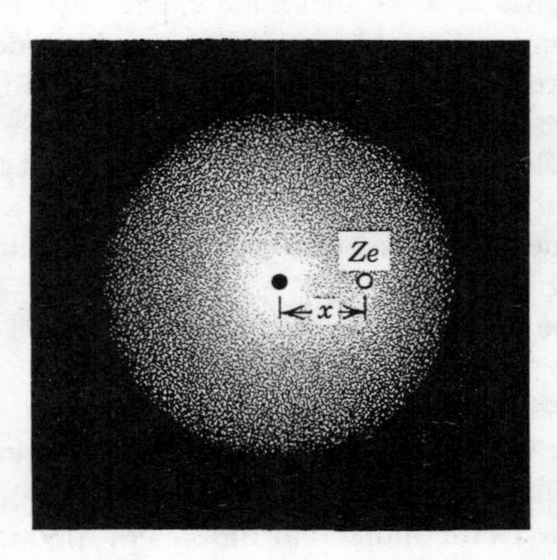

Figure 7. *A polarized molecule with electron cloud shifted relative to the nucleus.*

arrive at a rough estimate of the charge separation produced when the molecule is in a forcing field. Let us assume a monatomic molecule that is spherically symmetrical, that has a radius R, that has a positive point charge Ze at the center (Z is the atomic number, e the electronic charge) with a negative charge cloud of uniform negative charge density surrounding this central charge. We shall further make the approximation that the forcing field shifts the negative charge cloud with respect to the nucleus without changing its spherical shape. The nucleus is a distance x from the center of the negative charges. If the forces between the nucleus and the cloud are entirely electrical, the shift x will be such that the field at the nucleus due to the negative cloud will equal, and be oppositely directed to, the forcing field.

Gauss's law applied to a spherically symmetrical configuration, such as this negative cloud, shows that the field a distance x from the center is the same as if the charge within a sphere of radius x were concentrated at the center. The charge within this sphere is proportional to the volume, $(\frac{4}{3})\pi x^3$, and is given by

$$\frac{Ze(\frac{4}{3}\pi x^3)}{\frac{4}{3}\pi R^3} = Ze\left(\frac{x^3}{R^3}\right) \tag{8-1}$$

If E_L is the local forcing field, then

$$E_L = \frac{Ze(x/R)^3}{4\pi\epsilon_0 x^2} = \frac{Zex}{4\pi\epsilon_0 R^3} \tag{8-2}$$

giving for the charge separation,

$$x = \frac{4\pi\epsilon_0 R^3 E_L}{Ze} \tag{8-3}$$

Atomic radii are of the order of one angstrom or 10^{-10} meter. Using a reasonably high but not outlandish forcing field of 10^{+6} volts/meter, we get for $Z = 16$ (sulfur), $x \cong 3 \times 10^{-16}$ meter or 3×10^{-6} angstrom unit and for $Z = 53$ (iodine), $x \cong 1 \times 10^{-6}$ angstrom unit.

The approximations used here certainly introduce gross simplifications. It is, for example, doubtful that the entire electron cloud of a molecule where Z is not small shifts as a whole. More likely, only the more loosely bound outer electrons shift to produce the dipole moment. Conceivably, the further approximation of a spherically symmetrical charge continuum in place of the actual electron configuration introduces a less serious error. Even allowing, however, for an error in the computed dipole dimension of as much as several orders of magnitude, the results show that molecular separations are likely to be quite large compared to the dimensions of induced molecular dipoles and that thus the approximate rather than the exact equations for the dipole field may safely be used in computing the influence of a molecular dipole on even a very close neighbor.

9. Finding the Local Field at a Molecule

We now proceed with the details of finding the local field at a molecule. Each of the four parts of this field will be expressed in terms of the polarization, **P**, and the average electric field, **E**, within the dielectric.

(*a*) The field due to just the free charge density, σ, on the capacitor plates is σ/ϵ_0, which is D/ϵ_0, from the results of Problem 2 at the end of this chapter. Using the definition of **D** in terms of **E** and **P**, we get

$$\mathbf{E}_a = \mathbf{E} + \frac{\mathbf{P}}{\epsilon_0} \tag{9-1}$$

(*b*) The surface polarization charge density on the dielectric is the normal component of the polarization at the surface which, in this case, is **P** itself. The field due to these polarization charges is opposite in sense to the field due to the free charges.

$$\mathbf{E}_b = -\frac{\mathbf{P}}{\epsilon_0} \tag{9-2}$$

(*c*) Next, consider a spherical cavity of radius *a* about the field point (Figure 8) obtained by removing the dielectric material within the sphere without affecting the polarization, still *P*, at the surface. There is thus a polarization charge density on the surface of the cavity of $-P \cos \varphi$. The

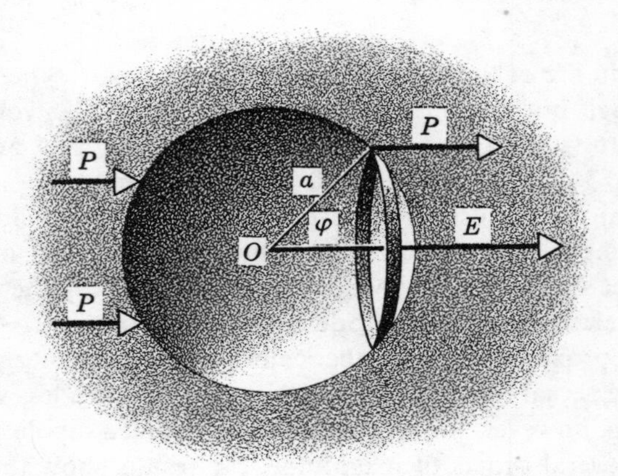

Figure 8. Electric field at O due to induced charges on surface of cavity.

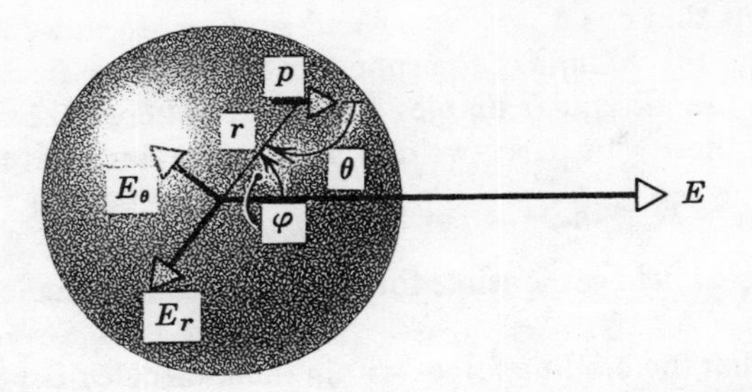

Figure 9. Electric field at O due to aligned dipoles in sphere of radius a.

charge on an annular ring of width $a d\varphi$ at φ produces a field at 0 of value

$$dE_c = \frac{2\pi a^2 P \cos^2 \varphi \sin \varphi \, d\varphi}{4\pi\epsilon_0 a^2} \tag{9-3}$$

Integrating Equation 9-3 over the sphere, we get the total contribution of the polarization charges on the cavity:

$$\mathbf{E}_c = \frac{\mathbf{P}}{3\epsilon_0} \tag{9-4}$$

(*d*) Lastly we find the field produced by the dipoles of the dielectric within the sphere. For a molecular dipole of moment p located at (r, φ), the field components in the r and θ directions (see Figure 9) were found in Section 8 of Chapter 2:

$$E_r = \frac{2p \cos \theta}{4\pi\epsilon_0 r^3} \tag{9-5}$$

$$E_\theta = \frac{p \sin \theta}{4\pi\epsilon_0 r^3}$$

The components in the x-direction (parallel to E) and in the y-direction (normal to E) are

$$\begin{aligned} E_{px} &= -E_r \cos \varphi - E_\theta \sin \varphi \\ E_{py} &= -E_r \sin \varphi + E_\theta \cos \varphi \end{aligned} \tag{9-6}$$

The dipoles in a volume $d\mathscr{V}$ have a dipole moment $P\, d\mathscr{V}$, which produces x and y components of the electric field at 0 of

$$dE_x = \frac{P}{4\pi\epsilon_0 r^3}(-2 \cos \theta \cos \varphi - \sin \theta \sin \varphi)\, d\mathscr{V} \tag{9-7}$$

$$dE_y = \frac{P}{4\pi\epsilon_0 r^3}(-2 \cos \theta \sin \varphi + \sin \theta \cos \varphi)\, d\mathscr{V}$$

Using the facts that $\cos\theta = -\cos\varphi$ and $\sin\theta = \sin\varphi$, we integrate each expression over the volume of the sphere to find that the electric field at the center due to the molecular dipoles within the sphere is zero.

Combining these four parts, we find that the total field at the center 0 is

$$\mathbf{E}_{\text{eff}} = \mathbf{E}_a + \mathbf{E}_b + \mathbf{E}_c = \mathbf{E} + \mathbf{P}/\epsilon_0 - \mathbf{P}/\epsilon_0 + \mathbf{P}/3\epsilon_0 = \mathbf{E} + \mathbf{P}/3\epsilon_0 \tag{9-8}$$

which is the local field responsible for polarizing the molecule at the center of the sphere.

Providing that the binding forces within the molecule obey Hooke's law, the induced dipole moment $\mathbf{p}$ is proportional to this effective field:

$$\mathbf{p} = \alpha\epsilon_0\mathbf{E}_{\text{eff}} \tag{9-9}$$

where α is the polarizability of the molecule. Then

$$\mathbf{P} = n\mathbf{p} = n\alpha\epsilon_0\mathbf{E}_{\text{eff}} \tag{9-10}$$

where n is the number of molecules per unit volume. Combining Equation 9-10 with

$$\mathbf{P} = (\kappa - 1)\epsilon_0\mathbf{E} \tag{9-11}$$

obtained from Equations 6-1 and 6-3, we get

$$\kappa = \frac{(1 + 2S)}{1 - S} \tag{9-12}$$

where $S = n\alpha/3$, relating the relative dielectric constant of the substance to the polarizability of a molecule. For gases under ordinary pressures, the dielectric constant differs from unity by only a small amount, in which case $S \ll 1$, and Equation 9-12 becomes

$$\kappa = (1 + 2S)(1 - S)^{-1} = (1 + 2S)(1 + S + \cdots) \tag{9-13}$$

where the binomial theorem has been used, with terms in S^2 and higher being dropped. Therefore κ differs from unity for such gases by

$$\kappa - 1 \approx 3S \approx n\alpha \tag{9-14}$$

10. Molecules with Permanent Dipole Moments

The previous theory predicts that all molecules in an electric field will develop an induced dipole moment which contributes to the polarization of the medium. An additional contribution is made by those molecules with permanent dipole moments because of the tendency of the permanent dipoles to align with the field.

A field $\mathbf{E}$ exerts a torque on a dipole $\mathbf{p}$ (Problem 12, Chapter 1) of magnitude $pE\sin\theta$, where θ is the angle between the direction of the dipole and the direction of the electric field. Such a dipole has a potential energy of $W = -pE\cos\theta$, if zero potential energy is assigned to an orientation perpendicular to the field. (Why is this a convenient zero?)

It is shown in statistical mechanics that under conditions of thermal equilibrium the number of molecules per unit volume whose dipole moments are oriented to make an angle with the electric field between the values θ and $\theta + d\theta$ is

$$dn = Me^{-W/kT}\, dW \tag{10-1}$$

where M is a constant of proportionality, k is Bolzmann's constant, and T is the absolute temperature.

Since $dW = pE\sin\theta\, d\theta$, and since each such molecule contributes a component of dipole moment in the direction of **E** of $p\cos\theta$, the total dipole moment per unit volume of the molecules so oriented is

$$dP = Mp^2Ee^{x\cos\theta}\cos\theta\sin\theta\, d\theta \qquad \text{where } x = pE/kT \tag{10-2}$$

The net dipole moment per unit volume for all orientations is the integral over θ of Equation 10-2:

$$\begin{aligned} P &= Mp^2E\int_1^{-1} e^{x\cos\theta}\cos\theta\, d(-\cos\theta) \\ &= (Mp^2E/x^2)[e^x(x-1) + e^{-x}(x+1)] \end{aligned} \tag{10-3}$$

The integral of Equation 10-1 over θ gives the total number of molecules per unit volume,

$$\begin{aligned} n &= MpE\int_1^{-1} e^{x\cos\theta} d(-\cos\theta) \\ &= (MpE/x)(e^x - e^{-x}) \end{aligned} \tag{10-4}$$

from which the constant M is found to be

$$M = \frac{nx}{pE(e^x - e^{-x})} \tag{10-5}$$

Equations 10-4 and 10-5 together give the polarization produced by the permanent molecular dipoles:

$$P = np\left(\coth x - \frac{1}{x}\right) \qquad \text{where } \coth x = \frac{e^x + e^{-x}}{e^x - e^{-x}} \tag{10-6}$$

For ordinary temperatures, where T is large, x is small and Equation 10-6 reduces very closely to

$$P \approx \frac{npx}{3} \approx \frac{np^2E}{3kT} \tag{10-7}$$

Equation 10-7 is that part of the polarization due to permanent dipoles and is combined with the polarization resulting from the induced dipole moments to get the total polarization. For a gas consisting of molecules with permanent dipole moments, Equation 9-14 thus becomes

$$\kappa - 1 = n\left(\alpha + \frac{p^2}{3\epsilon_0 kT}\right) \tag{10-8}$$

known as the Langevin-Debye formula. It predicts that the dielectric constant of such a gas is a function of the temperature. κ plotted against the quantity $1/T$ is a straight line with a slope of $np^2/3\epsilon_0 k$ and an intercept of $n\alpha + 1$. From the slope and the intercept of this graph the polarizability and the permanent dipole moment of the molecule can be obtained.

In an oscillatory forcing electric field, the contributions of the induced dipole moments and the permanent dipole moments to the dielectric constant are different functions of the frequency. In fact, the response of the permanent dipoles to orientation by an alternating field decreases rapidly as the frequency increases, thus permitting a further separation of the contributions of these two kinds of dipoles.

11. Conclusion

In this chapter is the first place where we have put emphasis on the molecular constitution of matter in the development of electromagnetic theory. The electrical behavior of dielectrics is explained on the assumption that molecular dipole moments are induced or pre-existing dipole moments are aligned by an electric field to produce the macroscopic polarization which is observed. This polarization is described in terms of a vector, **P**, which is a function of the applied electric field. Out of a linear combination of the electric field and the polarization, **P**, a new vector, **D**, is formed whose divergence depends only on the free charges present and not on the nature of the medium. In many materials, **D** is very closely proportional to *E*.

The electric field at a molecule in a polarized dielectric is not the same as the averaged field, **E**, used in the macroscopic theory. Rather we found that it must be obtained in terms of the influence of the surrounding molecular dipoles, in addition to the field imposed by the free charges. From this theory it is possible to infer the polarizability of molecules from macroscopic measurements on dielectric media.

In Chapter 7 we shall find that in many respects the analysis of the magnetic properties of a medium follows a similar pattern to that of the dielectric properties in this chapter. In reading Chapter 7 the student should watch for this parallelism; its recognition will help his understanding.

Discussion Questions

1. In the rough calculation of the molecular dipole moment of Section 8, suppose that, rather than the whole electron cloud shifting in one direction and the positive nucleus shifting in the other, actually the nucleus plus all electrons but one formed the positive element of the dipole and the remaining electron the negative element. What difference would this make in the calculation Would the conclusions from the calculations be significantly altered?

2. Suppose that intermolecular forces were such in a dielectric that the *induced* molecular dipole moment was formed at an angle θ to the electric field instead of parallel to it. How would this affect the arguments of Sections 2 through 8?

3. Two parallel plates are of infinite extent and uniformly charged with surface charge density σ and are separated by d meters. A disk of dielectric material of radius a and thickness d is located between the plates. How does the presence of the dielectric affect the electric field between the plates at distances greater than a from the center of the dielectric disk? Is it still true that $\mathbf{D}$ is independent of the presence of the dielectric? What about $\nabla \cdot \mathbf{D} = \rho$ (where ρ is only the free charge density)?

4. Taking the charge separation in a spherical polarized atom as x and the dipole moment as Zex, relate this dipole moment to the polarization vector, $\mathbf{P}$, and then to the average field in the dielectric, $\mathbf{E}$. What value of average field would need to be applied before the dipole size becomes too large for the use of the approximate dipole field equations? The relative dielectric constant for ordinary solids ranges from between 1 and 2 up to about 10. (The student is asked to justify a judgment, not to find a single right answer.)

Problems

1. Find the surface polarization charge density at an interface between two dielectrics. The values of the polarization vectors at this interface are $\mathbf{P}_1$ and $\mathbf{P}_2$. Establish your own convention as to the positive sense of the areas involved and test your answer against your intuition as to the sign of the surface charge to be expected.

2. Show that the magnitude of the displacement, $\mathbf{D}$, at the surface of a conductor is σ, where σ is the surface charge density on the conductor.

3. A coaxial cable has an inner conductor with circular cross section of radius a. The outer conductor is a circular cylinder of inner radius b. An insulating material of relative dielectric constant κ separates the two. What is the capacitance per unit length of this cable?

4. A point charge q is surrounded by a spherical shell of dielectric of inner radius a and outer radius b. Within a is empty space (except for q) and beyond b is empty space. Determine (1) the polarization within the dielectric, (2) the polarization charge density within the dielectric, (3) the surface polarization charge density on each surface of the dielectric, (4) the value of the electric field within and without the dielectric, and (5) the value of the displacement vector.

5. It is readily shown and is intuitively evident that a dipole experiences no net force (although it may experience a torque) in a uniform electric field. What force, however, will a dipole experience when the field is nonuniform?

6. A parallel plate capacitor carries a charge Q, the separation of the plates being d. A slab of dielectric of thickness t, where t is less than d, is between the plates, its surfaces being parallel to the plates. Obtain the potential difference between the plates, the field in the dielectric, and the field in the free space.

7. A parallel plate capacitor carries a charge Q, the separation of the plates being d. The space between the plates is occupied by two dielectrics of relative dielectric constants κ_1 and κ_2, the first having a thickness t and the second a thickness $d - t$. What is the difference of potential between the plates, the field in each dielectric, the surface charge density at the interface, and the capacitance?

8. Poisson's equation is $\nabla^2 V = -\rho/\epsilon_0$, where ρ is the total charge density both free and polarization. Thus, within a dielectric where the polarization charge density is $-\nabla \cdot \mathbf{P}$ and there are no free charges, this equation becomes:

$$\nabla^2 V = -\nabla \cdot \mathbf{P}/\epsilon_0$$

Show that, in a uniform, isotropic dielectric containing no free charges, the potential actually obeys Laplace's equation: $\nabla^2 V = 0$.

9. Find the potential and the field around and within an uncharged dielectric sphere of radius a located in an external field which would be uniform in the absence of the dielectric. Take the potential at the center of the sphere as zero. What boundary conditions at the surface of the sphere must hold? (Use an adaptation of the method followed in Section 14 of Chapter 2.)

10. What is the total induced dipole moment in the dielectric sphere of Problem 9? If it were a conducting sphere, what would be its total induced dipole moment?

11. An electric dipole of moment p is located inside a uniform, isotropic dielectric. At points far from the dipole, what will be the polarization charge density developed in the dielectric?

12. Suppose small metal spheres are dispersed throughout a region, their number being such that they occupy a small fraction F of the total volume of the region. If the space between the spheres has a relative dielectric constant of unity, what will be the relative dielectric constant of this assembly of spheres?

13. If the spheres of Problem 12 were of a dielectric material such as a plastic with a relative dielectric constant κ, what would be the over-all dielectric constant of the assemblage?

REFERENCES

Most of the references listed at the end of Chapter 2 will also make good collateral reading for this chapter. To them may be added:

1. Whitmer, *Electromagnetics*, 2nd ed., Prentice-Hall, 1962. Chapter **2.**

2. Reitz and Milford, *Foundations of Electromagnetic Theory*, Addison-Wesley, 1960. Chapters 4 and 5.

3. Harnwell, *Principles of Electricity and Electromagnetism*, McGraw-Hill, 1938. Chapters II and III.

4. Panofsky and Phillips, *Classical Electricity and Magnetism*, Addison-Wesley, 1955. Chapter 2.

4

Direct currents

In discussing the electrostatic behavior of matter, we made the assumption that all electrons were bound to atoms by quasi-elastic forces. This assumption is a good one in insulators, but is not justified in conductors where many electrons are free to move about without unique associations with particular atoms. In an electric field, such free electrons migrate against the field to produce surface concentrations of charge, the migration continuing until the field due to the surface charges is equal and opposite to the applied field. Under static conditions, the inside of a metal must always be a field-free region.

If the electrons which reach the surface are removed, and if those which move away from the opposite surface are replaced, there results a continuous movement of electrons, i.e., a current, through the conductor. We shall start by examining briefly and contrasting the classical and quantum mechanical theories of electrical conduction. Whereas the classical theory is simple and can be treated in some detail, unfortunately neither space nor the mathematical background expected of the reader allows other than a qualitative discussion of the quantum mechanical theory. Nevertheless, the reader should be able to appreciate some of the scope and beauty of the quantum mechanical analysis.

1. Classical Derivation of Ohm's Law

We imagine that the free electrons within a conductor form a kind of electron gas existing in the intermolecular spaces. Within a long cylindrical conductor of uniform cross-sectional area, $\mathscr{A}$, an electric field, E, constant in time, is established parallel to the axis (the conductor must be part of a complete circuit, so that charges are removed at one end and replaced at the other). There is then a force qE on the mobile carriers of charge q and mass m. By Newton's second law of motion the acceleration of these carriers is qE/m, provided they are subject to no other forces. Under the influence of this force the charges begin to move, but they soon collide with atoms and are slowed down or stopped. Our information is inadequate to analyze the motion of the carriers in detail, but we imagine a kind of averaged motion over many such stops and starts and describe it in terms of a pseudo-frictional force in addition to the force of the electric field. Based on macroscopic experience with frictional phenomena, we try an assumption that this force is proportional, and oppositely directed, to the average or smoothed out velocity. Including both forces on a charge carrier, Newton's second law takes the form

$$m \frac{dv}{dt} = qE - kv \tag{1-1}$$

where v is the velocity of the charge q parallel to the axis of the conductor and k is the "frictional" coefficient.

Figure 1 illustrates schematically the conductor in which the charges q at any instant are all moving with the averaged speed v. In a time dt, each charge moves a distance vdt, so that with n mobile charges per unit volume there are $nq\mathscr{A}v\,dt$ coulombs in that volume whose length is $v\,dt$ and cross section is $\mathscr{A}$. All of these charges will cross the plane $a \cdots a$, in a time dt so that the charge per unit time crossing the plane is $nq\mathscr{A}v$. This is defined as the current i in the conductor:

$$i = nq\mathscr{A}v \tag{1-2}$$

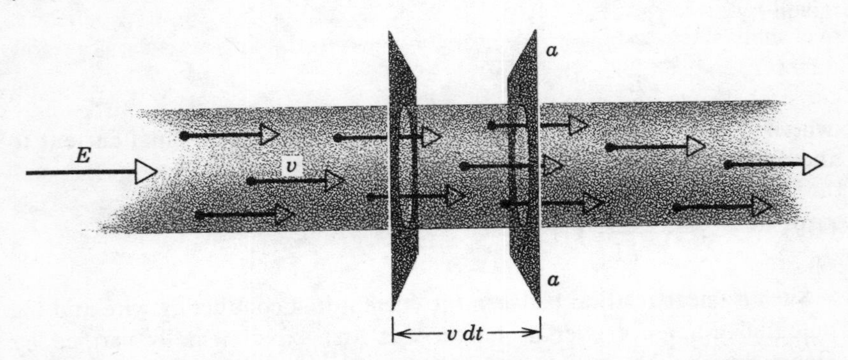

Figure 1. *Schematic representation of a conducting wire showing average motion of charge carriers in applied field E.*

Thus Equation 1-1 becomes

$$\frac{di}{dt} = \frac{nq^2\mathscr{A}E}{m} - \frac{ki}{m} \tag{1-3}$$

which can be rearranged as

$$\frac{di}{i - M} = -\frac{k\,dt}{m} \tag{1-4}$$

where $M = nq^2\mathscr{A}E/k$. Integrating both sides of Equation 1-4 gives:

$$\ln(i - M) = -\frac{k}{m}t + \ln C \tag{1-5}$$

where $\ln C$ is a constant of integration, which is evaluated by counting time from the instant the field is turned on, at which time, $i = 0$, so that $C = -M$, giving finally:

$$i = \frac{nq^2\mathscr{A}E}{k}(1 - e^{-kt/m}) \tag{1-6}$$

The current starts from zero and rises exponentially, approaching the final value

$$I = \frac{nq^2\mathscr{A}E}{k} \tag{1-7}$$

Experimentally, it is found that the current reaches its final value almost instantaneously, suggesting that k/m, the ratio of the "friction" to the mass of the charge carriers, is relatively large. The electric field, E, is, in terms of the potential difference, V, between the ends of the conductor, given by

$$E = \frac{V}{L} \tag{1-8}$$

where L is the length of the conductor, thus relating the final current to the potential difference by

$$I = \left(\frac{nq^2\mathscr{A}}{kL}\right)V \tag{1-9}$$

Such a linear relation between the current in a conducting wire and the potential difference over its length was first experimentally verified by Georg Simon Ohm in Germany in 1827. The quantity in the parentheses involves constants which are characteristic of the conductor and is called the conductance, usually symbolized by G. The reciprocal of G is the resistance, R,

$$R = \frac{1}{G} = \left(\frac{k}{nq^2}\right)\left(\frac{L}{\mathscr{A}}\right) \tag{1-10}$$

which is, in a sense, a measure of the opposition to the flow of charges in the conductor; it is directly proportional to the length, inversely proportional to the cross sectional area. The proportionality constant (k/nq^2), the resistivity, is a property of the material of the conductor. It is hardly surprising that the resistivity should be inversely proportional to the number of the free charges within the conductor and it is interesting that it is not dependent on the sign of those charges.

The neatness with which this theory fits the observed facts of metallic conductors makes it very appealing. It correctly relates electrical resistance to the cross section and length of a conductor; it properly relates the resistance to the number of free charge carriers; and, at least qualitatively, it seems to give the appropriate dependence of resistance on temperature, inasmuch as we should expect the "frictional" effect to be greater with greater thermal vibration of the atoms within the metal. But results contrary to this theory are readily found in the behavior of many materials which show resistances decreasing instead of increasing with temperature. Carbon is a familiar material of this type but there are many others. As a class such substances are called semiconductors, since usually their resistivities, though much less than insulators, are several orders of magnitude greater than metallic conductors. It is true that the theory might be modified by postulating that the number of charge carriers is a function of the temperature in such a way that the resistance decreases with temperature, but at best this would be arbitrary, since the theory provides no suggestion why it should be so.

It is disappointing that an attractively simple theory such as the foregoing should not be more satisfactory. The simple elegance with which it predicts Ohm's law leads us to hope that it might have even broader validity. But it is always well for the student of physics to be reminded that agreement with the results of some experiments can never conclusively "prove" a theory, although disagreement with an experiment does "disprove" a theory. No matter how much experimental confirmation there is for a theory, we can never unambiguously demonstrate that an experiment will never be performed whose results will disagree with the predictions of the theory. Experimental confirmation can only increase the probability that a theory is a correct description of the working of nature. For the foregoing classical theory of electrical conduction we, of course, have adequate evidence contrary to its predictions. So we now look at a theory whose predictions are consistent with a wider range of experimental evidence.

2. The Band Theory of Electrical Conduction

The reader has no doubt already had some acquaintance with the theory of the hydrogen atom. The single electron associated with the hydrogen nucleus, existing as it does in a region in which the potential energy varies inversely with the distance from the nucleus, can have only certain sharply defined energies. These energies are related to each other inversely as the squares of whole numbers. On the other hand, an electron which is far enough away from the nucleus to be effectively free of the attraction of the positive charge can have any energy; or, in other words, the allowed energies form a continuum. The restriction of the energies of an electron to specific values occurs where an electron is confined through a potential function to a restricted region.

If two atoms are close together, the potential energy in the region between them is lowered, permitting electrons at some levels to exist in the proximity of either nucleus. In addition, this modification of the potential function alters the energy level pattern. The Pauli Exclusion Principle requires that no two electrons of a system can have precisely the same energy. Since there are now in the dual system twice as many electrons, what were single levels in the one atom system, now split up into two, with the separation of the doublets being determined by the proximity of the atoms, the closer together the atoms the greater the separation of the doublets formed. The effect is shown schematically in Figure 2, where energy is plotted vertically, and distance horizontally.

When three atoms are bound together, the energy levels split into triplets; for four atoms, quadruplets, etc. A metallic crystal may be thought of as a macromolecule of many atoms bound together in an orderly array. We visualize the energy levels of the individual atoms as

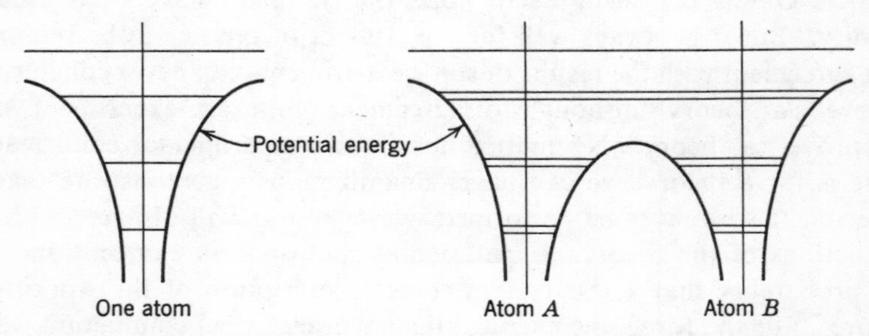

Figure 2. *Energy levels in an isolated atom and in a coupled pair of atoms.*

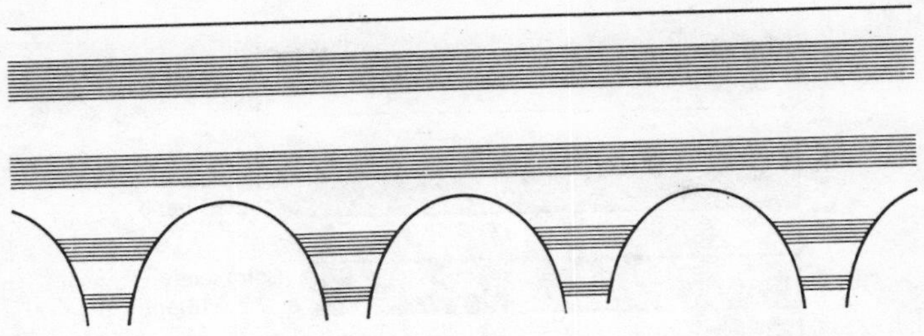

Figure 3. Energy bands in a crystal of many coupled atoms.

being expanded into a very large number of very closely spaced levels, the spacing being so close as to form effectively a continuum of levels over a limited energy range. The energy level pattern is suggested schematically in Figure 3. Although the separation of the energy levels within a band is very small, for an electron they are still distinct levels and the Pauli Exclusion Principle, limiting each energy level to one electron, controls the energy distribution of the electrons. There are, in general, many more possible levels than there are electrons to fill them, and the pattern of occupied and unoccupied levels is essential in explaining the different behavior of good conductors, semiconductors, and insulators.

A possible occupation pattern for a good conductor is shown in Figure 4 where three bands of energy levels are illustrated, the lower one being completely occupied, while the middle one is only partially occupied. The effect of an external electric field is to impart energy to the electrons. An electron can accept such energy, provided there is a level corresponding to its new energy vacant for it to change to. Since there is an effectively continuous range of energy levels above those in the partially filled band, these electrons readily accept kinetic energy from the electric field, and a current can exist in the metal. The greater the electric field, the more electrons shift to higher levels and the greater the current.

It must be remembered that in quantum mechanical analyses the

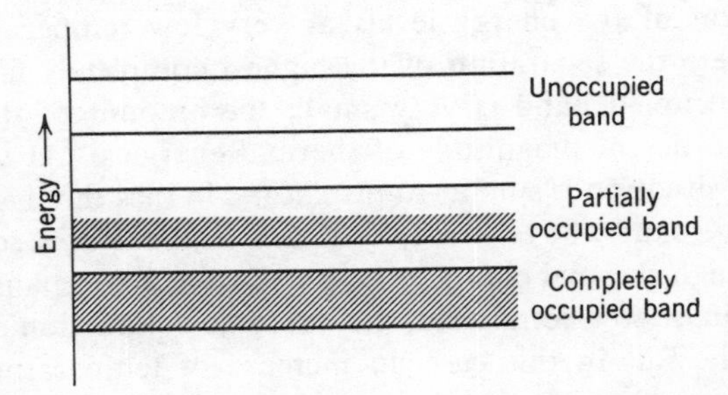

Figure 4. Occupied and unoccupied energy levels in a conductor.

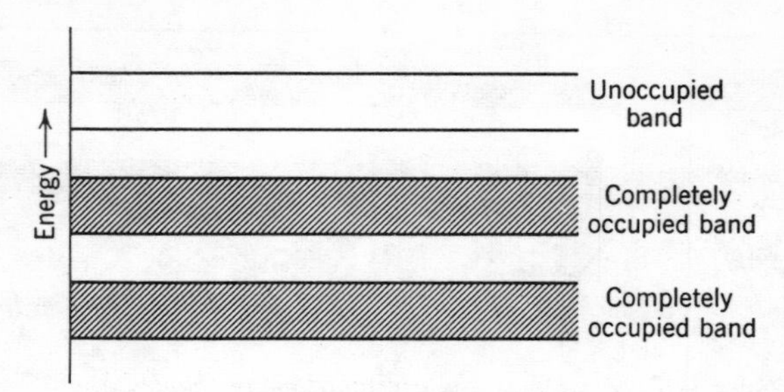

Figure 5. Energy bands in an insulator.

electrons are treated as waves which move through the crystal. It can be shown that if the crystal is an accurately repetitive structure of stationary atoms, these electron waves move freely through the structure. If there are irregularities in the structure, there are reflections of the waves at these irregularities, which may very roughly be compared to the frictional effect used in the classical theory. Such irregularities are partly the result of imperfections in the crystals and partly the result of thermal vibrations of the atoms. The greater the thermal vibrations, the greater will be the reflection of the electron waves and the greater is the electrical resistance, consistent with the observed variation of resistance with temperature in metals.

From the above concept of the energy level occupation in a metal it is a simple step to the explanation of the behavior of an insulator. Here, too, there are bands of energy levels, but the highest occupied band is completely filled (Figure 5). Thus, even though an electric field is applied, no electron can accept energy from the field because there are no energy levels available for it to change to. Should the electric field be high enough to be able to impart energy corresponding to the jump to the next higher band, the insulator suffers electrical breakdown and becomes a conductor.

An interesting prediction of this band theory of conductors appears in the explanation of semiconductor behavior. For a pure semiconductor, the occupation of the energy levels at very low temperatures is as in Figure 6, where the separation of the highest completely filled band and the next unoccupied band is very small, the amount of this separation being of the order of magnitude of thermal energies. At very low temperatures the situation is similar to insulators, in that there is a filled band with an empty band above it. At ordinary temperatures some electrons will have enough thermal energy to appear in the upper band where, with the application of an electric field, they become conduction electrons, just as in a metal. But in this case an increase of temperature introduces more conduction electrons, while at the same time increasing the thermal

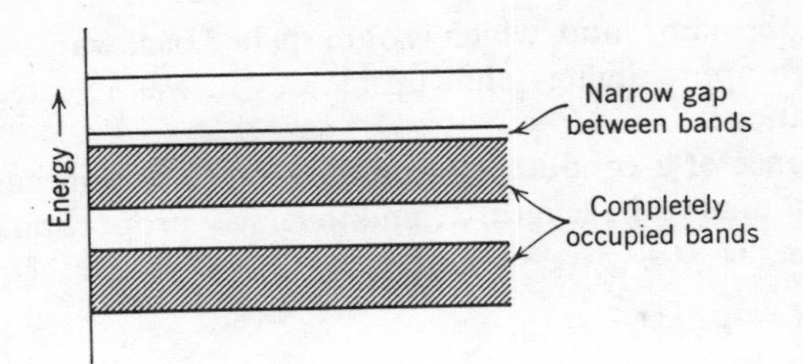

Figure 6. Energy bands in a semi-conductor.

vibrations of the atoms to interfere with their movement. The former tends to increase the electrical conductance, while the latter tends to decrease it; the result turns out to be a net increase in conductance with temperature, as is experimentally observed in semiconductors.

It is also possible to make a semiconductor out of what would normally be considered an insulating material by introducing controlled amounts of impurities which modify the energy level structure of the crystal. The details of how this is done are treated in books on physical electronics* which the student may wish to consult.

3. Metallic Conductors—Ohm's Law

In 1827 Georg Simon Ohm, who was then a professor of mathematics in a university in Cologne, published a paper in which he described experiments designed to relate the electric current in a conducting wire to the potential difference between the ends of the wire. He found a linear relation which is now usually given in the form:

$$V = RI \tag{3-1}$$

where V is the potential difference, I the current, and R is a constant of proportionality, the resistance, which depends on the dimensions of the conductor, the material of which it consists, and the temperature. It is measured in units appropriately called ohms, one ohm being one volt per ampere, or one volt-second per coulomb. If the temperature is held constant, Equation 3-1 is found to hold quite accurately over a very great range of current values.

Sometimes it is convenient to display Ohm's law in the form

$$I = GV \tag{3-2}$$

where G, the conductance, is the reciprocal of the resistance. The units of G are reciprocal ohms which have been given, perhaps in a moment of

* Hemenway, Henry, Caulton, *Physical Electronics* (Wiley).

desperation, the name mho, which is ohm spelled backward. (Conceivably "oyw," which approximates ohm upside-down, was rejected for reasons of pronunciation!)

The resistance of a conductor of uniform cross section has been found to be proportional to its length, L, and inversely proportional to its cross sectional area, $\mathscr{A}$,

$$R = \rho L/\mathscr{A} \tag{3-3}$$

where ρ, a constant characteristic of the material of the conductor and the temperature, is the resistivity, which from Equation 3-3 has the dimensions of ohm-meters. It follows that the conductance, G, takes the form:

$$G = \frac{\sigma\mathscr{A}}{L} \tag{3-4}$$

where σ is the reciprocal of the resistivity, called the conductivity, and is measured in mhos per meter.

For some purposes it is convenient to express Ohm's law in a point form. For this we need to define the concept of current density or current per unit area. Consider the current at a small area, $\Delta\mathscr{A}$, such as in Figure 7, where charges moving with a velocity $\mathbf{v}$ pass through a surface A–B, a small portion of which is $\Delta\mathscr{A}$. All the charges within the cylinder of slant height $\mathbf{v}\,dt$ and base area $\Delta\mathscr{A}$ pass through $\Delta\mathscr{A}$ during a time dt. The volume of this cylinder is $v \cos\theta\, dt\, \Delta\mathscr{A}$ and the total charge passing through $\Delta\mathscr{A}$ in a time dt is $dQ = nq\mathbf{v}\, dt \cdot \Delta\mathscr{A}$. Thus, the current over the area $\Delta\mathscr{A}$ is

$$\Delta I = dQ/dt = nq\mathbf{v} \cdot \Delta\mathscr{A} \tag{3-5}$$

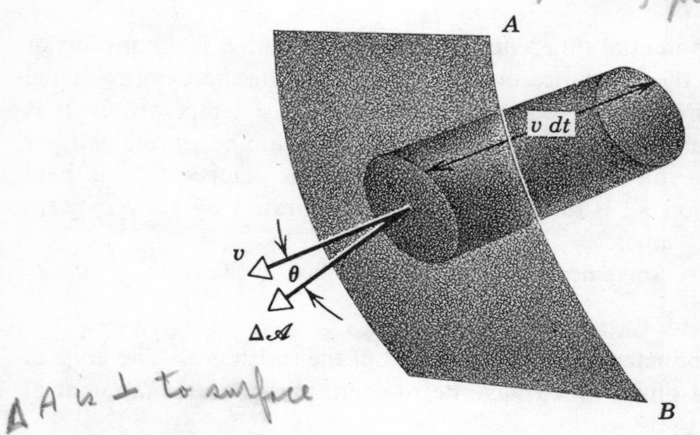

Figure 7. Charges in cylinder pass through surface A-B in a time dt.

where n is the number of charge carriers per unit volume and q is the charge on each carrier. Thus this vector $nq\mathbf{v}$ is the current per unit area, the current density $\mathbf{J}$ (the "conduction" current density, when it is necessary to distinguish it from the displacement current density), so that

$$\Delta I = \mathbf{J} \cdot \Delta\mathscr{A} \tag{3-6}$$

and, after passing to the limit, the current over a finite area becomes

$$I = \int \mathbf{J} \cdot d\mathscr{A} \tag{3-7}$$

It is readily seen that the dimensions of $\mathbf{J}$ are amperes per square meter. Note that since $\mathbf{J}$ involves the product of the charge on the carriers and the velocity $\mathbf{v}$, it is immaterial whether we consider the current density as resulting from positive charges moving in one direction or negative charges moving in the opposite direction. Conventionally, we think of $\mathbf{J}$ as having the direction in which positive charges move.

We now return to Figure 7 and imagine that the velocity of the charges, $\mathbf{v}$, is the result of an electric field $\mathbf{E}$ in the direction of $\mathbf{v}$. Let the slant height of the cylinder be $d\mathbf{l} = \mathbf{v}\,dt$, and the potential difference along the cylinder be $dV = \mathbf{E} \cdot d\mathbf{l} = E\,dl$, where the second form results because $\mathbf{E}$ and $d\mathbf{l}$ are parallel vectors. Since the cross-sectional area of this cylinder is $\Delta\mathscr{A} \cos\theta$, its conductance is, from Equation 3-4,

$$G = \frac{\sigma\,\Delta\mathscr{A} \cos\theta}{dl} \tag{3-8}$$

and Ohm's law from Equation 3-2 becomes

$$J\,\Delta\mathscr{A} \cos\theta = \left(\frac{\sigma\,\Delta\mathscr{A} \cos\theta}{dl}\right)(E\,dl) \tag{3-9}$$

or

$$\mathbf{J} = \sigma\mathbf{E} \tag{3-10}$$

where the return to the vector notation merely reaffirms what has been assumed earlier, that $\mathbf{J}$ and $\mathbf{E}$ are parallel. Since $\mathbf{J}$ and $\mathbf{E}$ are quantities measured at the same point in the medium, this is a point form of Ohm's law, and will be found useful from time to time as a theoretical tool.

4. Combinations of Resistances

Where the same charges pass in turn through two or more conductors, the conductors are said to be connected in series. If the individual conductors have resistances of R_1, R_2, $\cdots$, it is always possible to

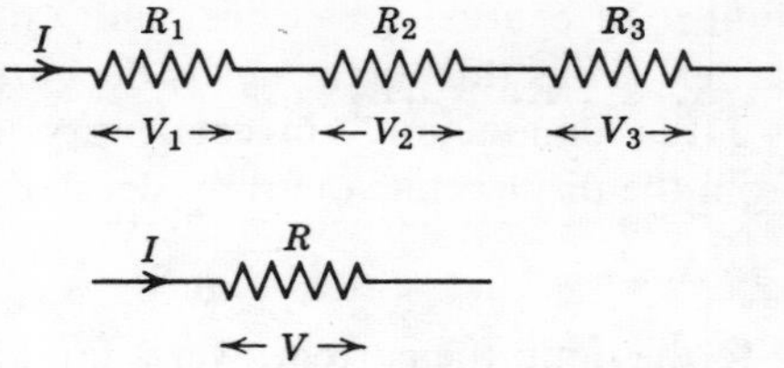

Figure 8. Resistances in series and equivalent resistance.

compute the value of an equivalent single resistance in which the ratio of potential difference to current will be the same as in the series combination. Thus in Figure 8 a current I is common to the three resistances, the arrow showing the direction of motion of positive charges, and V_1, V_2, and V_3 are the potential differences across the respective resistances. The total potential difference, V, across the series combination is

$$V = V_1 + V_2 + V_3 \tag{4-1}$$

From Ohm's law, Equation 4-1 becomes

$$V = IR = IR_1 + IR_2 + IR_3 = I(R_1 + R_2 + R_3) \tag{4-2}$$

from which the equivalent resistance R is

$$R = R_1 + R_2 + R_3 \tag{4-3}$$

When conductors are connected so that the same potential difference exists across each, the resistances are said to be connected in parallel, and it is similarly possible to determine an equivalent resistance in which the ratio of potential difference to current is the same as for the parallel combination. In Figure 9 the resistances R_1, R_2, and R_3 are connected in parallel and the total current is the sum of the individual currents:

$$I = I_1 + I_2 + I_3 \tag{4-4}$$

Again introducing Ohm's law, we find that

$$V/R = V/R_1 + V/R_2 + V/R_3 \tag{4-5}$$

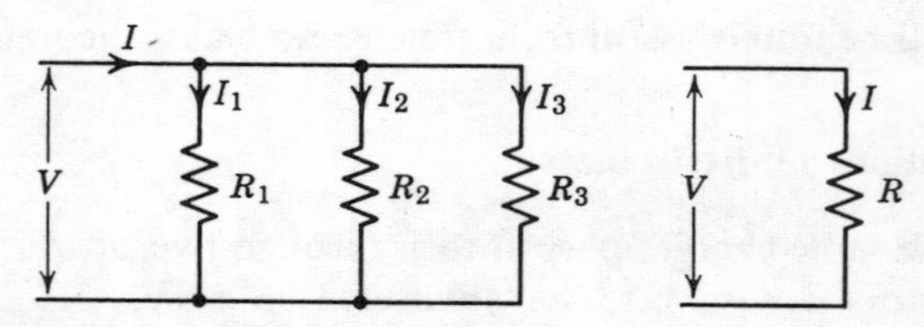

Figure 9. Resistances in parallel and equivalent resistance.

so that the equivalent resistance of a parallel combination is found from

$$1/R = 1/R_1 + 1/R_2 + 1/R_3 \tag{4-6}$$

Evidently the equivalent resistance of a number of resistors in series is always greater than the largest resistance in the group, whereas the equivalent resistance of a parallel combination is always smaller than the smallest resistance in the group. Note that capacitors may also be combined in series and in parallel by replacing the resistances in Equations 4-3 and 4-6 with reciprocal capacitances.

Many simple kinds of electrical circuits can be analyzed into combinations of series and parallel resistances for which a single equivalent resistance can be computed. However, expecially in circuits in which there is more than one source of power, such an analysis is often difficult. A more elegant method of general applicability developed by Kirchhoff will be discussed later for the solving of any kind of circuit problem.

5. Energy and Power Conversion

In Chapter 2, in the discussion of the energy of a charge in an electric field, it was shown that for a charge q moving from a point A where the potential is V_A to a point B where the potential is V_B, the following energy balance must hold:

$$\int \mathbf{F} \cdot ds + q(V_A - V_B) = (m/2)(v_B{}^2 - v_A{}^2) \tag{5-1}$$

where $\mathbf{F}$ is the resultant of any nonelectrostatic forces, m is the mass carrying the charge q, and v_A and v_B are the velocities of m at A and B, respectively.

When the path A–B is in a vacuum and no nonelectrical forces act on m, the loss in electrical potential energy of the charge q is equal to the gain in kinetic energy of the mass m. An example is the acceleration of ions through a potential difference in an electrostatic field.

In a conducting wire in which there is a steady current, the charge carriers on the average suffer no change in kinetic energy. The Equation 5-1 becomes

$$W + q\,\Delta V = 0 \tag{5-2}$$

where W is the work of nonelectrical forces and ΔV is the difference of potential, $V_A - V_B$.

The nonelectrical forces acting on the charge may be of several kinds, of which the "pseudo-frictional" force has already been discussed. Like frictional forces in mechanics it is associated with the development of heat in the conductor. A charge may also be affected by other types of forces, by means of which its potential energy is converted into mechanical energy, into chemical energy, or into radiant energy.

In any case when a charge dq decreases its potential energy by an amount $V\,dq$, an amount of energy of some other form, dW, appears. (In analyzing electric circuits, we shall not have occasion to use the concept of absolute potential relative to a point at infinity, and shall therefore drop the somewhat clumsy notation, ΔV. We shall henceforth use the symbol V for potential difference.) The power is

$$P = dW/dt = V(dq/dt) = VI \tag{5-3}$$

where dt is the time in which the energy dW evolves and I is the current, dq/dt, in the conductor. Equation 5-3 applies to any kind of energy conversion.

In that case where only "pseudo-frictional" forces are acting on the charge in a conductor, all the potential energy is converted into heat. Combining Equations 5-3 with Ohm's law, we get alternative forms:

$$P = IV = I^2R = V^2/R \tag{5-4}$$

James P. Joule in 1840 carried out the original experiment investigating the heat developed in a conductor carrying a steady current, and the last two forms in Equation 5-4 are commonly called Joule's law.

6. Sources of Emf

In a strict sense, we do not use electrical energy directly. We use forms of mechanical energy, heat energy, sound energy, and manufactured radiant energy. We get this energy from chemical energy in fuels, potential energy in water, radiant energy from the sun, etc. Electrical energy represents an intermediate step in the conversion of energy as it is found in nature to the forms useful to us. Those places in an electrical system where such energy conversions take place—either into electrical energy or from electrical energy to other kinds—we call, with an exception to be noted, sources of emf.

A positive source of emf is a region in an electrical system where another form of energy is converted to electrical energy. Examples are (*a*) electric generators, where mechanical energy is converted, (*b*) electric cells, where chemical energy is converted, (*c*) thermocouples, where heat energy is converted, and (*d*) solar cells, where radiant energy is converted. A negative source of emf, or a "sink" of emf, on the other hand, is a point where electrical energy is converted to some other form. Examples are (*a*) electric motors, where the conversion is to mechanical energy, (*b*) storage batteries being charged, with a conversion to chemical energy, and (*c*) thermocouples used in reverse where, through the Peltier effect, heat energy is developed.

A characteristic of a source of emf is that the energy conversion which takes place is in principle reversible. A motor driven in reverse becomes a generator; a charged battery becomes a source of electrical energy; and a thermocouple may convert energy in either direction. One form of energy conversion—the conversion of electrical energy into heat in a resistance—is not reversible, since it is not possible, even in principle, to put heat back into a conductor and obtain electrical energy. A resistance is never regarded as a negative source of emf.

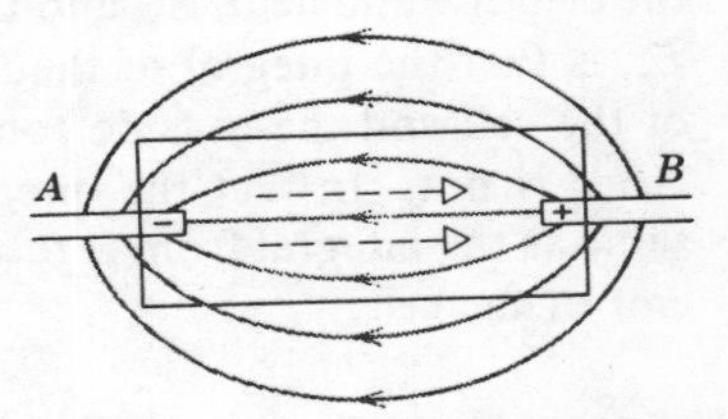

Figure 10. Schematic of a cell showing field due to chemical forces and electrostatic field of accumulated charges at the poles.

Historically, the term emf is an abbreviation of the words "electromotive force," a misnomer for which there should long ago have been a replacement. Common usage, however, has made replacement difficult and we compromise by using the abbreviation to suppress the misleading implications of the word "force." There are, of course, forces acting on charges within a source of emf, but the measure of such a source is in terms of energy, not force.

An emf is measured in terms of the amount of energy reversibly converted per unit of charge passing through the device. The unit, energy per unit charge, is the volt, and we speak of an emf of $\mathscr{E}$ volts—a 6-volt battery, a 100-volt generator, a 5-millivolt thermocouple emf. Thus a 6-volt battery contributes 6 joules of electrical energy for each coulomb of charge passing through it (although some of this energy is immediately converted to heat within the battery because of the resistance of the materials through which the charge must move). When the battery is being charged, each coulomb passing through contributes 6 joules of energy to be stored in the form of chemical energy (but more than 6 joules are required to move 1 coulomb through the battery because of the heat conversion, which is in addition to the stored chemical energy).

The forces which act on charges that pass through a cell are, of course, associated with an electric field (electric field strength is defined as force per unit charge, without specification of the origin or nature of the force). In Figure 10 is shown schematically a cell with negative and positive terminals, *A* and *B*, respectively. The field associated with the chemical forces of the cell is suggested by the dashed arrows; it tends to move positive charges toward the positive terminal where, if the cell is open-circuited, they collect, as do their counterparts at the negative terminal, until their electrostatic field equals and nullifies the field of the chemical forces. This electrostatic field is indicated by the heavier lines, and exists not only within but also outside the cell.

It is this electrostatic field which is responsible for the movement of charges along a conductor placed across the terminals of the cell while the field associated with the chemical forces acts to compensate for the depletion of the charge at the terminals. An important difference between the electrostatic field, E_s, and the field associated with the chemical forces, E_c, is that the integral of the former around the closed circuit consisting of the cell and the outside conductor is zero, whereas the integral of the latter is not. In fact the integral of E_c around the circuit, which is the same as the integral from A to B, since E_c is zero outside the cell, is just the emf of the cell:

$$\oint \mathbf{E}_c \cdot d\mathbf{l} = \int_A^B \mathbf{E}_c \cdot d\mathbf{l} = \mathscr{E} \tag{6-1}$$

where $\mathscr{E}$ is the emf, the work done by the cell per unit positive charge. When the circuit contains only the cell and resistance, with no negative emfs, all the energy supplied to charges by the cell is converted to heat in the resistance. The energy supplied per second by the chemical action is, from the definition of emf, $\mathscr{E}I$, when a current I exists in the circuit. From Joule's law, the heat developed per second in a resistance, R, outside the cell is I^2R and in the internal resistance, r, of the cell is I^2r. Consequently, conservation of energy requires that

$$\mathscr{E}I = I^2R + I^2r \tag{6-2}$$

from which, dividing by I,

$$\mathscr{E} = I(R + r) \tag{6-3}$$

By Ohm's law, IR is the potential difference, V, between the terminals of the cell, so that

$$\mathscr{E} = V + Ir \tag{6-4}$$

showing that when the current is zero, the terminal potential difference equals the emf of the cell. Since the emf is a constant of the cell (before depletion of the chemical energy), it is the terminal voltage which must decrease as the current increases, the rate of decrease being a function of the internal resistance of the cell.

The above discussion has used a cell as an example of a source of positive emf. The same reasoning, however, applies to any other type of emf, the only difference being the internal mechanism whereby energy is imparted to the charges.

7. Matching of Load to Internal Resistance

Figure 11 is a schematic diagram of an electrical circuit containing an emf, symbolized by the circle (it may be a generator, a battery, a solar cell, etc.) with a resistance r in series with the emf, and a resistance R in the external circuit. From Equations 5-4 and 6-3 the energy converted to heat per second in the external resistance is

$$P = I^2R = \mathscr{E}^2R/(R + r)^2 \qquad (7\text{-}1)$$

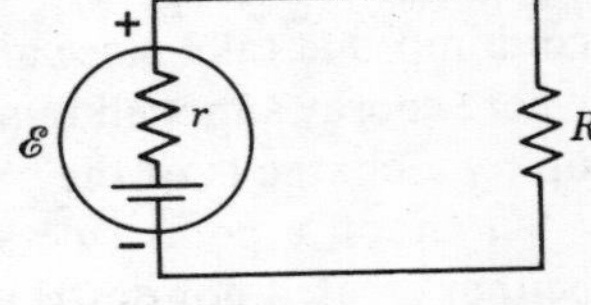

Figure 11. A source of emf and a load R.

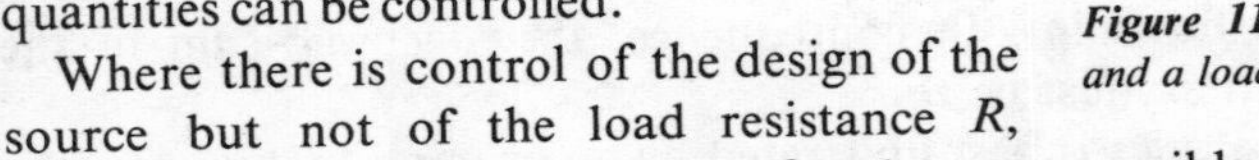

It is often important to determine how the greatest power can be obtained from the source. The answer depends on what quantities can be controlled.

Where there is control of the design of the source but not of the load resistance R, Equation 7-1 shows that making $\mathscr{E}$ as large as possible and r as small as possible maximizes the power in the load. The power which is wasted as internal heating of the generator is I^2r and clearly the delivery of maximum power to the load includes the minimizing of losses within the source.

In other circumstances, particularly in communications circuitry, there is little choice or control of the source, but the load resistance R can be chosen so as to obtain maximum power from the source. By differentiating Equation 7-1 with respect to R, it is readily found that the condition of maximum power supplied to R when $\mathscr{E}$ and r are fixed, is

$$R = r \qquad (7\text{-}2)$$

which leads to the interesting result that the maximum power is supplied to the load when an equal amount of power is wasted within the generator, that is, when the electrical efficiency of the generator is 50%!

8. Kirchhoff's Rules

An electrical circuit may be said to be solved when all the currents, resistances, and potential differences in all the branches are known. A common type of problem is for the emfs and the resistances to be given and from them to determine the currents (and possibly the potential differences) in every part of the system. In other problems resistances may be asked for that will lead to specified currents for specified emfs.

It was pointed out in Section 4 that certain simple types of circuits containing a single emf and combinations of series and parallel resistances can be solved by a direct use of Ohm's law and the equations for obtaining the equivalents of series and parallel combinations. But particularly when several emfs exist in different branches of a circuit, this simple approach is not possible; even with a single emf, there can be combinations of resistors which are cumbersome to combine into series and parallel groupings. G. R. Kirchhoff, in 1847, formulated two rules of circuit analysis which are tools for the solving of systems of any complexity. These rules are statements of the conservation of energy and the conservation of charge as they apply to an electrical circuit.

Kirchhoff's loop rule states that the sum of the energy changes per coulomb that take place around any closed path in a system must be zero, where energy supplied to a charge is taken as of one sign and energy given up by a charge is of the opposite sign.

Kirchhoff's point rule states that at a junction in a circuit, charge is neither created nor destroyed, so that as much flows away from a junction of conductors as flows in. In consequence, the algebraic sum of the currents at a junction must be zero.

The use of these rules is best illustrated by an example. The system in Figure 12 shows two sources of emf, $\mathscr{E}_1$ and $\mathscr{E}_2$, with internal resistances r_1 and r_2, respectively, simultaneously supplying energy to a load resistance R. Consider the emfs and the resistances as known, and let it be required to find the current in each element. These currents are indicated in Figure 12 as i_1, i_2, and i_3.

The first step in proceeding to a solution of this circuit is to assign directions to the currents. These directions are assigned arbitrarily. Should a negative value for a current be obtained in the final solution, the sense of this current is opposite to that originally assigned. Arrows in Figure 12 indicate an arbitrary choice of directions, in which the flow of charge is toward junction A for each current. Application of the point rule provides the following relation among the currents:

(8-1) $$i_1 + i_2 + i_3 = 0$$

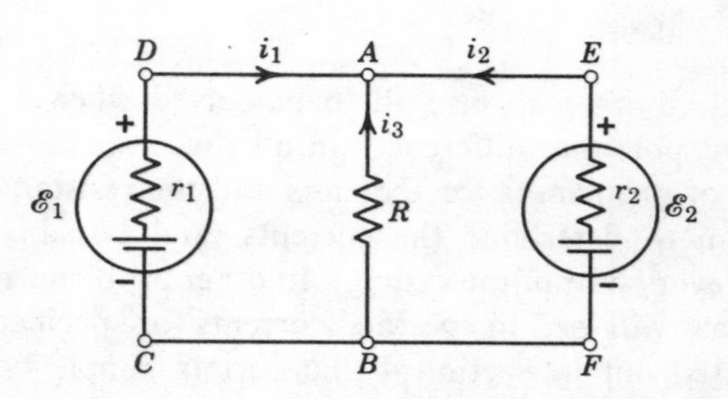

Figure 12. A network with two sources of emf.

in which current directed toward the junction has been taken as positive. Clearly, no new information is obtained by applying the point rule to junction *B*.

We now apply the loop rule to the various closed paths that can be traced in this system. (Connecting wires are taken to have zero resistance.) For one closed path let us choose the loop which includes $\mathscr{E}_1$ and the load *R*. Starting at *C* and moving upward through $\mathscr{E}_1$ to *D*, there is an increase of potential from the negative to the positive terminal of $\mathscr{E}_1$ volts less the drop of potential resulting from the current in the internal resistance r_1. From *D* to *A* there is no change in potential, and from *A* to *B* there is a rise in potential through the resistance *R* since, we are traversing the circuit in this branch in a direction opposite to that of the current. The loop rule requires that the sum of these potential changes be zero:

$$\mathscr{E}_1 - i_1 r_1 + i_3 R = 0 \tag{8-2}$$

The same reasoning applied to the loop *FEAB* gives

$$\mathscr{E}_2 - i_2 r_2 + i_3 R = 0 \tag{8-3}$$

Since there are three unknown currents to be found, the three equations, 8-1, 8-2, and 8-3, give sufficient information for the solution. The loop rule could have also been applied to the path *CDEF*, but the information so obtained is redundant unless either 8-2 or 8-3 is not used.

The currents in this system are found from the simultaneous solution of the above three equations, a simple task so long as the resistances are not functions of the currents. The reader may readily check that the currents are:

$$i_1 = \frac{\mathscr{E}_1(R + r_2) - \mathscr{E}_2 R}{R(r_1 + r_2) + r_1 r_2} \tag{8-4a}$$

$$i_2 = \frac{\mathscr{E}_2(R + r_1) - \mathscr{E}_1 R}{R(r_1 + r_2) + r_1 r_2} \tag{8-4b}$$

$$i_3 = -\frac{\mathscr{E}_1 r_2 + \mathscr{E}_2 r_1}{R(r_1 + r_2) + r_1 r_2} \tag{8-4c}$$

Evidently, whatever the values of the circuit constants, the direction of i_3 was originally chosen wrong. Unless $\mathscr{E}_1 = \mathscr{E}_2$, moreover, it is possible for either i_1 or i_2 to be reversed.

The solving of circuit problems by the use of Kirchhoff's rules is usually simple and straight-forward, but nevertheless experience and an orderly technique can be of considerable value. The following are suggestions on procedure:

(*a*) Make a clear circuit diagram, marking the polarities of the emfs and the assumed directions of the currents.

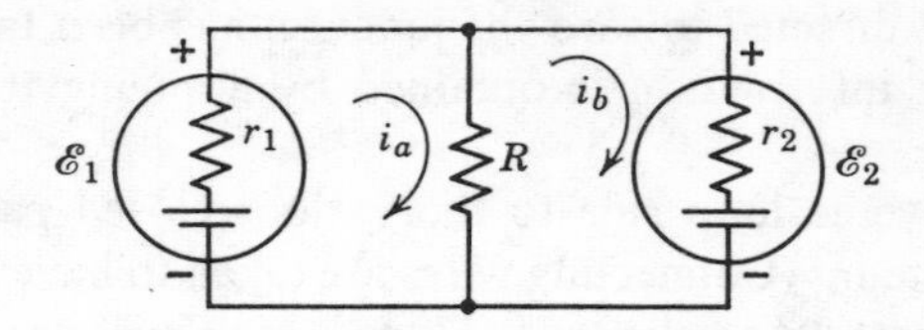

Figure 13. Network of Figure 12 with loop currents indicated.

(*b*) Write equations based on the point rule for enough junctions to include all the variable currents, avoiding redundant junctions.

(*c*) Write equations based on the loop rule, choosing loops so as to include each unknown quantity at least once (there will be rare exceptions), and writing enough loop equations so that the total number of equations equals the number of unknown quantities.

(*d*) Arrange the equations in an orderly fashion and solve simultaneously, preferably by the method of successive substitutions. Except in the simplest cases, the temptation to use determinants should, in the author's opinion, be resisted. The method of successive substitution permits a dimensional check at all times (if the problem is done in symbols) and, if done in an orderly fashion, the work can be traced backward to search out an error.

A modified procedure, which derives from Kirchhoff's rules, is that which uses loop currents. Using the same circuit as in Figure 12, we assign a current to each loop (see Figure 13). Here i_a is considered to be a current associated with charges moving clockwise around the left-hand loop, and i_b with charges moving clockwise around the right-hand loop. Thus i_a corresponds to i_1 in Figure 12, i_b to $-i_2$, and $i_b - i_a$ to i_3. Because there are now only as many unknowns as loops, fewer equations are required. We write for the left-hand loop:

$$\mathscr{E}_1 - i_a r_1 - (i_a - i_b)R = 0 \tag{8-5a}$$

and for the right-hand loop:

$$\mathscr{E}_2 + i_b r_2 + (i_b - i_a)R = 0 \tag{8-5b}$$

The reader can readily show that the solution of Equations 8-5 is consistent with that of Equations 8-4. For most problems the method employing loop currents is a worthwhile simplification.

9. Temperature Dependence and Nonlinear Elements

The foregoing treatment of d-c circuits is, in its fundamentals, completely general, although the discussion and the examples have assumed that the resistances are constants, independent of the potential differences across them or the currents in them. Such may not always be even a good approximation because resistances of conductors and semiconductors are functions of temperature and of the applied voltage, sometimes rather complicated functions. Because heat is generated by current in a resistor, it follows that resistance is likely to be a function of the current unless this heat is removed at the same rate at which it is produced.

The dependence of resistance upon temperature is expressed in terms of the fractional change in resistance per degree change in temperature, α, a quantity which, for most metallic conductors, has a very small value and changes very slowly with temperature. It is defined by

$$\alpha = \frac{1}{R}\frac{dR}{dt} \tag{9-1}$$

where t is the temperature. If the resistance at temperature t_0 is R_0, integration of Equation 9-1 gives

$$R = R_0 e^{\alpha(t-t_0)} \tag{9-2}$$

where R is the resistance at the temperature t. Since α is a small quantity, Equation 9-2 can be simplified for moderate temperatures by expanding $e^{\alpha(t-t_0)}$ in terms of a power series in $\alpha(t - t_0)$ and dropping higher powers than the first, giving

$$R = R_0[1 + \alpha(t - t_0)] \tag{9-3}$$

For pure metals and most alloys, α is positive with a value of the order of 10^{-3} per degree centigrade or less; for a few alloys and for most nonmetals which come under the class of semiconductors it is negative. Some semiconductors, called thermisters, are specially designed with a relatively high negative temperature coefficient and, when calibrated, can be used in connection with appropriate circuits for temperature measurements or temperature control.

Certain combinations of semiconductors can be constructed to have resistances which are a function of the polarity of the applied voltage, with the resistance in one direction being considerably higher than for the other. By analogy with the similar properties of diode vacuum tubes, such elements are called crystal diodes, and find use where alternating currents are to be rectified. Usually their resistance is also a function of the magnitude of the applied voltage. In fact, certain types often employed in connection with microwave detectors (Chapters 9 and 12) have the property that, for small applied voltages, the current varies with the square of the voltage.

A comparatively new development in the past decade has been the transistor, which is a layered combination of semiconductors in which the currents are functions of the voltages at the layers. The most common type of transistor contains three elements and behaves in many ways similar to the three-element vacuum tube. Because of their small size, their long life, and the fact that they do not depend on thermal emission to function, and therefore evolve very little heat, they have found wide application in miniaturized equipment, in computers, and in other uses where space or power requirements are limited.

Discussion Questions

1. In Section 6 a distinction was made between electrical energy and mechanical, thermal, acoustical, and radiant energy. Is such a distinction fundamental?

2. Discuss some possibilities for conversion from and to electrical energy. Should all of them be classed as emf's? Or should some come under the class of resistance? Can such a distinction always be made rigorously? Is it a necessary one?

3. The internal resistance of a cell can be measured in at least two ways: (*a*) by measurement of terminal voltage as a function of resistance; or (*b*) by finding that resistance in which the maximum power is dissipated. Discuss what might be the expected accuracy of the result in each case in terms of the instruments available in your laboratory.

4. When magnitudes are to be determined from a graph, it is convenient when possible to display the relation in such a way that it appears as a straight line. After determining the emf and the internal resistance of a cell by finding the terminal potential difference as a function of load resistance, how can the data be plotted to produce a straight-line graph? What features of the graph determine r and the emf?

5. The circuit in Figure Q5 is essentially the same as the one discussed in Section 8. What must be the relation among the emf's, $\mathscr{E}_2$ and $\mathscr{E}_1$, and the

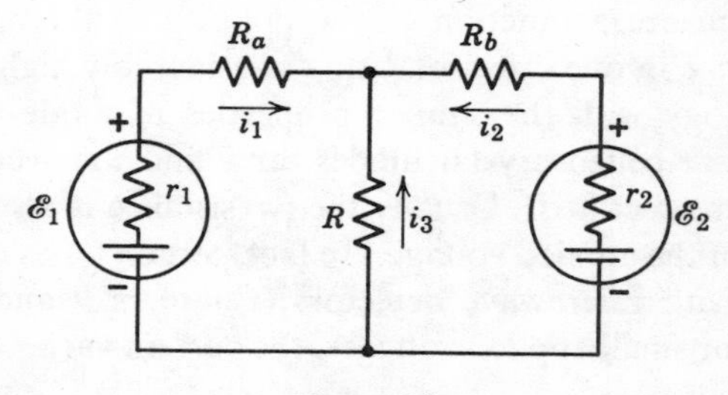

Figure Q5.

resistances of the system if the current i_2 is to be zero? (This is the basic structure of a potentiometer.) If the current i_3 is 1 ma, what does this say about the value of R? Suggest how the potentiometer may be made direct reading. Develop other modifications that make it more useful and convenient as a laboratory instrument.

6. In the theory developed in Section 1 it was assumed that the frictional force on the electrons was proportional to the velocity. Develop the theory of electrical conduction on the assumption that the frictional force is proportional to the second power of the velocity, and determine the relation predicted between the current and the voltage.

7. Show that the circuit of Figure 12 may be solved by the following steps: (*a*) find the currents in each branch on the assumption that $\mathscr{E}_2 = 0$ (but r_2 unchanged); (*b*) find the currents in each branch on the assumption that $\mathscr{E}_1 = 0$ (but r_1 unchanged); (*c*) sum for each branch the currents found in (*a*) and (*b*) to obtain the actual current when both emf's are present. Compare with the results already given, and then generalize to obtain what is called the Superposition Theorem.

8. Thevenin's theorem says that the current in the ith branch of a network can be found by the following steps: (*a*) Open the ith branch and find the potential difference, V_i, at the opened terminals; (*b*) find the resistance, r_i, of the rest of the network as measured at the opened terminals when all the emf's are assumed zero (without changing their internal resistances); (*c*) treat the rest of the network as a single emf of V_i and internal resistance r_i which is connected to the resistance in the ith branch, R_i, and thus find the current in the ith branch. Test this theorem on the circuit of Figure 12. This theorem requires a lengthy statement, but it is often relatively simple to apply and is of considerable importance in circuit analysis.

Problems

1. In copper the density of charge carriers is approximately 10^{29} per cubic meter. Assuming each carrier has one electronic charge, find the average drift velocity of a carrier in a cylindrical copper conductor of radius 1 mm for a current of one ampere.

2. A cylinder of carbon has an outer radius of b and an axial hole of radius a. A copper conductor fills the inner hole and a copper sleave makes contact with the outer surface. If the resistivity of carbon is ρ, find the resistance per unit length of the cylinder of carbon between the inner and the outer conductors. (Take the conductivity of copper to be effectively infinite.)

3. The circuit in Figure P3 is to be used to measure the internal resistance and the emf of a cell. The current is measured first with R_1 and then with R_2 in the circuit. Express the emf, $\mathscr{E}$, and the internal resistance, r, of the cell in terms of I_1, I_2, R_1, and R_2.

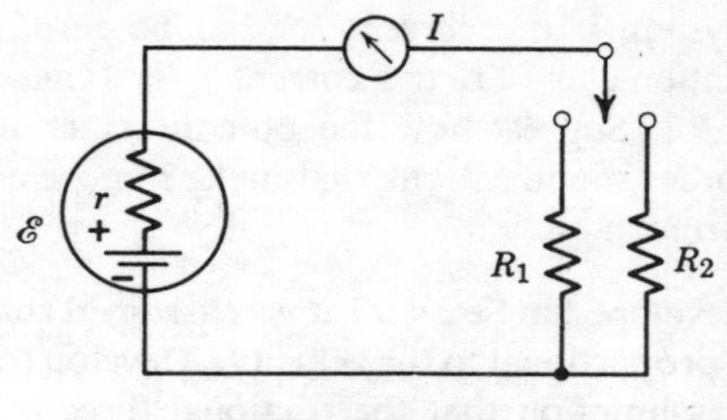

Figure P3

4. An Ayrton shunt is a divided resistor across the terminals of a current meter as in Figure P4(*a*). The fraction s/R can be varied from zero to one. Show that the current, i, in the meter is directly proportional to the current, I, which is to be measured, and that for a given I, the meter current is directly proportional to s. What advantages does the Ayrton shunt have for changing meter sensitivities over the direct changing of shunts such as in the circuit in Figure P4(*b*)?

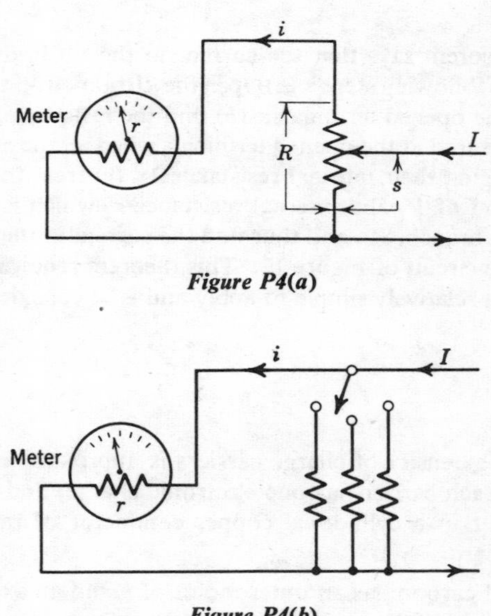

Figure P4(a)

Figure P4(b)

5. A source of emf, ℰ, and negligible internal resistance is in series with a resistance R and a parallel group of n resistances, each of resistance r.

(*a*) In terms of r and R, what must be the number n, in order that the total rate of heat production in the parallel group be a maximum?

(*b*) Suppose that the value of r is variable. Find the value of r for which the heat production in the parallel group when $n = N$ is the same as when $n = 1$.

Plot P versus r when $n = 1$ and when $n = 16$, with $R = 3$ ohms, and indicate the range of values of r for which $P_{n=1}$ is greater than $P_{n=16}$ and the range for which $P_{n=1}$ is less than $P_{n=16}$.

6. A generator at a power station has an internal resistance of R' and a terminal voltage of $\mathscr{E}$ when it is supplying no current. It is connected by means of a two-wire transmission line (total resistance r) to a factory whose net resistance is R. Answers to the following questions should be expressed in terms of $\mathscr{E}$, R', r, and R.

(*a*) What is the total power developed by the generator?

(*b*) What is the power supplied to the transmission line at the terminals of the generator?

(*c*) What is the power delivered to the load R?

(*d*) What is the power lost in the transmission line?

(*e*) With the power delivered to the load R considered as useful output, what is the electrical efficiency of the system?

(*f*) R and $\mathscr{E}$ are to be increased in such a way as to double the voltage at the factory terminals and yet keep the power delivered to the factory the same. What are the new values of R and $\mathscr{E}$ and how is the power loss in the transmission line affected?

7. N identical cells are available, each with an emf $\mathscr{E}$ and an internal resistance r. These cells are to be grouped into p parallel groups, each of s cells in series, the resulting battery to supply power to a load resistor R. Determine the values of p and s which will result in the maximum heat developed in R. (Is this an example of the theorem developed in Section 7?)

8. Figure P8 is a Wheatstone's bridge circuit. The five bridge currents and the battery current are labeled. By means of Kirchhoff's rules, determine the current, i_5, in terms of the resistances and the battery voltage. Check your results by finding the condition on the resistances R_1, R_2, R_3, and R_4 for which i_5 is zero.

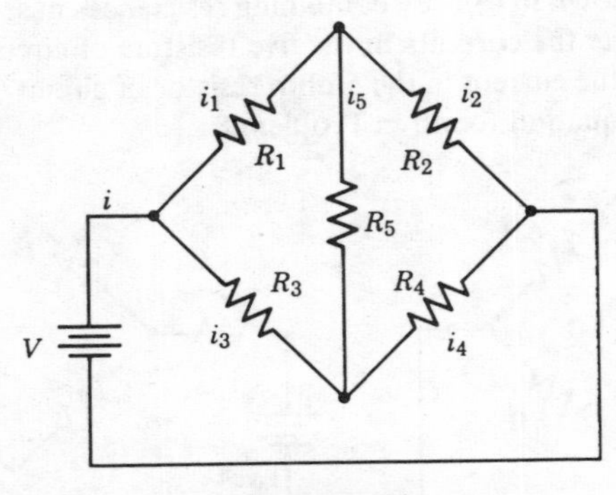

Figure P8

9. Three resistances a, b, and c are connected as in Figure P9(*a*). Three other resistances x, y, and z are connected as in Figure P9(*b*). Show that if the following relations hold:

$$a = \frac{xy}{(x + y + z)} \qquad b = \frac{xz}{(x + y + z)}$$

$$c = \frac{yz}{(x + y + z)}$$

it is impossible to distinguish electrically between these two circuits by measurements made at the terminals *A*, *B*, and *C*.

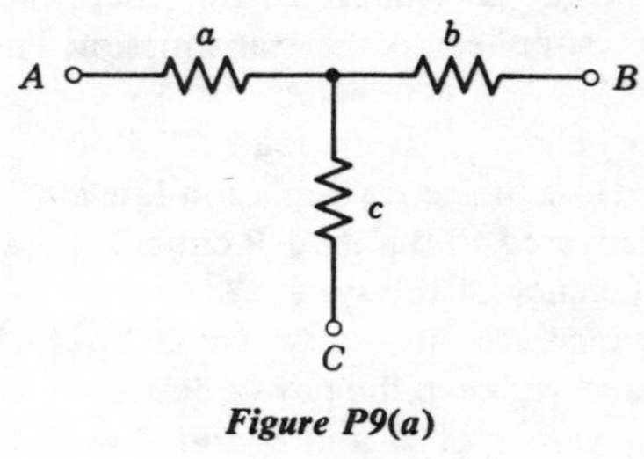

Figure P9(a)

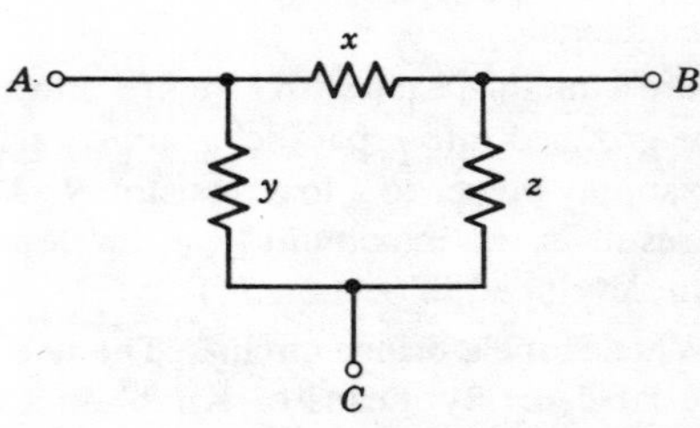

Figure P9(b)

10. Using the results of Problem 9, transform the bridge circuit in (*a*) of Figure P10 into the circuit shown in (*b*). By combining resistances in series and parallel combinations, compute the currents in the five resistors of circuit (*b*), and from this information find the current in the 5-ohm resistor of circuit (*a*). Check your results by using the equation found in Problem 8.

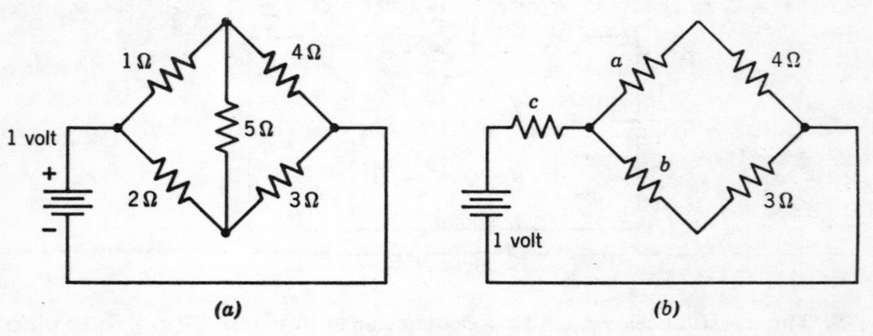

Figure P10

11. Twelve identical wires are connected together so as to form the edges of a cube. If the resistance of each wire is r, what is the resistance between two corners of the cube which are at the ends of a diagonal?

12. On the cube of Problem 11, what is the resistance between the two ends of any one wire?

13. In the circuit in Figure P13, what is the effect on the currents in the various resistances of connecting a 6-ohm resistor across the points AB?

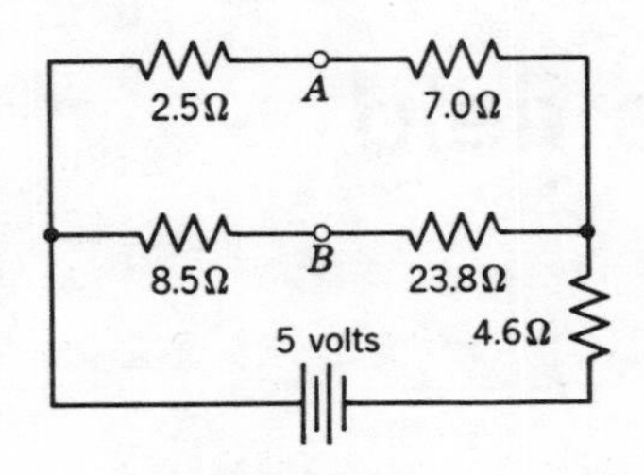

Figure P13

14. What is the current in the battery in the circuit in Figure P14? To what value must the 4-ohm resistor be changed so that there will be no current in the 3-ohm resistor?

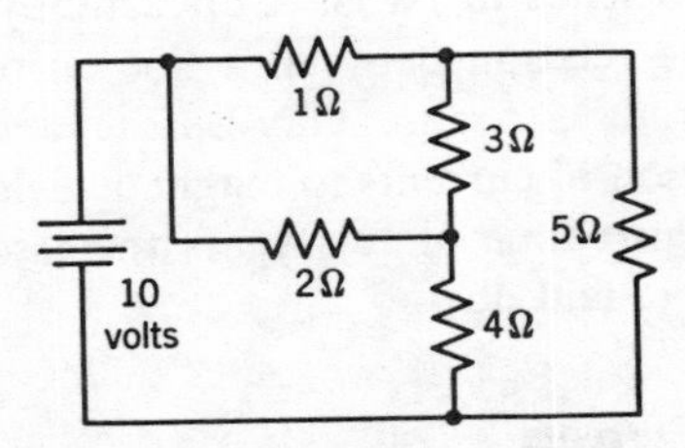

Figure P14

15. There is a need for constructing a resistor whose resistance is temperature-independent within the limits of the approximation of Equation 9-3. The magnitude of the resistance is to be 90 ohms and it is proposed to use carbon with a temperature coefficient of $-0.0005/°C$ and nichrome with a temperature coefficient of $+0.0004/°C$. What must be the magnitude of the resistance of each part and how should they be connected?

REFERENCES

There are many good treatments of direct current circuits. The following two are good examples:

1. Scott, *The Physics of Electricity and Magnetism*, Wiley, New York, 1959, Chapter 5.
2. Harnwell, *Principles of Electricity and Electromagnetism*, McGraw-Hill, New York, 1938, Chapter IV and parts of Chapter V.

5

Currents and the magnetic field

On the 11th of September, 1820, André Marie Ampère got the news of the discovery by Hans Christian Oersted that a current in a wire can produce a deflection of a compass needle. Within one week he presented the first of his papers on the subject of the relation of magnetic fields to currents to the Academy of Science in Paris. Concurrently, Jean Biot and Felix Savart developed the relation between a line current and its associated magnetic field. Faraday, Arago, Gay-Lussac and others also eagerly investigated the relation of currents to magnetic fields, with the result that in the remarkably short time of two years the essential foundations for electromagnetism were laid down.

1. Forces Between Currents

Recall that with electrostatics we started with the effect of one point charge on another point charge, and therefrom predicted the interactions of aggregates of charges. Similarly with currents, we start with an idealization of a "point" current element and its influence on another such element; then, by summing over all the current elements in interacting circuits, we obtain the force between whole circuits.

In Figure 1, two such current elements are shown, $d\mathbf{l}_1$ and $d\mathbf{l}_2$, separated by a distance $\mathbf{r}$. Both $d\mathbf{l}_1$ and $d\mathbf{l}_2$ are infinitesimal lengths that are parts of closed circuits in which currents I_1 and I_2 exist. Each element is a vector whose positive direction is taken as the direction in which positive charges are flowing in its circuit. The distance $\mathbf{r}$ is also a vector whose positive

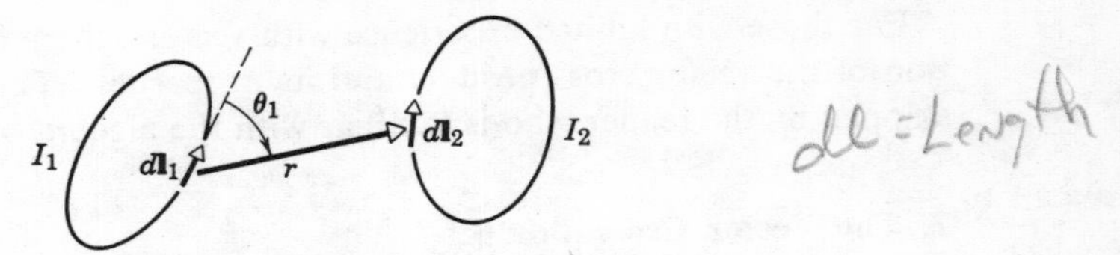

Figure 1. $d\mathbf{l}_1$ *and* $d\mathbf{l}_2$ *are elements of closed circuits.*

direction shall be taken from $d\mathbf{l}_1$ to $d\mathbf{l}_2$. The vector $\mathbf{r}$ makes an angle θ_1 with the positive direction of $d\mathbf{l}_1$. Let these two vectors fix a plane in space whose normal is directed in the sense given by the thumb of the right hand when the fingers curl from $d\mathbf{l}_1$ to $\mathbf{r}$. The element $d\mathbf{l}_2$ makes an angle θ_2 with this normal. These are the geometrical relationships we need to describe the interaction between the two current elements.

The force, $d^2\mathbf{F}_2$, acting on the element $d\mathbf{l}_2$ because of the current in $d\mathbf{l}_1$, is proportional to both currents and to the inverse square of their separation, as well as to the sines of θ_1 and θ_2.

$$d^2F_2 = \frac{\mu_0 I_1 I_2}{4\pi}\frac{dl_1\,dl_2}{r^2}\sin\theta_1 \sin\theta_2 \tag{1-1}$$

Here, $\mu_0/4\pi$ is a constant of proportionality. The direction of the force on dl_2 is always perpendicular to the directions of dl_2 and to the normal to the plane of dl_1 and r. In contrast to electrostatics where the theory begins with a force between scalar point charges, the theory of the interaction of currents begins with the force between vector current elements. In both cases, however, the starting point is an inverse square law.

The constant μ_0 plays a role analogous to the reciprocal of ϵ_0 in electrostatics. μ_0 is called the absolute permeability of free space and its value is $4\pi \times 10^{-7}$ newton per ampere² in the MKS system, for which a justification must await further development of the theory.

Equation 1-1 is not a relation which can be tested directly by experiment. (Strictly speaking, neither could Coulomb's law be tested directly, applying as it does to point charges.) It is not possible to isolate the effect of an element of a circuit from the rest of the circuit of which it is a part. The test of the validity of Equation 1-1 lies in its ability to predict the force which one complete current system exerts on another complete system, the prediction being made by integrating Equation 1-1 over circuit 1 to find the total force on one element of circuit 2 and then integrating this force over the whole of circuit 2. Where geometry is relatively simple and there is reasonable symmetry, such integrations, for which there will be illustrations later, are readily accomplished. For the time being we shall regard Equation 1-1 as a fundamental relation which has been adequately tested.

For those with limited experience with vectors, there follows a description of the vector cross product and its properties. This section may be skipped by the reader who is familiar with the algebra of vectors.

2. The Vector Cross Product

In the algebra of vectors, the dot product is a scalar quantity which, for two vectors, **R** and **P**, is defined by $\mathbf{R} \cdot \mathbf{P} = RP \cos \theta$, where θ is the smaller angle between the positive directions of **R** and **P**.

A second kind of vector product, called the cross product, is one in which the result is a vector. This product is written $\mathbf{R} \times \mathbf{P}$, and is defined as a vector perpendicular to the plane containing **R** and **P**, whose positive sense is indicated by the thumb of the right hand when the fingers curl from **R** toward **P** through the smaller angle between their positive directions. In Figure 2, the positive sense of the cross product of **R** and **P** is into the plane of the paper. This rule for the positive sense of the cross product is also expressed as the direction that a conventional screw would progress when turned from **R** to **P** through the smaller angle.

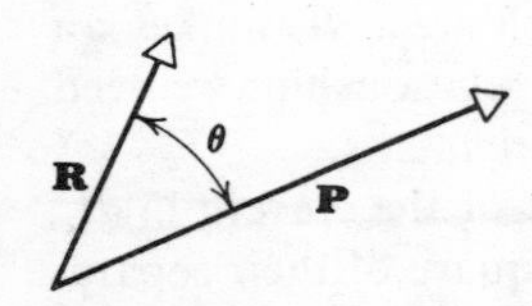

Figure 2. The cross product of **R** *and* **P** *is directed into the paper.*

The magnitude of the cross product is defined as $RP \sin \theta$. Whereas the dot product of two vectors is zero when the two vectors are normal to each other, the cross product is zero when the two vectors are parallel. And, whereas the value of the scalar product is not changed by changing the order of the vectors, the cross product changes sign. Algebraically, this is evident from the fact that $\sin \theta = -\sin(-\theta)$; geometrically, from the right-hand rule, $\mathbf{R} \times \mathbf{P}$ is a vector directed into the paper, whereas $\mathbf{P} \times \mathbf{R}$ is directed out from the paper for the vectors shown in Figure 2. Another geometrical interpretation of the magnitude of the cross product is that it is the area of the parallelogram of which **R** and **P** are two sides.

Evidently the cross product of two perpendicular vectors is, in magnitude, the simple product of the individual magnitudes. If these magnitudes are each unity, the cross product is unity, and the result is a unit vector normal to both. Thus if **i** and **j** are vectors of unit length in the x and y directions, respectively, then $\mathbf{i} \times \mathbf{j} = \mathbf{k}$, where **k** is a unit vector in the z direction, similarly, $\mathbf{j} \times \mathbf{k} = \mathbf{i}$ and $\mathbf{k} \times \mathbf{i} = \mathbf{j}$; whereas $\mathbf{i} \times \mathbf{i} = \mathbf{j} \times \mathbf{j} = \mathbf{k} \times \mathbf{k} = 0$. Consequently the nine terms in the cross product of **R** and **P**, when each is written in component form,

$$(2\text{-}1) \qquad \begin{aligned} &(\mathbf{i}R_x + \mathbf{j}R_y + \mathbf{k}R_z) \times (\mathbf{i}P_x + \mathbf{j}P_y + \mathbf{k}P_z) \\ &= \mathbf{i} \times \mathbf{i}R_xP_x + \mathbf{i} \times \mathbf{j}R_xP_y + \mathbf{i} \times \mathbf{k}R_xP_z \\ &\quad + \mathbf{j} \times \mathbf{i}R_yP_x + \mathbf{j} \times \mathbf{j}R_yP_y + \mathbf{j} \times \mathbf{k}R_yP_z \\ &\quad + \mathbf{k} \times \mathbf{i}R_zP_x + \mathbf{k} \times \mathbf{j}R_zP_y + \mathbf{k} \times \mathbf{k}R_zP_z \end{aligned}$$

$$d\vec{F}_2 = \left(\frac{\mu_0}{4\pi}\right) I_1 I_2 \left(d\vec{l}_2 \times \frac{d\vec{l}_1 \times \vec{r}}{r^3}\right)$$

reduce to the six terms:

$$\mathbf{R} \times \mathbf{P} = \mathbf{i}(R_yP_z - R_zP_y) + \mathbf{j}(R_zP_x - R_xP_z) + \mathbf{k}(R_xP_y - R_yP_x) \tag{2-2}$$

a result which for some is easier to remember when displayed in the form of a determinant:

$$\mathbf{R} \times \mathbf{P} = \begin{vmatrix} \mathbf{i} & \mathbf{j} & \mathbf{k} \\ R_x & R_y & R_z \\ P_x & P_y & P_z \end{vmatrix} \tag{2-3}$$

An appreciation of the geometrical significance of the vector cross product will be obtained from a consideration of a few standard operations. As mentioned, the cross product itself is the vector representation of the area of a parallelogram of which the constituents of the product are the sides. Now let us perform a scalar multiplication of this cross product with a third vector, **Q**. Such an operation is written $\mathbf{Q} \cdot \mathbf{R} \times \mathbf{P}$. It is not necessary to write $\mathbf{Q} \cdot (\mathbf{R} \times \mathbf{P})$, since there can be no ambiguity about the order in which the scalar and the vector operations are to be performed. The statement $(\mathbf{Q} \cdot \mathbf{R}) \times \mathbf{P}$ has no meaning because the vector product is defined only between two vectors.

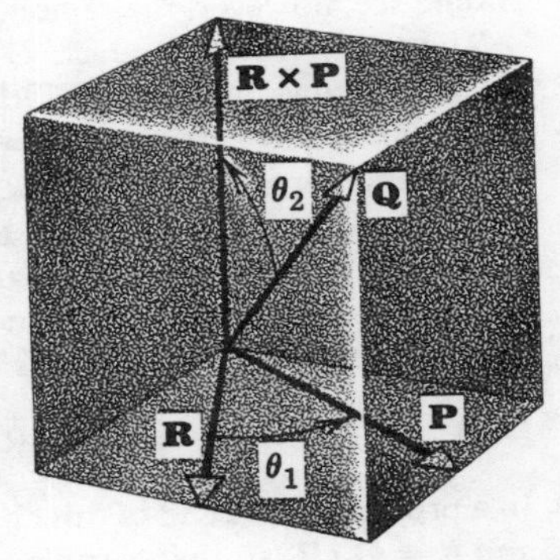

Figure 3. Illustrating the formation of a triple scalar product.

Figure 3 illustrates the three vectors, **Q**, **R**, and **P**, and indicates relevant angles. The cross product of **R** and **P** is drawn normal to the plane of **R** and **P**, and **Q** makes an angle θ_2 with this normal. The scalar product of **Q** with $\mathbf{R} \times \mathbf{P}$ is the projection of **Q** on this normal, $Q \cos \theta_2$, times the magnitude of $\mathbf{R} \times \mathbf{P}$, which is seen to be the volume of the parallelepiped determined by the three vectors i.e., $QRP \sin \theta_1 \cos \theta_2$.

Since the volume of a parallelepiped is the area of the base times the height, it is immaterial which face is taken as the base. We can thus equally well consider the base as being the face formed by the edges **Q** and **P**, with the projection of **R** on the normal to this face as the height. Then the volume would be $\mathbf{R} \cdot \mathbf{P} \times \mathbf{Q}$. An extension of this reasoning will show that all of the following are equivalent statements for the volume:

$$\begin{aligned} \mathbf{Q} \cdot \mathbf{R} \times \mathbf{P} &= \mathbf{R} \cdot \mathbf{P} \times \mathbf{Q} = \mathbf{P} \cdot \mathbf{Q} \times \mathbf{R} = \mathbf{R} \times \mathbf{P} \cdot \mathbf{Q} \\ &= \mathbf{P} \times \mathbf{Q} \cdot \mathbf{R} = \mathbf{Q} \times \mathbf{R} \cdot \mathbf{P} \end{aligned} \tag{2-4}$$

From Equations 2-4 we obtain the general rule that, in such a triple scalar product, the result is unchanged when the locations of the dot and the cross are interchanged; and that, for fixed positions of dot and cross, the result is not

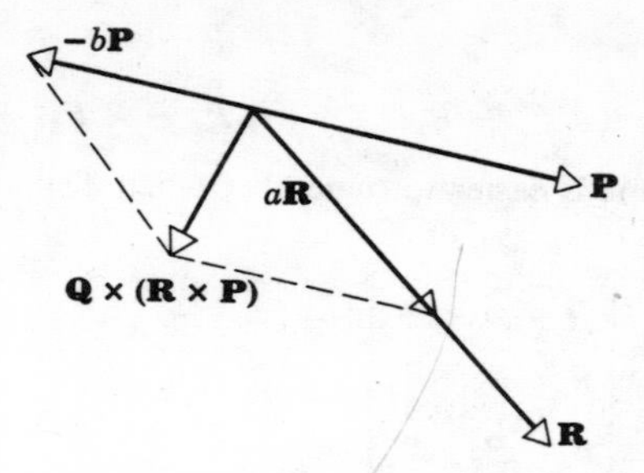

Figure 4. A triple vector product.

affected by a cyclic change in the order of the three constituents. The reader should see, however, that a noncyclic permutation changes the sign of the result.

The triple vector product, $\mathbf{Q} \times (\mathbf{R} \times \mathbf{P})$ is also useful. An examination of Figure 3 shows that this quantity is a vector which must lie in the plane of **R** and **P**, since it is perpendicular to both **Q** and $\mathbf{R} \times \mathbf{P}$. Here the parentheses are important, because, $(\mathbf{Q} \times \mathbf{R}) \times \mathbf{P}$ is a quite different vector, being in the plane of **Q** and **R**. Since $\mathbf{Q} \times (\mathbf{R} \times \mathbf{P})$ is a vector in the plane of **R** and **P**, it can be regarded as the diagonal of a parallelogram whose sides are parallel to **R** and **P**, respectively, as shown in Figure 4. That is, one side is, say, $a\mathbf{R}$, where a is some scalar number, and the other is $-b\mathbf{P}$, where b is another scalar number. Thus

$$\mathbf{Q} \times (\mathbf{R} \times \mathbf{P}) = a\mathbf{R} - b\mathbf{P} \tag{2-5}$$

In a problem at the end of this chapter the reader is asked to show that $a = \mathbf{Q} \cdot \mathbf{P}$ and $b = \mathbf{Q} \cdot \mathbf{R}$, so that a triple vector product can be expanded into the form:

$$\mathbf{Q} \times (\mathbf{R} \times \mathbf{P}) = (\mathbf{Q} \cdot \mathbf{P})\mathbf{R} - (\mathbf{Q} \cdot \mathbf{R})\mathbf{P} \tag{2-6}$$

In terms of the angles shown in Figure 3, the magnitude of the triple vector product is $QRP \sin \theta_1 \sin \theta_2$.

3. The Vector Form of the Force Law

We now return to Equation 1-1, writing the current elements as vectors, $d\mathbf{l}_1$ and $d\mathbf{l}_2$, each with the direction of the conventional current in its own loop, and writing $\mathbf{r}$ as a vector with a direction from $d\mathbf{l}_1$ to $d\mathbf{l}_2$. The force law then takes the form:

$$d^2\mathbf{F}_2 = \left(\frac{\mu_0}{4\pi}\right) I_1 I_2 \left(d\mathbf{l}_2 \times \frac{d\mathbf{l}_1 \times \mathbf{r}}{r^3}\right) \quad (136) \tag{3-1}$$

$$= \left(\frac{\mu_0}{4\pi}\right) I_1 I_2 \left(d\mathbf{l}_2 \times \frac{d\mathbf{l}_1 \times \mathbf{r}_u}{r^2}\right) \quad (137)$$

where $\mathbf{r}_u$ is a vector of unit length in the direction of $\mathbf{r}$. It is left to the reader to show that Equation 3-1 is the proper translation of Equation 1-1 into vector notation.

The total force which circuit 1 exerts on circuit 2 is the double integration of 3-1 over the two circuits:

$$\mathbf{F}_2 = \left(\frac{\mu_0}{4\pi}\right) I_1 I_2 \oint_2 d\mathbf{l}_2 \times \oint_1 \frac{d\mathbf{l}_1 \times \mathbf{r}}{r^3} \tag{3-2}$$

where the subscripts on the integration signs indicate the circuit over which the integration is to be performed.

The position of $d\mathbf{l}_1$ relative to a set of Cartesian axes will be given in terms of variables x', y', z'; and the position of $d\mathbf{l}_2$, in terms of x, y, and z. Then

$$\mathbf{r} = \mathbf{i}(x - x') + \mathbf{j}(y - y') + \mathbf{k}(z - z') \tag{3-3}$$

and

$$r^2 = (x - x')^2 + (y - y')^2 + (z - z')^2 \tag{3-4}$$

so that only $\mathbf{r}$ is a function of both sets of variables, a fact that becomes important in later considerations.

Equation 3-2 involves vector integrations, which can be tedious unless advantage can be taken of regularities and symmetries. But its significance does not entirely rest with the forces between circuits that can be computed with it. It also plays a fundamental role in the formulation of further electromagnetic theory, and we shall find that often important deductions can be drawn from such theory without the necessity for computations of specific cases.

4. The Force Between Parallel Wires

An example of a direct computation with Equation 3-2 is the determination of the force which a straight wire of length L, carrying a current I_2, experiences when it is a distance x from a parallel wire of effectively infinite length carrying a current I_1. The situation is illustrated in Figure 5.

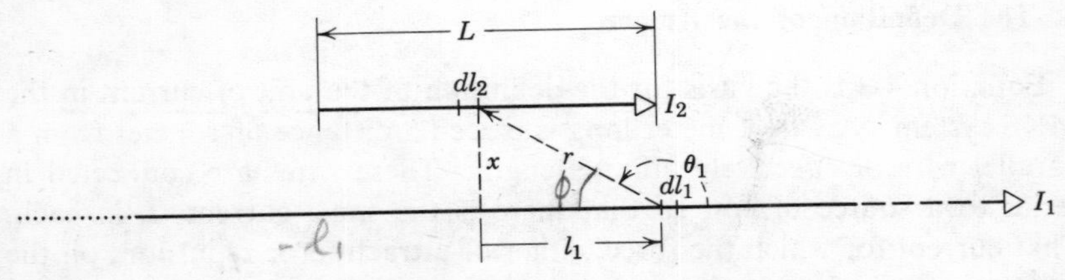

Figure 5. Computing the force between parallel wires.

It may properly be asked how it is that we propose to use Equation 3-2 to compute the force on a straight wire, when Equation 3-2 clearly requires integration over a complete current loop. The answer is in two parts. We propose, first, to find only that part of the total force on the circuit of I_2 which acts on the length L, ignoring those forces that act on the rest of the system. Second, we postulate that the current I_1 exists in a system of which only the straight wire illustrated is close enough to I_2 to have significant effect. Here we are making use of the inverse square law by putting the return loop effectively at infinity.

We first compute the integral over circuit 1 in Equation 3-2. Let the element $d\mathbf{l}_1$ be a distance $\mathbf{l}_1$ from the foot of the perpendicular, x. Then

$$l_1 = -x \cot \theta_1, \qquad dl_1 = x \csc^2 \theta_1 \, d\theta_1, \qquad \text{and} \qquad r = x \csc \theta_1 \tag{4-1}$$

Moreover, the direction of the vector integrand is the same—out from the paper—for all the elements on I_1; consequently, the integration becomes an algebraic operation, and the second integral in Equation 3-2 is

$$\begin{aligned} \int_{-\infty}^{\infty} \frac{d\mathbf{l}_1 \times \mathbf{r}}{r^3} &= \int_0^{\pi} \frac{x \sin \theta_1 \csc^2 \theta_1 \, d\theta_1}{x^2 \csc^2 \theta_1} \\ &= \frac{1}{x} \int_0^{\pi} \sin \theta_1 \, d\theta_1 = \frac{2}{x} \end{aligned} \tag{4-2}$$

Clearly this integral has the same value at all points on the length L, so that the total force on L is

$$\begin{aligned} F_L &= \frac{\mu_0}{4\pi} I_1 I_2 \int_0^L \left(\frac{2}{x}\right) dl_2 \\ &= \frac{\mu_0 I_1 I_2 L}{2\pi x} \end{aligned} \tag{4-3}$$

The direction of the force is one of attraction between the currents I_1 and I_2. The reader should satisfy himself that, with a reversal of one of the currents, the force becomes a repulsion without change in magnitude.

5. The Definition of the Ampere

Equation 4-3 is the basis for the definition of the unit of current in the MKS system. A wire 1 meter long is placed a distance of 1 meter from a parallel wire of effectively infinite length. These wires are connected in series to a source of emf so that there is the same current, I, in each. That current for which the force, either of attraction or repulsion, on the 1-meter wire is 2×10^{-7} newtons is defined as a current of 1 ampere.

The value of $\mu_0 = 4\pi \times 10^{-7}$ newton/ampere^2 was chosen with foresight, not only to simplify certain much used electromagnetic equations, but also to permit the unit of current to correspond with the unit which has been traditionally used in the laboratory.

In our discussion of Coulomb's law in Chapter 1, the coulomb as a unit of charge was arbitrarily introduced, because its definition had to wait until this point. Charge has already been related to current by

$$I = dQ/dt \tag{5-1}$$

Thus a coulomb of charge is that amount of charge which passes a cross section of a circuit in 1 second when there is a steady current of 1 ampere in the circuit.

We shall see in later work that the dielectric constant of free space cannot be chosen arbitrarily, but is fixed by its relation to μ_0 and the speed of light, c:

$$\epsilon_0\mu_0 = 1/c^2 \tag{5-2}$$

In the way in which we have developed the definition of the ampere above, the quantity which is arbitrarily chosen—in addition, of course, to the units of mass, length, and time—is the value of μ_0. It is only a small change of point of view to say that the size of the ampere is arbitrarily chosen, thereby fixing the value of the permeability of free space. In mechanics, three fundamental units of measurement, mass, length, and time, are sufficient for the description of the mechanical behavior of matter. One more fundamental unit is required for the description of electrical behavior. The ampere has been chosen to be this fundamental unit, but it is also possible to choose to define charge as a fundamental unit through Coulomb's law, after which current becomes a derived unit through Equation 5-1. This choice is not a matter of logic but of convenience. It is not possible to store either a unit of current or a unit of charge in a vault as the unit of mass and the unit of length are stored, but apparatus is devised which makes it possible to reproduce a known current when needed. A similar kind of device for reproducing a known charge can be built, but we do not commonly use charge meters, whereas ammeters are in wide use and require calibration against a standard. Such a standardizing instrument is at the Bureau of Standards in Washington, D.C.

6. The Magnetic Flux Density

At this point in the consideration of the interactions of currents we are at the stage where we were in the treatment of Coulomb's law when we were computing the force which one charge exerts on another and had not yet developed the concept of a field of force. The electric field was introduced after it was seen that the force on a point charge is always proportional to the magnitude of the charge, and we defined the electric field as this proportionality factor, making it a vector parallel to the force on a positive charge.

An examination of Equation 3-1 shows that the force on a current element $I_2 d\mathbf{l}_2$ is proportional to the magnitude of this element, and we are led to a definition of a field vector, somewhat analogous to **E**, which relates the force on a current element to the current element itself:

$$d\mathbf{F}_2 = I_2\, d\mathbf{l}_2 \times \mathbf{B} \tag{6-1}$$

where **B** is called the magnetic flux density or the magnetic induction. Equation 6-1 suggests in principle an experimental way of measuring **B**, somewhat like that discussed in Section 3 of Chapter 1 for **E**. That is, we do not need to know what gives rise to the **B**-field in order to measure it. We measure the force on a test circuit element at a point. Even ignoring the fact that determining the force on an *element* of a circuit has its difficulties, the measurement is not so simple as for the electric field because of the vector cross product in the definition.

The definition 6-1 leads to the force on a complete circuit:

$$\mathbf{F}_2 = I_2 \oint_2 d\mathbf{l}_2 \times \mathbf{B} \tag{6-2}$$

where the integration is over the contour of circuit 2.

When the current systems that give rise to the **B**-field are known, Equation 3-1 shows that this field can be computed from those currents by

$$\mathbf{B} = \frac{\mu_0}{4\pi} I_1 \oint_1 \frac{d\mathbf{l}_1 \times \mathbf{r}}{r^3} \tag{6-3}$$

A **B**-field that is due to more than one current system is the vector sum of computations like Equation 6-3 over each system. Equation 6-3 is variously called Ampère's law, Biot's law, or a generalization of the Biot-Savart law.

From Equation 6-1 it is seen that the direction of the force on a current element is perpendicular to the element as well as perpendicular to **B**. The order of the components in this vector product should be carefully noted—reversal will give the wrong direction of the resulting force.

A comparison of Equations 4-2 and 4-3 with 6-3 shows that the **B**-field associated with the current in the infinite wire of Section 4 is

$$B = \frac{\mu_0 I_1}{2\pi x} \quad \text{(infinite straight wire)} \tag{6-4}$$

and thus the force on the second wire can be written

$$F_2 = BI_2L \tag{6-5}$$

The magnetic flux density associated with a long straight wire drops off with the inverse first power of the distance from the wire in the same way that the electric field dropped off with distance from an uniformly charged infinite wire. But attention should be called to the difference in the direction of the field. If we draw a continuous line so that the **B**-field at every point is tangent to this line, it forms a circle about the wire. Although such a circular field is a special case arising out of the simple symmetry of the straight wire, it is nevertheless a general fact that there are no beginnings or ends to such lines of the **B**-field, that any line so drawn will either close in on itself or continue indefinitely. Especially in the simpler cases, it is useful to visualize the **B**-field in terms of flux lines, just as for the **E**-field.

7. The Flux Density on the Axis of a Circular Loop

Figure 6 shows schematically a circular loop conductor of radius a, with the current I in the loop entering the paper at the bottom and coming out from the paper at the top. We shall compute the magnetic flux density, **B**, at a point on the axis of the loop a distance x from the center.

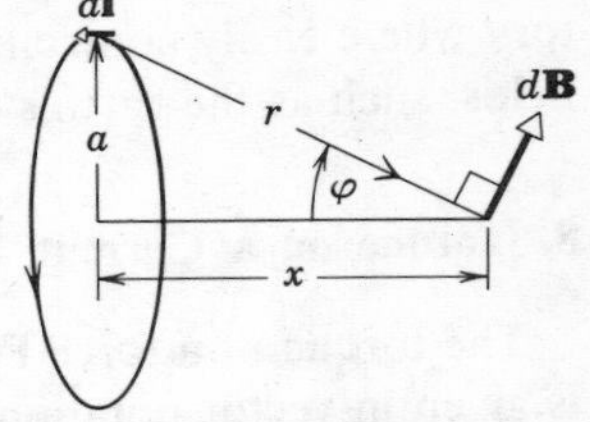

Figure 6. Field on the axis of a circular loop.

An infinitesimal element at the top of the loop produces a field strength $d\mathbf{B}$ at the point on the axis given by

$$d\mathbf{B} = \frac{\mu_0 I}{4\pi}\frac{d\mathbf{l} \times \mathbf{r}}{r^3} \tag{7-1}$$

where **r** is shown on the diagram. The vector $d\mathbf{B}$ is normal to both **r** and $d\mathbf{l}$. A corresponding element of the loop at the bottom produces an element of field symmetrically opposite. Since the loop as a whole consists of pairs of such elements, the components of the field perpendicular to the axis cancel, while the axial components add. The axial component of Equation 7-1 is

$$dB_x = dB \sin\varphi = \frac{\mu_0 I}{4\pi}\frac{dl \sin\varphi}{r^2} \tag{7-2}$$

where φ and r are constants in the integration over the current loop. The total **B**-field on the axis is thus

$$B_x = \frac{\mu_0 I a \sin \varphi}{2r^2} \tag{7-3}$$

But $\sin \varphi = a/r$ and $r^2 = a^2 + x^2$. Consequently,

$$B_x = \frac{\mu_0 I a^2}{2(a^2 + x^2)^{3/2}} \tag{7-4}$$

An examination of the nature of the variation of this axial field with x suggests an interesting application of such a current loop. Evidently, the field **B** decreases monotonically with x. The slope of the curve of B versus x is zero at the center of the loop and approaches zero for very large x. Its curvature is negative near the coil, becomes positive for larger x, and must be zero at an intermediate point. Therefore, in the vicinity of the point of zero curvature the graph approximates a straight line over a limited region. Call this point of zero curvature x_1. If another identical current loop is placed at a point $2x_1$ from the first, the field of this second loop will be increasing approximately linearly at x_1, and within a limited region around x_1 the field will be very closely constant with x. We may intuitively expect this also to be approximately true at points off but near the axis in this region. There is thus a small region of uniform magnetic flux density between the coils.

Coils so arranged are called Helmholtz coils and are used in the laboratory where easily controlled uniform fields are needed or where unwanted fields, such as the earth's field, are to be canceled.

8. Torque on a Current Loop

The torque of a force **F** acting at a distance **r** from a center of rotation is given in vector notation by

$$\mathbf{T} = \mathbf{r} \times \mathbf{F} \tag{8-1}$$

The direction of the vector torque is the direction of the axis of rotation.

Consider a plane closed current loop of arbitrary shape in a region where there is a uniform magnetic flux density **B** in the x direction. This loop carries a current I, and we fix our attention on an element of this loop, $d\mathbf{l}$. From Equation 6-2 the force on this element is

$$d\mathbf{F} = I(d\mathbf{l} \times \mathbf{B}) \tag{8-2}$$

and the torque which this force produces about the origin of coordinates is

$$d\mathbf{T} = I[\mathbf{r} \times (d\mathbf{l} \times \mathbf{B})] \tag{8-3}$$

which, from Equation 2-6, becomes

$$dT = I[(\mathbf{r} \cdot \mathbf{B})\, d\mathbf{l} - (\mathbf{r} \cdot d\mathbf{l})\mathbf{B}] \tag{8-4}$$

Since we have assumed that

$$\mathbf{B} = \mathbf{i}B + \mathbf{j}0 + \mathbf{k}0 \tag{8-5}$$

and since

$$\mathbf{r} = \mathbf{i}x + \mathbf{j}y + \mathbf{k}z \tag{8-6}$$
$$d\mathbf{l} = \mathbf{i}\, dx + \mathbf{j}\, dy + \mathbf{k}\, dz$$

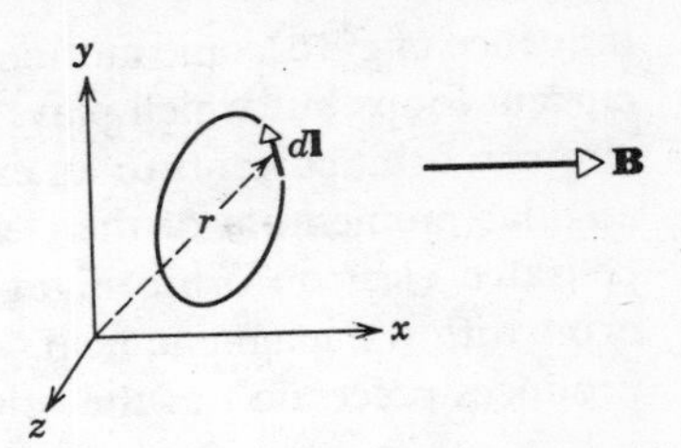

Figure 7. Computing the torque on a current loop.

Equation 8-4 becomes

$$d\mathbf{T} = IB[\mathbf{i}(-y\, dy - z\, dz) + \mathbf{j}(x\, dy) + \mathbf{k}(x\, dz)] \tag{8-7}$$

The torque on the whole loop is the integral of Equation 8-7 over the loop. Consequently the three components of the torque on the current loop are

$$T_x = -IB \oint (y\, dy + z\, dz) \tag{8-8a}$$

$$T_y = IB \oint x\, dy \tag{8-8b}$$

$$T_z = IB \oint x\, dz \tag{8-8c}$$

The integral in Equation 8-8*a* is zero, while that in 8-8*b* is the projection of the area of the loop on the *xy*-plane, and in 8-8*c* is the projection of the area on the *xz*-plane. Therefore the Equations 8-8 can be combined into the vector relation:

$$\mathbf{T} = I\mathscr{A} \times \mathbf{B} \tag{8-9}$$

where $\mathscr{A}$ is the vector area determined by the perimeter of the loop. Evidently, a current loop in a uniform **B**-field experiences a torque only about axes perpendicular to the field. A galvanometer coil, for example, is acted upon by a torque which is proportional to the current in the coil and proportional to the sine of the angle between the **B**-field and the vector area of the coil.

By analogy with the behavior of the electric dipole in an electric field (review Problem 12 of Chapter 1, and express the results in the form of a vector product), the quantity $I\mathscr{A}$ is called the magnetic dipole moment or, more briefly and more commonly, the magnetic moment, **M**, of the current loop. A current loop in a magnetic field behaves as an electric dipole in an electric field, except that there is an angular momentum associated with the circulating current. This introduces the complicating influence of gyroscopic motion, an effect which is negligible in macroscopic current loops but which plays an important role in the reaction of atomic electron orbit currents to an external magnetic field. If it were not for the angular momentum of the electron in its orbit, an atom with an uncompensated electron orbit might be expected to align the area vector of its orbit with the magnetic field. Instead, the torque from the magnetic field produces precession of the orbit about the direction of the field.

9. The Force on a Single Moving Charge

A charge Q moving with a velocity $\mathbf{v}$ in a magnetic flux density $\mathbf{B}$ experiences a force given by

$$\mathbf{F} = Q\mathbf{v} \times \mathbf{B} \tag{9-1}$$

This relation follows from Equation 6-5, which may be written without the subscripts and in vector form:

$$\mathbf{F} = I\mathbf{L} \times \mathbf{B} \tag{9-2}$$

where $\mathbf{L}$ is the length of the wire (including the direction) and $\mathbf{B}$ is the field at the wire. We have seen in Section 1 of Chapter 4 that in terms of the charge carriers, the current in a wire can be expressed as

$$I = nq\mathscr{A}v = nq\mathscr{A} \cdot \mathbf{v} \tag{9-3}$$

where in the vector form we make use of the fact that $\mathbf{v}$ and the area vector, $\mathscr{A}$, are parallel. Thus we get

$$I\mathbf{L} = \rho(\mathscr{A} \cdot \mathbf{v})\mathbf{L} = \rho(\mathscr{A} \cdot \mathbf{L})\mathbf{v} \tag{9-4}$$

where $\rho = nq$, n being the number of charge carriers per unit volume. The form of Equation 9-4 is possible because $\mathbf{L}$ and $\mathbf{v}$ are parallel.

Now consider a charged body of cross section $\mathscr{A}$ and length $\mathbf{L}$ moving with a velocity $\mathbf{v}$. If the density of charge on the body is ρ, then the total charge is $Q = \rho(\mathscr{A} \cdot \mathbf{L})$, so that for this charged body, the quantity $Q\mathbf{v}$ plays the same role as $I\mathbf{L}$ in a wire. If the charge is not distributed uniformly over the carrier, Equation 9-1 nevertheless holds true for each element of volume, $d\mathscr{V}$, of the body; and since each element is moving with the same velocity, it holds for the whole body. Equation 9-1 is usually applied to cases where the moving charge can be considered as a point charge.

Instead of deducing Equation 9-1 from Equation 6-5, it is also possible to take 9-1 as a definition of $\mathbf{B}$, rather than 6-1. Equation 6-1 is then a consequence of 9-1 and the rest of the development follows.

Since the force on a moving charge involves the cross product of the velocity and the flux density, so that this force is perpendicular to both, it cannot act to change the magnitude of the velocity, only its direction. In a region where the *B*-field is uniform and a charge Q has a velocity normal to **B**, the path of the charge is a circle of radius r such that

$$\frac{mv^2}{r} = QvB \tag{9-5}$$

and

$$r = \frac{mv}{QB} \tag{9-6}$$

where m is the mass of the charge carrier. The radius of the circle is thus a measure of the momentum of the carrier. Where **B** and **v** are known, the ratio of mass to charge can be computed.

The kinetic energy of the carrier is $mv^2/2$. Where **v** itself may not be known a knowledge of the kinetic energy and the momentum (from the radius of the path) is sufficient information to compute the ratio m/Q. The kinetic energy is particularly simple to determine when that energy is acquired as a result of the charges having passed through a known difference of potential.

In general, if any two of the three quantities, momentum, kinetic energy, and velocity, are known, the ratio of mass to charge can be computed.

A mass spectrograph is an instrument with which this ratio is measured for atoms or molecules for which the charge is assumed to be known. Many types of mass spectrographs have been designed and built, the various kinds differing in respect to which pair of quantities are used to compute the mass, as well as in the way in which they make use of certain geometrical arrangements to focus the ion beams. One common type is illustrated in Figure 8, where an ion is formed near the positive plate of a parallel plate capacitor, is accelerated to a speed v as it moves toward the negative plate where it passes through a slit into a region of uniform flux density **B**, normal to the velocity. The ion travels through a semicircle and is registered on a photographic plate.

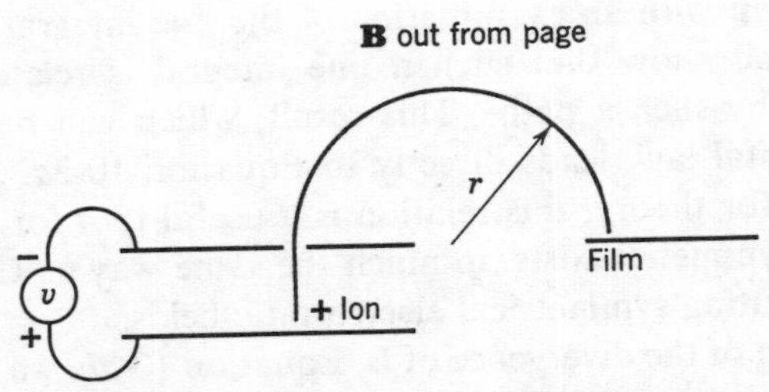

Figure 8. A simple mass spectrograph.

Mass spectrographs employing other principles are to be investigated in connection with a discussion question at the end of this chapter.

When the velocity of the ion is not normal to **B**, it may be resolved into a component parallel to **B** and a component normal to **B**. The parallel component is unaffected by **B**, whereas the normal component is changed so that the ion circles with the **B** direction as an axis. The complete motion is thus a helix along the **B** direction; the greater the value of the flux density, the tighter the helix. Protons approaching the earth from the sun tend to follow a helical path along the magnetic field of the earth, the helices becoming tighter as they approach the poles.

10. Introduction to Ampère's Circuital Law

We have found two fundamental laws for the electrostatic field

$$\nabla \cdot \mathbf{E} = \frac{\rho + \rho_P}{\epsilon_0} \quad \text{or} \quad \nabla \cdot \mathbf{D} = \rho \tag{10-1a}$$

$$\oint \mathbf{E} \cdot d\mathbf{l} = 0 \tag{10-1b}$$

and we now proceed to find comparable laws involving the magnetic flux density. Before the end of this chapter, we shall have shown that

$$\oint \mathbf{B} \cdot d\mathbf{l} = \mu_0 \int \mathbf{J} \cdot d\mathscr{A} \tag{10-2a}$$

$$\nabla \cdot \mathbf{B} = 0 \tag{10-2b}$$

where **J** is the current density.

Thus the electrostatic field is related to the density of its sources through a divergence, whereas the magnetic flux density is related to the density of its sources through a line integral. Moreover, the fact that the line integral of the electrostatic field is zero means that there are no field lines which form closed contours and such a field is called irrotational. The vanishing of the divergence of **B**, on the other hand, means that **B**-field lines have no beginnings or ends, for which property the **B**-field is referred to as being solenoidal.

We shall begin with an examination of the line integral of **B** around a closed path, and show that such a line integral is related to the total current linked by such a path. This result, which will be referred to as Ampère's circuital law, leads directly to Equation 10-2*a*. In addition to its importance for theory, this relation is a useful tool for computing the **B**-field where symmetry exists, in much the same way as Gauss's law was an aid in computing symmetrical electrostatic fields.

The vanishing of the divergence of **B**, Equation 10-2*b*, we shall find to be a necessary consequence of Equation 6-3.

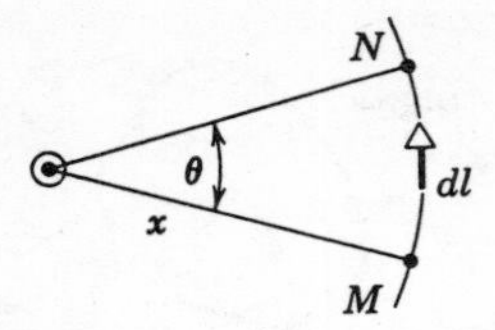

Figure 9. Line integral of **B** *along an arc around an infinite straight wire.*

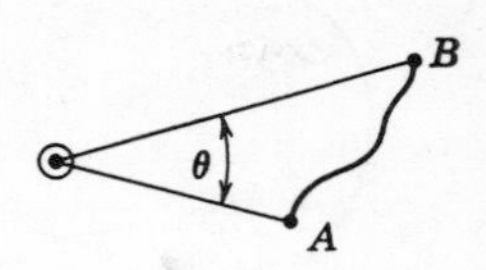

Figure 10. Line integral of **B** *on an arbitrary path.*

Before discussing a general derivation of Ampère's circuital law, let us begin by considering a special case and one or two applications to get a feeling for this law.

The **B**-field associated with the current in an infinite straight wire was found in Section 6 to be

$$B = \frac{\mu_0 I}{2\pi x} \tag{10-3}$$

where I is the current in the wire and x is the perpendicular distance from the wire to the point where the field is being determined. The direction of **B** is perpendicular to both I and x, following the right-hand rule. The line integral of **B** over an arc of a circle of radius x, which intercepts an angle θ at the wire, is illustrated in Figure 9, where the wire is shown in cross section with the current directed out from the paper. Since $dl = x\,d\theta$ and $d\mathbf{l}$ is parallel to **B**,

$$\int_M^N \mathbf{B} \cdot d\mathbf{l} = \int_M^N Bx\,d\theta = \frac{\mu_0 I \theta}{2\pi} \tag{10-4}$$

showing that this integral is a function only of θ and not of the distance x from the wire. For an arbitrary path covering an angle θ, as in Figure 10, the result is the same because the path can be approximated as closely as we please by alternate radial and arc segments. Along the radial segments, $d\mathbf{l}$ is normal to **B** and the integral over these segments is zero, while the integral over the arcs is the same as in Equation 10-4.

Clearly, then, if the path of integration is a closed one that completely encircles the current, as in (*a*) of Figure 11, the value of the integral is $\mu_0 I_0$ whereas if it does not encircle the current as in (*b*) of Figure 11, the value of the integral is zero. Later we shall show that this relation,

$$\oint \mathbf{B} \cdot d\mathbf{l} = \mu_0 I \tag{10-5}$$

is true in general, where I is the total current encircled by the path of

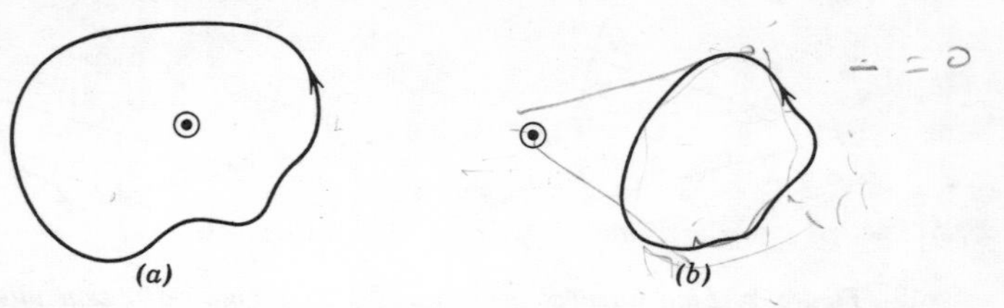

Figure 11. *Closed paths encircling a current (a) and not encircling a current (b).*

integration, regardless of the shape of the conductor giving rise to the field.

But before demonstrating this general theorem, we shall assume its truth and look at two cases where it is useful in computing the **B** field.

11. The Magnetic Flux Density within a Solenoid

A solenoid is formed by a uniformly wound helical conductor of constant radius, the length of the winding being large compared to the radius. In such a solenoid, shown schematically in cross section in Figure 12, the intensity of the field just outside the solenoid —along the line DC, for example—is negligible so long as the points on this line are far from the ends of the solenoid; the intensity within the solenoid —for points on the line AB, for example —is directed parallel to the axis of the solenoid and is constant over the length of AB.

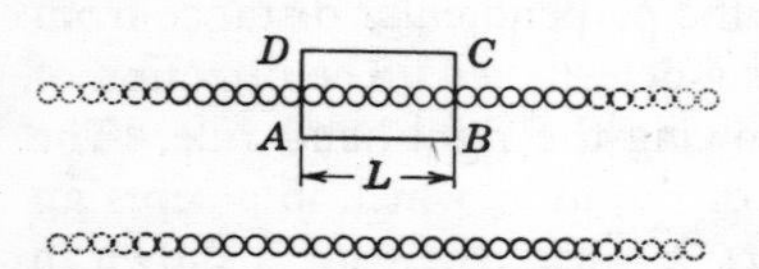

Figure 12. *Circuital path of integration in a solenoid.*

The line integral of B over the closed path $ABCD$ consists of four parts:

$$\oint \mathbf{B} \cdot d\mathbf{l} = \int_A^B \mathbf{B} \cdot d\mathbf{l} + \int_B^C \mathbf{B} \cdot d\mathbf{l} + \int_C^D \mathbf{B} \cdot d\mathbf{l} + \int_D^A \mathbf{B} \cdot d\mathbf{l} \tag{11-1}$$

The integral from A to B is BL, since **B** is constant; the integrals over BC and over DA are zero since **B** is normal to the path of integration; and the integral over CD is zero, because **B** is zero. Hence

$$\oint \mathbf{B} \cdot d\mathbf{l} = BL \tag{11-2}$$

With n conducting turns per unit length and a current I in the turns, the total current encircled by the path $ABCD$ is $(nL)I$. Consequently, the magnetic flux density within the solenoid at points far from the ends is given by

$$B = \mu_0 nI \tag{11-3}$$

independent of where the path AB is located over the cross section of the solenoid. Such long solenoids have been employed when accurately uniform flux densities were required over a limited volume.

12. The Magnetic Flux Density within a Toroid

A toroid is a solenoid which has been bent into a circle, as illustrated in Figure 13. Ampère's circuital law can be used to determine the flux density at a point a distance r from the center. From the symmetry of the configuration it is seen that **B** has the same magnitude at all points on a circle of radius r, and the right-hand rule shows that **B** is directed along the arc of this circle in a counterclockwise direction when the current is in the sense of the arrows in Figure 13. The line integral of **B** around this circle is thus

$$\oint \mathbf{B} \cdot d\mathbf{l} = 2\pi r B \tag{12-1}$$

With a total of N turns carrying current I, the circuital law requires that

$$2\pi r B = \mu_0 N I \tag{12-2}$$

and hence

$$B = \frac{\mu_0 N I}{2\pi r} \tag{12-3}$$

This flux density is not uniform over the cross section of the toroid, but if r is large compared with the cross-sectional dimensions of the coil itself, then very nearly $N = 2\pi r n$ where n is the turns per unit length and B within the toroid becomes very closely

$$B \cong \mu_0 n I \tag{12-4}$$

the same expression as for the long solenoid. A long solenoid may be regarded as the limiting case of a toroid as the radius of the toroid is allowed to become very large.

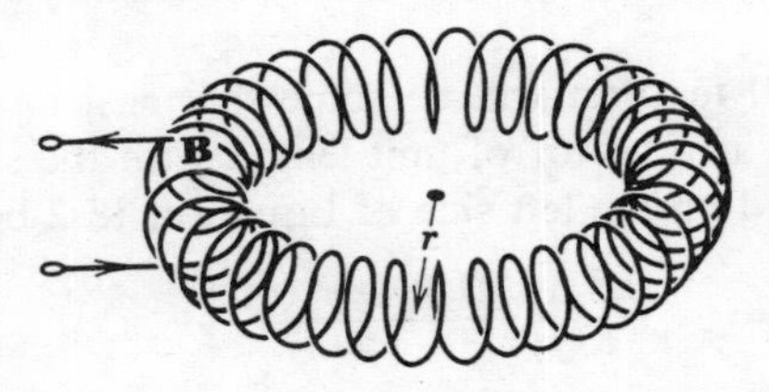

Figure 13. The **B***-field within a toroid is uniform on a circle of radius r.*

13. Ampère's Circuital Law: A General Proof

A general demonstration of the validity of Ampère's circuital law is fundamental to electromagnetic theory, and as a by-product of such a demonstration the reader will gain experience in the manipulation of vectors as well as in their geometrical interpretation.

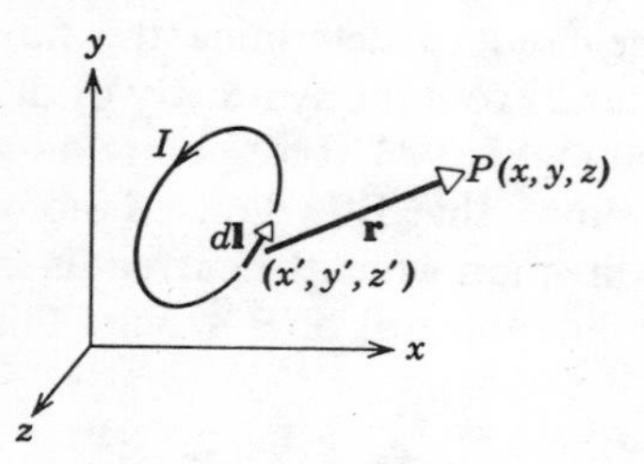

Figure 14. Current loop giving rise to **B**-*field at P.*

The proof starts with the defining relation for B given in Equation 6-3:

$$\mathbf{B} = \frac{\mu_0 I}{4\pi} \oint \frac{d\mathbf{l} \times \mathbf{r}}{r^3} \tag{13-1}$$

where the subscripts on I and $d\mathbf{l}$ used in Section 6 are no longer necessary. The integration in Equation 13-1 is to be carried out over the complete current loop that gives rise to **B**. Thus in Figure 14, the flux density at the point P, with coordinates x, y, and z, is the summation of the contributions of the elements $d\mathbf{l}$, with coordinates x', y', and z'. Having computed **B** at the general point P, we are then required by Ampère's circuital law to find the line integral of **B** over a closed path such as illustrated in Figure 15. Let an element of this closed path be designated by $\delta\mathbf{r}$. The complete expression for Ampère's circuital law is

$$\oint \mathbf{B} \cdot \delta\mathbf{r} = \left(\frac{\mu_0 I}{4\pi}\right) \oint_P \oint_I \frac{d\mathbf{l} \times \mathbf{r} \cdot \delta\mathbf{r}}{r^3} \tag{13-2}$$

where the integral labeled P is over the closed path involving the coordinates x, y, z, and the integral labeled I is over the current loop involving the coordinates x', y', and z'. It will be convenient initially to evaluate the product $\mathbf{B} \cdot \delta\mathbf{r}$, leaving its integration around the closed path until later.

The relation between the point P and the current loop is the same, whether we move the point P through a vector distance $\delta\mathbf{r}$ or, keeping P fixed, move each point on the current loop by an amount $\delta\mathbf{r}'$, where $\delta\mathbf{r} = -\delta\mathbf{r}'$. Figure 16 illustrates the complementary character of the two alternatives.

The geometrical features will be somewhat easier to understand if we define a vector $\mathbf{r}_u$ as a vector of unit length with the same direction as $\mathbf{r}$. Then the integrand on the left side of Equation 13-2 becomes:

$$\mathbf{B} \cdot \delta\mathbf{r} = -\frac{\mu_0 I}{4\pi} \oint_I \frac{[d\mathbf{l} \times \mathbf{r}_u] \cdot (\delta\mathbf{r}')}{r^2} \tag{13-3}$$

the integration being only over the primed coordinates. The numerator

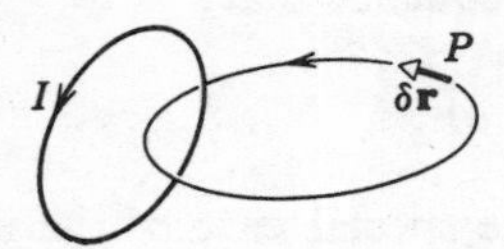

Figure 15. Path of integration of **B** *linking current loop.*

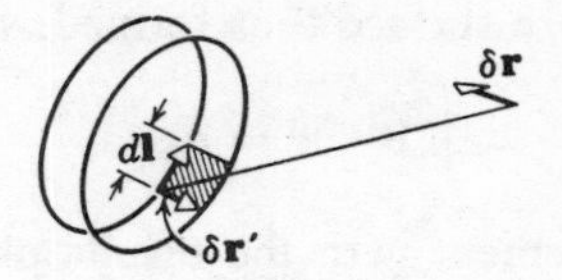

Figure 16. $\delta\mathbf{r}' = \delta'\mathbf{r}$.

in the integrand is a triple scalar product such as was described in Section 2 of this chapter. In such a product the dot and the cross may be interchanged or the vectors may be permuted in cyclic order without changing the value of the product, but a permutation not in cyclic order reverses the sign of the product. The following manipulations of the numerator can thus be made:

$$-d\mathbf{l} \times \mathbf{r}_u \cdot \delta\mathbf{r}' = \mathbf{r}_u \times d\mathbf{l} \cdot \delta\mathbf{r}' = \mathbf{r}_u \cdot d\mathbf{l} \times \delta\mathbf{r}' \tag{13-4}$$

so that Equation 13-3 becomes

$$\mathbf{B} \cdot \delta\mathbf{r} = \frac{\mu_0 I}{4\pi} \oint \frac{\mathbf{r}_u \cdot d\mathbf{l} \times \delta\mathbf{r}'}{r^2} \tag{13-5}$$

A geometrical interpretation of the integrand in Equation 13-5 is now possible. From Figure 16 it is seen that the element $d\mathbf{l}$ in moving through the distance $\delta\mathbf{r}'$ sweeps out the area given by $d\mathbf{l} \times \delta\mathbf{r}'$, the direction of the vector area being outward from the band-like surface swept out by the whole current loop. The dot product of $d\mathbf{l} \times \delta\mathbf{r}'$ with the unit vector $\mathbf{r}_u$ can be interpreted either as the component of this area vector along the direction of $\mathbf{r}$, or as the projection of the area itself on a plane normal to $\mathbf{r}$. This area projection divided by r^2 is the solid angle which $d\mathbf{l} \times \delta\mathbf{r}'$ intercepts at the point P, and hence the integral on the right of Equation 13-5 is the solid angle which the band-like surface swept out by the whole current loop intercepts at P. Writing this solid angle as $d\Omega$, Equation 13-5 becomes

$$\mathbf{B} \cdot \delta\mathbf{r} = \frac{\mu_0 I}{4\pi}\, d\Omega \tag{13-6}$$

We now let the current loop move through that series of displacements $\delta\mathbf{r}'$ which are step by step complementary to the elements $\delta\mathbf{r}$ of the closed path of P. In this process the loop sweeps out a closed surface. If the original path of the point P threads through the current loop, then in the complementary case the point P is inside the surface swept out by the loop. The integration of Equation 13-6 is an integration of the solid angle over the entire surface thus formed, with the consequence that

$$\oint \mathbf{B} \cdot \delta\mathbf{r} = \mu_0 I \tag{13-7}$$

in agreement with the results obtained in the special case of the infinite straight wire. If the path of the point P does not thread through the current loop, the complementary surface does not enclose the point P and the integrated solid angle is zero.

If, instead of a filamentary loop of current, the current is distributed in space so that there is a current density **J** which is a function of position, then the total current encircled by a closed path will be

$$I = \int \mathbf{J} \cdot d\mathcal{A} \tag{13-8}$$

where the integration is over an area of which the path is a perimeter. Consequently, a more general form of Equation 13-7 is

$$\oint \mathbf{B} \cdot \delta\mathbf{r} = \mu_0 \int \mathbf{J} \cdot d\mathcal{A} \tag{13-9}$$

It is worthy of note that the area of integration on the right of Equation 13-9 is not uniquely defined but can be any area whose perimeter is the path of integration on the left.

14. The Curl of a Vector and Stokes's Theorem

In order to proceed further with a discussion of the properties of the magnetic flux density, it is desirable to develop another feature of vector calculus.

The operator del has been used to form the gradient of a scalar function of space and the divergence of a vector function of space. We now examine the interpretation of the cross product of ∇ with a vector function. This is called the curl—sometimes, in European texts, the rotation—and is written variously as $\nabla \times \mathbf{R}$, curl **R**, or rot **R**, where **R** is a general vector function. In this book the first form is used. Its expansion in terms of the rectangular components of the vector **R** is formed just as the cross product of two vectors:

$$\nabla \times \mathbf{R} = \mathbf{i}\left(\frac{\partial R_z}{\partial y} - \frac{\partial R_y}{\partial z}\right) + \mathbf{j}\left(\frac{\partial R_x}{\partial z} - \frac{\partial R_z}{\partial x}\right) + \mathbf{k}\left(\frac{\partial R_y}{\partial x} - \frac{\partial R_x}{\partial y}\right) \tag{14-1}$$

$$= \begin{vmatrix} \mathbf{i} & \mathbf{j} & \mathbf{k} \\ \partial/\partial x & \partial/\partial y & \partial/\partial z \\ R_x & R_y & R_z \end{vmatrix}$$

The curl function is another type of space differentiation of a vector and is itself a vector.

Stokes's theorem (*not* Stokes's law in physics for a body falling in a viscous medium) is a relation between the line integral of a vector around a closed path and the integral of the curl of that vector over an area of which the closed path is a perimeter. Thus

$$\oint \mathbf{R} \cdot d\mathbf{l} = \int \nabla \times \mathbf{R} \cdot d\mathcal{A} \tag{14-2}$$

As a first step in demonstrating Stokes's theorem we need to generalize the concept of the gradient. In the differential calculus it is shown that when a scalar function has the value φ at a point x, then at a point $x + dx$ its value to first order differentials is

$$\phi + \frac{\partial \phi}{\partial x} dx \tag{14-3}$$

In three dimensions, on the other hand, the value of ϕ at $x + dx, y + dy, z + dz$ is given by

$$\phi + \frac{\partial \phi}{\partial x} dx + \frac{\partial \phi}{\partial y} dy + \frac{\partial \phi}{\partial z} dz = \phi + \nabla\phi \cdot d\mathbf{r} \tag{14-4}$$

where $\mathbf{r} = \mathbf{i}x + \mathbf{j}y + \mathbf{k}z$ and $d\mathbf{r} = \mathbf{i}\, dx + \mathbf{j}\, dy + \mathbf{k}\, dz$

By an analogous operation, when the value of a vector function at a point fixed by the position vector $\mathbf{r}$ is $\mathbf{R}$, then its value at the point $\mathbf{r} + d\mathbf{r}$ is

$$\mathbf{R} + \nabla\mathbf{R} \cdot d\mathbf{r} \tag{14-5}$$

where now the second part, representing the change in the function $\mathbf{R}$, has nine terms when expanded in Cartesian coordinates:

$$\begin{aligned} \nabla\mathbf{R} \cdot d\mathbf{r} = {} & \mathbf{i}\left[\frac{\partial R_x}{\partial x} dx + \frac{\partial R_x}{\partial y} dy + \frac{\partial R_x}{\partial z} dz\right] \\ & + \mathbf{j}\left[\frac{\partial R_y}{\partial x} dx + \frac{\partial R_y}{\partial y} dy + \frac{\partial R_y}{\partial z} dz\right] \\ & + \mathbf{k}\left[\frac{\partial R_z}{\partial x} dx + \frac{\partial R_z}{\partial y} dy + \frac{\partial R_z}{\partial z} dz\right] \end{aligned} \tag{14-6}$$

Figure 17 shows a plane figure whose sides are the infinitesimal vectors $d\mathbf{r}_1$ and $d\mathbf{r}_2$. Let $\mathbf{R}$ be a vector function of position whose value at the center of the figure is $\mathbf{R}$ and whose values along the sides of the figure are given by expressions of the form of Equation 14-5. We form the line integral of the vector $\mathbf{R}$ along the path *MNOPM* and show its relation to the curl of $\mathbf{R}$.

From Equation 14-5 the value of R along the four parts of this path are

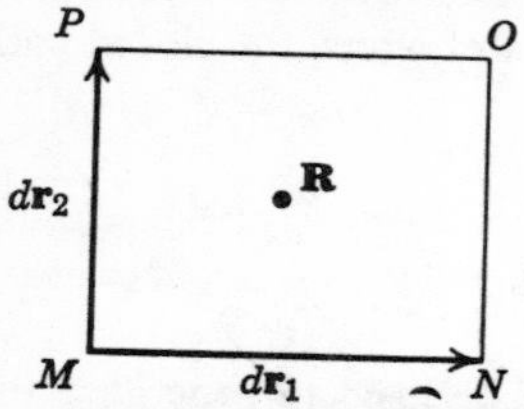

Figure 17. The line integral of R around the perimeter is related to the $\nabla \times \mathbf{R}$ *at the middle.*

$$\text{(14-7)} \qquad \text{from } M \text{ to } N \qquad \mathbf{R} - \nabla\mathbf{R}\cdot\frac{d\mathbf{r}_2}{2}$$

$$\text{from } N \text{ to } O \qquad \mathbf{R} + \nabla\mathbf{R}\cdot\frac{d\mathbf{r}_1}{2}$$

$$\text{from } O \text{ to } P \qquad \mathbf{R} + \nabla\mathbf{R}\cdot\frac{d\mathbf{r}_2}{2}$$

$$\text{from } P \text{ to } M \qquad \mathbf{R} - \nabla\mathbf{R}\cdot\frac{d\mathbf{r}_1}{2}$$

Thus the line integral about this closed path in the counterclockwise sense is

$$\text{(14-8)} \qquad \oint \mathbf{R}\cdot d\mathbf{l} = (\mathbf{R} - \tfrac{1}{2}\nabla\mathbf{R}\cdot d\mathbf{r}_2)\cdot d\mathbf{r}_1 + (\mathbf{R} + \tfrac{1}{2}\nabla\mathbf{R}\cdot d\mathbf{r}_1)\cdot d\mathbf{r}_2 + (\mathbf{R} + \tfrac{1}{2}\nabla\mathbf{R}\cdot d\mathbf{r}_2)\cdot(-d\mathbf{r}_1) + (\mathbf{R} - \tfrac{1}{2}\nabla\mathbf{R}\cdot d\mathbf{r}_1)\cdot(-d\mathbf{r}_2)$$

We further note that the vector area of $MNOP$ is

$$\text{(14-9)} \qquad d\mathbf{r}_1 \times d\mathbf{r}_2 = \mathbf{i}(dy_1\,dz_2 - dy_2\,dz_1) + \mathbf{j}(dz_1\,dx_2 - dz_2\,dx_1) + \mathbf{k}(dx_1\,dy_2 - dx_2\,dy_1) = \mathbf{i}\,d\mathscr{A}_x + \mathbf{j}\,d\mathscr{A}_y + \mathbf{k}\,d\mathscr{A}_z$$

where $d\mathscr{A}_x$, $d\mathscr{A}_y$, and $d\mathscr{A}_z$ are the components of the vector representing the area of $MNOPM$. Using Equations 14-6 and 14-9 to expand 14-8 we get after collecting terms:

$$\text{(14-10)} \qquad \oint \mathbf{R}\cdot d\mathbf{l} = \left(\frac{\partial R_z}{\partial y} - \frac{\partial R_y}{\partial z}\right) d\mathscr{A}_x + \left(\frac{\partial R_x}{\partial z} - \frac{\partial R_z}{\partial x}\right) d\mathscr{A}_y + \left(\frac{\partial R_y}{\partial x} - \frac{\partial R_x}{\partial y}\right) d\mathscr{A}_z = \nabla \times \mathbf{R}\cdot d\boldsymbol{\mathscr{A}}$$

as the value of the line integral around this infinitesimal area.

The curl of a vector is thus related to the line integral of that vector over an infinitesimal closed path. Its relation to a *finite* closed path is indicated in Figure 18. Here the area within the closed path has been subdivided into small regions, the circular arrows indicating the direction in which to perform the line integral of the vector field over the boundary of each region. The summation of these line integrals over all the boundaries will equal the line integral over the outer boundary, since any line common to two subregions is traversed in opposite directions by integrations around the two regions and thus drops out of the final sum. Since each line integral about a subregion is given by an expression of the form of Equation 14-10, the line integral about the whole region becomes:

$$\oint \mathbf{R} \cdot d\mathbf{l} = \int \nabla \times \mathbf{R} \cdot d\mathscr{A} \tag{14-11}$$

where the area involved in the integral on the right has the path of integration on the left as a perimeter. There are an infinite number of such areas, however, and any of them can be used in Equation 14-11, a fact that will be important in later considerations.

The relation given in Equation 14-11 is known in mathematics as Stokes's theorem. Its derivation involved no restrictions on the nature of the function R (other than the usual ones of continuity and finiteness) and hence the theorem is general. We shall have considerable occasion to use it in ways comparable to the way we used Gauss's theorem. Gauss's theorem converts a volume integral to an area integral, or the reverse, when the volumes and areas are properly related. Stokes's theorem similarly converts an area integral to a line integral, or the reverse, when the area and the line are properly related.

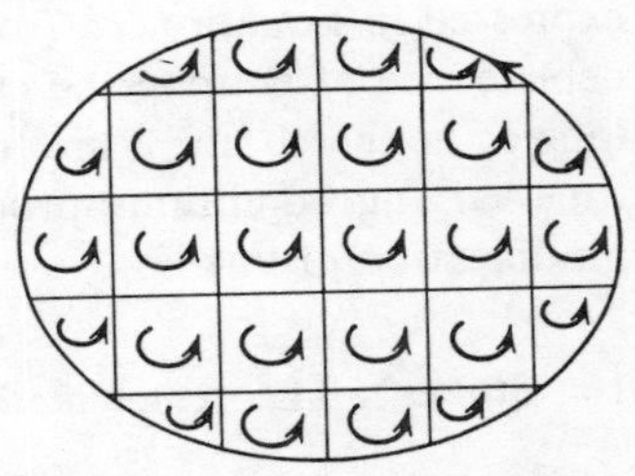

Figure 18. *Integrating the curl over a finite area.*

15. The Point Form of Ampère's Law

The left side of Equation 13-9 is a line integral around a closed path of the vector field **B**. By Stoke's theorem this integral can be converted to an area integral involving the curl of **B**, so that Equation 13-9 becomes

$$\int \nabla \times \mathbf{B} \cdot d\mathscr{A} = \mu_0 \int \mathbf{J} \cdot d\mathscr{A} \tag{15-1}$$

another integral form of Ampère's circuital law. The law applies to any area, however small; let us write it for an infinitesimal area, in effect by removing the integral signs:

$$\nabla \times \mathbf{B} \cdot d\mathscr{A} = \mu_0 \mathbf{J} \cdot d\mathscr{A} \tag{15-2}$$

In general, it does not follow that if two scalar products are equal, such as

$$\mathbf{R} \cdot \mathbf{P} = \mathbf{Q} \cdot \mathbf{P} \tag{15-3}$$

that necessarily **R** must equal **Q**. If, however, the equality holds for any orientation of the vector **P**, then clearly **R** and **Q** must be equal. Since

Equation 15-2 holds for any area, $d\mathscr{A}$, it therefore follows that

$$\nabla \times \mathbf{B} = \mu_0 \mathbf{J} \tag{15-4}$$

which is the point form of Ampère's circuital law, relating the rates of change of the magnetic flux density at a point in space with the current density at the same point. Should it be possible to find an infinitesimal area for which Equation 15-4 does not hold or an integrated area for which 15-1 does not hold, then something is wrong with Ampère's law as expressed in Equation 13-7 or 13-9 and we shall need to reinvestigate the dependence of **B** upon the current. No such difficulty is encountered so long as neither **B** nor **J** are functions of time. But we shall find later that time-varying conditions produce contradictions that are resolved by a modification of this law.

16. Stokes's Theorem and the Electric Field

It was demonstrated in Chapter 2 that conservation of energy required that the line integral around a closed path of the electrostatic field must vanish. From Stoke's theorem this line integral may be converted to an area integral over the area for which the line of integration is the perimeter:

$$\oint \mathbf{E} \cdot d\mathbf{l} = \int \nabla \times \mathbf{E} \cdot d\mathscr{A} = 0 \tag{16-1}$$

from which, by the same reasoning that lead to Equation 15-4, we get

$$\nabla \times \mathbf{E} = 0 \tag{16-2}$$

An identity in vector algebra says that the curl of the gradient of any scalar function is always zero:

$$\nabla \times \nabla \phi = 0 \tag{16-3}$$

a fact that the reader can readily prove to himself by performing the indicated operations in Cartesian coordinates (although the validity of the relation does not depend on the choice of coordinate system).

A comparison of Equations 16-2 and 16-3 shows that it is always possible to express the vector **E** as the gradient of some scalar function, say $-V$:

$$\mathbf{E} = -\nabla V \tag{16-4}$$

where the negative sign is chosen in order that V may be interpreted as work done by an outside agent, as in Chapter 2.

Equation 15-4, relating the magnetic flux density to the current, shows that in such regions where the current density is zero the magnetic flux density may also be expressed in terms of a scalar potential function, but the integral form of Ampère's circuital law requires that the region over which such a scalar potential is used must be so limited that it is not possible to find a line integral of **B** which encircles any current. In certain special cases such a scalar magnetic potential has applications.

17. The Magnetic Vector Potential

It is, however, possible to find a vector function from which **B** can always be found by differentiation, comparable to the scalar electrostatic potential whose differentiation gives **E**. That is, just as the electrostatic field, **E**, can be expressed as the gradient of a scalar function, V, so we shall now show that the magnetic flux density, **B**, can be expressed as the curl of a vector function, **A**:

$$\mathbf{B} = \nabla \times \mathbf{A} \tag{17-1}$$

This function is called the vector potential. The name is chosen by analogy with V, but "potential" in this case should not be directly associated with the concept of work. There is an indirect association, however, for the reader can easily show that the time derivative of **A**, like the space derivative of V, has the dimensions of volts per meter.

To develop the relation 17-1 we again go back to the Equation 6-3:

$$\mathbf{B} = \frac{\mu_0 I}{4\pi} \oint \frac{d\mathbf{l} \times \mathbf{r}}{r^3} \tag{17-2}$$

and refer to Figure 14. It is again to be emphasized that the integration and the vector $d\mathbf{l}$ involve only the variables x', y', and z', that the point P is fixed by the coordinates x, y, and z, whereas **r** involves both sets of coordinates. Thus

$$\mathbf{r} = \mathbf{i}(x - x') + \mathbf{j}(y - y') + \mathbf{k}(z - z') \tag{17-3}$$
$$r = \sqrt{(x - x')^2 + (y - y')^2 + (z - z')^2}$$

It is left as a problem for the reader to show that

$$\nabla\left(\frac{1}{r}\right) = -\frac{\mathbf{r}}{r^3} \tag{17-4}$$

by means of which Equation 17-2 takes the form:

$$\mathbf{B} = -\frac{\mu_0 I}{4\pi} \oint d\mathbf{l} \times \nabla\left(\frac{1}{r}\right) \tag{17-5}$$

Another vector identity which the reader can readily verify by performing the indicated operations is

$$\nabla \times (q\mathbf{R}) = q\nabla \times \mathbf{R} - \mathbf{R} \times \nabla q \tag{17-6}$$

where **R** is a vector function and q a scalar function of position. Thus the integrand of Equation 17-5 may be written as

$$d\mathbf{l} \times \nabla(1/r) = (1/r)\nabla \times d\mathbf{l} - \nabla \times (d\mathbf{l}/r) \tag{17-7}$$

But because the differentiations in the curl are with respect to the unprimed variables, the first term on the right of Equation 17-7 is zero, and 17-5 takes the form

$$\mathbf{B} = \frac{\mu_0 I}{4\pi} \oint \nabla \times \left(\frac{d\mathbf{l}}{r}\right) \tag{17-8}$$

Integration and differentiation operations can be interchanged, being with respect to different variables, so that

$$\mathbf{B} = \frac{\mu_0 I}{4\pi} \nabla \times \oint \frac{d\mathbf{l}}{r} \tag{17-9}$$

which shows that **B** can always be written as the curl of a vector **A**, where **A** is defined by

$$\mathbf{A} = \frac{\mu_0 I}{4\pi} \oint \frac{d\mathbf{l}}{r} \tag{17-10}$$

Neither **B** nor **A** are necessarily easy to compute for an arbitrary current distribution, but in most cases **A** may be expected to be somewhat simpler than **B**. **A** is still a vector function and, therefore, the simplification is not as great as that introduced by using the scalar function V to arrive at **E**. Nevertheless, the vector potential has important theoretical implications for electromagnetic theory beyond its occasional usefulness in computing **B**.

One such implication can be shown very simply by using the vector identity that the divergence of the curl of any vector is always identically zero:

$$\nabla \cdot \nabla \times \mathbf{R} = 0 \tag{17-11}$$

where **R** is any vector function. A comparison of Equation 17-11 with 17-9 shows that the relation

$$\nabla \cdot \mathbf{B} = 0 \tag{17-12}$$

is a completely general relation for the magnetic flux density, analogous to Gauss's law for the electrostatic field. Equation 17-12 says that there

are no charges analogous to electric charges from which the **B** field arises, and that for any volume in space as much flux of **B** enters the volume as leaves it.

18. Further Comments on the Vector Potential

Although the curl of the function given in Equation 17-10 does give the **B** field associated with the current I, it is not the only function whose curl is **B**. Just as the scalar potential is defined to within an arbitrary constant, so clearly is the vector potential. But the indefiniteness of **A** goes beyond a mere additive constant, for Equation 16-3 shows that a function **A′** given by

$$\mathbf{A}' = \mathbf{A} + \nabla\phi \tag{18-1}$$

where ϕ is any scalar function whatsoever, will have the same curl as the function **A**, and hence will produce the same magnetic flux density, **B**. For reasons not connected with electromagnetic theory, the formation of a function such as **A′** from a function **A** by the addition of the gradient of a scalar is called a gauge transformation. Essentially, this indefiniteness of the function **A′** arises because, for a vector function to be defined to within only an additive constant, it is necessary to specify *both* its curl and its divergence. Only the curl of **A** is defined, through Equation 17-1.

We shall find it helpful on occasion to place a further restriction on the vector potential by arbitrarily fixing its divergence. This does not change the relation between the vector potential and the magnetic flux density, which only involves the curl. For example, suppose we insert Equation 17-1 into the point form of Ampère's circuital law:

$$\nabla \times \mathbf{B} = \nabla \times (\nabla \times \mathbf{A}) = \mu_0\mathbf{J} \tag{18-2}$$

The curl of the curl of a vector is the gradient of the divergence less the Laplacian:

$$\nabla \times (\nabla \times \mathbf{A}) = \nabla(\nabla \cdot \mathbf{A}) - \nabla^2\mathbf{A} \tag{18-3}$$

so that Equation 18-2 becomes

$$\nabla(\nabla \cdot \mathbf{A}) - \nabla^2\mathbf{A} = \mu_0\mathbf{J} \tag{18-4}$$

Since the divergence of **A** may be chosen at our convenience, we can choose it to be zero, thus reducing Equation 18-4 to

$$\nabla^2\mathbf{A} = -\mu_0\mathbf{J} \tag{18-5}$$

which has the form of Poisson's equation, discussed in Chapter 2. Vector equations of the form of 18-5 will be discussed in more detail in Chapter

10, where a different choice for the gauge transformation will be found convenient.

19. Summary

The differential relation giving the force between current elements (Equation 3-1) is an hypothesis whose integral over complete current loops can be experimentally verified and which provides us with an expression for the force between "point" elements of current, such as Coulomb's law represents for point charges. Like Coulomb's law, it is an "action-at-a-distance" relation and, again like Coulomb's law, we separate it into two parts by introducing the concept of a field, such that the force on a current element is determined by the **B**-field where the current element resides, and the **B**-field, in turn, is determined by current elements elsewhere.

The integral and differential properties of the **B**-field, the magnetic flux density, were then worked out in terms of the curl and the divergence. The line integral of **B** around a closed path of integration was found to be proportional to the total current linked by the integration path, from which it follows that the curl of **B** is proportional to the current density. It was shown that the divergence of **B** is alway zero, and hence that the integral of **B** over any closed surface is also zero. The vector **B**, moreover, can always be replaced by the curl of another vector called the vector potential, where this equivalence is analogous to the equivalence between the electrostatic field and the gradient of the static potential.

Discussion Questions

1. Prepare a report on mass spectrographs, discussing the various types of instruments, the problems of focusing in each, their advantages and disadvantages. An old but still excellent summary of the principles and designs of such instruments is by Bleakney, *American Physics Teacher*, **4**, 12, (1936).

2. Discuss the instrument described in Problem 20. This instrument is called a "velocity selector," since it will sort out from an ion beam of heterogeneous velocities that particular velocity determined by the values of the fields. It is interesting that the velocity selected is not affected by the magnitude of the charge on the ion. What effect does the ion charge have on the velocity selecting function of the instrument? In particular, could it be used to sort out velocities in a beam of neutral particles?

3. Discuss the principles of the cyclotron and its operation in terms of the concepts of Section 9. Show that the synchronism of the circulating ions with the alternating electric field is independent of the energy of the ions and that the energy of the ions is determined by their radius of curvature of their path at a given instant, provided the energy is not so great that the mass is affected by relativistic considerations. [Ref: Goble and Baker, *Element of Modern Physics*, p. 366, Ronald, 1962. Scott, *The Physics of Elec. and Mag*., p. 256–58.]

4. The vector form of the force law between two point elements of current is given in Equation 3-1 which expresses the force on element 2, the separation, **r**, being defined as a vector terminating on 2. By interchanging the subscripts on the element of length, $d\mathbf{l}_1$ and $d\mathbf{l}_2$, and reversing the direction of **r** by introducing a negative sign, the resulting expression is the force on element 1. From Newton's law, these two forces should be equal in magnitude and opposite in direction. Are they?

Consider the corresponding question with regard to the integrated form of Equation 3-1. Is the force on the complete current loop 1 equal and opposite to the force on 2? [Ref: Jackson, *Classical Electrodynamics*, p. 136, Wiley. Panofsky and Phillips *Classical Elec. and Mag*., 1st ed., p. 112, Addison-Wesley.]

5. A length of metallic ribbon of width b carries a longitudinal current I as a result of an electric field E in the ribbon associated with a difference of potential across its ends. Normal to the flat face of the ribbon is a uniform flux density B, from which the charge carriers in the ribbon experience a force toward one edge of the ribbon, thereby producing a concentration of charge carriers near this edge. Find the potential difference across the ribbon. Can a measurement of this potential difference be used to decide whether the charge carriers are positive or negative? This effect is known as the Hall effect. Discuss various uses to which it might be put. [Ref: Ham and Slemon, *Scientific Bases of Elec. Eng*., p. 205, Wiley. Duckworth, *Elec. and Magnetism*, p. 224ff, Holt, Rinehart, Winston.]

6. At the end of Section 17 it is demonstrated that $\nabla \cdot \mathbf{B} = 0$ is a general law, and it is interpreted in words as saying that for any volume in space as much flux of **B** enters the volume as leaves it. Another interpretation that has long been popular says that lines of **B**-flux are continuous and always form closed curves. This latter interpretation has been subjected to close scrutiny and the discussions are worth examining with care. One of the earliest references in this discussion is J. Slepian, *Am. J. Phys*., **19**, 87 (1951); and one of the latest is Mario Iona, *Am. J. Phys*., **31**, 398 (1963). The reader will find footnotes in these references leading him to others in an interesting exercise in careful thinking.

Problems

1. Show that

$$\mathbf{Q} \cdot \mathbf{R} \times \mathbf{P} = \begin{vmatrix} Q_x & Q_y & Q_z \\ R_x & R_y & R_z \\ P_x & P_y & P_z \end{vmatrix}$$

2. Using Cartesian coordinate notation, show that

$$\mathbf{Q} \times (\mathbf{R} \times \mathbf{P}) = (\mathbf{Q} \cdot \mathbf{P})\mathbf{R} - (\mathbf{Q} \cdot \mathbf{R})\mathbf{P}$$

3. Show that: $(\mathbf{Q} \times \mathbf{P}) \times \mathbf{R} = (\mathbf{R} \cdot \mathbf{Q})\mathbf{P} - (\mathbf{R} \cdot \mathbf{P})\mathbf{Q}$

4. Show that $\nabla \times \nabla\phi = 0$, where ϕ is any scalar function of position.

5. Show that $\nabla \cdot \nabla \times \mathbf{R} = 0$, where $\mathbf{R}$ is any vector function of position.

6. Show that $\nabla \times (\nabla \times \mathbf{R}) = \nabla(\nabla \cdot \mathbf{R}) - \nabla^2\mathbf{R}$

7. What is the magnetic flux density within a coaxial cable in which the inner and the outer conductors carry equal currents in opposite directions? The inner conductor has a radius a and the inner radius of the outer conductor is b. Assume that the current in the inner conductor is only on the surface.

8. What is the magnetic flux density in the plane containing two parallel wires separated by a distance d and carrying oppositely directed currents, I?

9. What is the magnetic flux density on the normal through the center of a rectangular conducting loop of dimensions a and b, carrying a current I?

10. A thin flat ribbon carries a current I uniformly distributed across its width, b. Find the magnetic flux density at points in the plane of the ribbon and at points in a plane normal to the plane of the ribbon through its center line.

11. In terms of the radius, a, of a coil, find the separation, $2x_1$, of a pair of Helmholz coils so that there is a limited region of approximately uniform magnetic flux density between them, as discussed in Section 7.

12. A tangent galvanometer is an instrument formerly used to measure current. It consists of a circular coil mounted with its plane in the earth's magnetic meridian. At the center of the coil is a compass needle whose angle with the plane of the coil is determined by the combined effect of the earth's field and the field of the current in the coil. In one such instrument, the coil is wound in two equal parts, with a lead brought out at the half-way point. When one-half of the coil is connected in a circuit containing a source of constant emf and a resistance, the needle makes an angle of arctan (0.6) with the plane of the coil. When both halves are connected in series in the same circuit, the angle is arctan (1.0). What is the angle when the two halves are connected in parallel?

13. A solenoid has a length L, has n turns per unit length, and carries a current I. Its radius is a. Consider a length dx of this solenoid, containing $n\,dx$ turns, as a current loop, as in Section 7, and find the axial magnetic flux density at a point a distance x from the plane of dx. Then integrate this flux density over the length of the solenoid to find the total flux density at this point. Apply this result to a point within the solenoid, far from either end. Compare with the results of Section 11. Also find the flux density on the axis in the plane of either end. How does this value compare with the flux density well within the solenoid?

14. Use a closed path similar to that used in Section 11 to find the magnetic flux density within the toroid discussed in Section 12.

15. The rectangular loop of Problem 9 is in the same plane as an infinite straight wire carrying a current I_0, the wire being parallel to the sides b, with the nearest side being a distance x from the straight wire. What is the force on the rectangular loop? Show in a diagram the directions of the currents in order for the force to be an attraction.

16. A circular loop of wire of radius a is in the same plane with an infinite straight wire, the center of the loop being a distance x from the wire. If the currents in the loop and the wire are I and I_0, respectively, what is the force on the loop? Show in a diagram the directions of the currents so that this force is an attraction.

17. A Faraday disk motor consists of a solid conducting disk of radius a, turning on an axle through the center and normal to the plane of the disk. One terminal of a source of emf is connected to the axle and the other to a brush riding on the rim of the disk. Normal to the face is a uniform flux density B. Show that the torque on the disk is $(BIa^2)/2$, where I is the current between the rim and the axle.

18. A solenoid has 8 turns per centimeter and carries 4 amperes. A small coil of 20 turns, average radius of 1 centimeter, and carrying a current of 100 ma, is located within the solenoid, its vector area making an angle of 30° with the magnetic flux density. What is the torque on the coil?

19. From Equation 7-4, show that the magnetic flux density on the axis of a circular loop at points far from the loop compared to its radius is given by

$$B_x = \frac{\mu_0}{4\pi} \frac{2m}{r^3}$$

where m is the magnetic moment of the loop. Compare this result with the corresponding electric field on the axis of a dipole as found from Equation 8-5, Chapter 2.

20. A beam of ions enters a region in which an electric field **E** and a magnetic flux density **B** are normal to each other and both normal to the velocity of the ions. Show that if the velocity of the ions is related to the fields by

$$\mathbf{v} = \frac{\mathbf{E} \times \mathbf{B}}{B^2}$$

the ions will pass through the region undeflected, but that ions of other velocities will be bent out of the beam. Draw a diagram of the vectors representing the electric field and the magnetic flux density for a beam of positive ions directed out from the paper. (See also Discussion Question 2.)

21. Figure P21 shows one type of mass spectrograph, in which an ion of charge Q and mass m traveling with a velocity $\mathbf{v}_0 = \mathbf{i}v_{0x} + \mathbf{j}v_{0y}$ enters a region in which a uniform electric field is in the y direction and a uniform magnetic flux density is in the z direction. Show that the path of the ion is a cycloid and discuss the nature of the cycloidal path for different values of $\theta = \arctan \dfrac{v_{0x}}{v_{0y}}$.

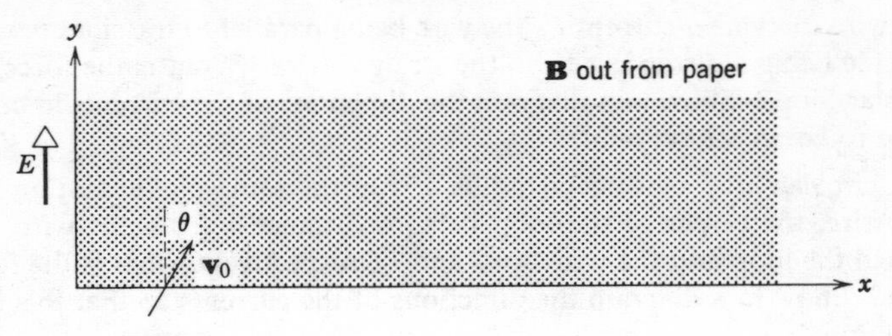

Figure P21

22. Show that the ions in Problem 21 will in general pass through the $x - z$ plane (the plane of the entering slit) in a direction parallel to $\mathbf{v}_0$ at points whose separation is both proportional to m/Q and independent of the direction or the magnitude of $\mathbf{v}_0$.

23. Using Equation 13-7, find the magnetic flux density at points within the cross section of a cylindrical conductor of radius a which carries a current I, where the current is uniformly distributed over the cross section. Sketch a graph of the flux density versus radial distance from the center of the conductor, and extend the graph to include points outside the conductor.

24. The magnetic flux density **B** is given in Equation 6-1 in terms of a filamentary current I. Show that **B** may be expressed in terms of the current density **J** by

$$\mathbf{B} = \frac{\mu_0}{4\pi} \int \frac{\mathbf{J} \times \mathbf{r}}{r^3}\, d\mathscr{V}$$

25. Equation 17-10 expresses the vector potential in terms of the filamentary current I. Show, from the result of Problem 24, that in terms of a current density **J**, the vector potential is given by

$$\mathbf{A} = \frac{\mu_0}{4\pi} \int \frac{\mathbf{J}}{r}\, d\mathscr{V}$$

26. Two infinite parallel wires carry equal currents in opposite directions. As illustrated in cross section in Fig. P26, a point P is located a distance $\mathbf{r}_1$ from the left-hand wire and $\mathbf{r}_2$ from the right-hand wire. Show that the vector potential **A** at the point P is given by: $\mathbf{A} = \mathbf{k}\left(\frac{\mu_0 I}{2\pi}\right) \ln \frac{r_2}{r_1}$ where **k** is a unit vector

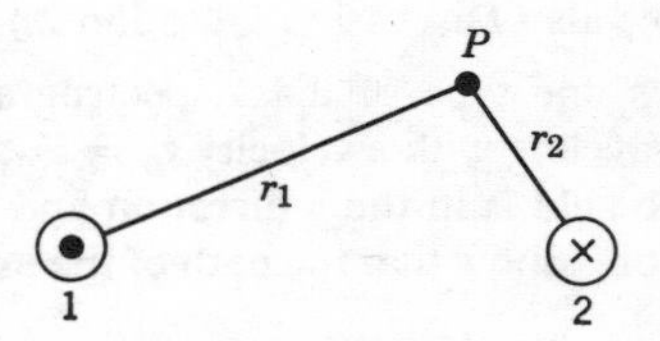

Figure P26

in the direction of the current in wire 1. Do the results of Problem 8 follow from this?

27. From considerations of symmetry, what must be the value of the vector potential on the axis of a circular loop of radius a at a point a distance x from the plane of the loop? In view of the value of B_x at this point (Equation 7-4), what is the value of the rate of change of the z component of A in the y direction?

REFERENCES

The references of previous chapters will, in most cases, be profitable reading on the subjects of this chapter as well. In addition, "Ampère as a Contemporary Physicist," by R. A. R. Tricker, in *Contemporary Physics* (London), August, 1962, is highly recommended.

6

Faraday's law

Two of the field relations which were discussed in the foregoing chapters,

$$\nabla \times \mathbf{E} = 0$$
$$\nabla \times \mathbf{B} = \mu_0 \mathbf{J}$$

apply to steady-state conditions and require further generalization if they are to apply to time-dependent fields. In this chapter we study the relation of electric fields to *changing* magnetic flux densities. In Chapter 12 we find that magnetic flux densities are related to *changing* electric fields.

1. Faraday's Law

In the years around 1830 Michael Faraday, an English physicist and chemist, combined remarkable intuitive insight with experimental skill to show that changing the magnetic conditions in the region of a conducting medium resulted in charges moving while the change was taking place and that the resulting currents became greater with increases in the rate at which the magnetic conditions were changed.

The vague expression, "magnetic conditions," has been chosen intentionally because the precise explanation of Faraday's results requires the defining of a new quantity, the total magnetic flux, Φ, over an area. It is defined by

$$\Phi = \int_c \mathbf{B} \cdot d\mathscr{A} \tag{1-1}$$

where the integration is to be carried out over an area having a closed curve such as c for a perimeter, as in Figure 1. The curve c will often, but not necessarily, be fixed by the location of a conducting loop. It is evident from Equation 1-1 why **B** has been called the magnetic flux *density*, since it appears in the form of flux per unit area. It will be left for the student to show that all areas having the curve c in Figure 1 as a perimeter are associated with the same total flux.

Faraday was able to synthesize the results of many seemingly different experiments* into a remarkably simple law which, expressed in our terms, can be stated as follows:

Around any closed path that may be traced in space there exists an emf which is the negative of the time rate of change of the magnetic flux over an area of which this closed path is a perimeter.

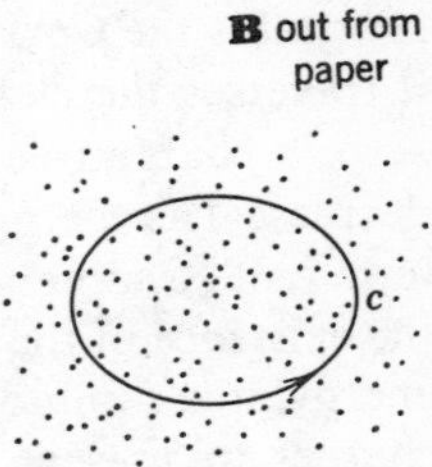

Figure 1. Total flux Φ *over area of curve c is integral of* **B**.

Emf has already been identified as the line integral of the electric field. According to Faraday's law, even though there is no battery, thermocouple, generator, or other obvious source of emf in the closed path, the line integral of **E** need not be zero. Using Equation 1-1, Faraday's law becomes:

$$\oint \mathbf{E} \cdot d\mathbf{l} = -\frac{d\Phi}{dt} = -\frac{d}{dt}\int \mathbf{B} \cdot d\mathscr{A} \tag{1-2}$$

When the magnetic flux is constant in time, the line integral of **E** around the closed path is zero, and Faraday's law reduces to the familiar relation for static electric fields.

The integral on the right side of Equation 1-2 can be a function of time either because **B** changes in magnitude or direction or because the curve c changes shape or orientation. When the curve itself changes, the limits of integration are functions of the time. Various applications of Faraday's law involve all of these possibilities. A loop rotates in a flux field; the flux field itself rotates while the loop remains fixed; the magnitude of the flux density changes; or under less common circumstances, the loop may expand or contract. In each of these cases, an emf is induced in the loop (and there is a current if the loop is a conductor).

For the above reasons the most general form of Faraday's law is that in which the derivative operator stands outside the integral sign, where it can operate on the limits of integration as well as on the integrand. But we

---

* Shamos, *Great Experiments in Physics*, Chap. 10 and 11. A. I. Sharlin, "From Faraday to the Dynamo," *Scientific American*, May, 1961.

shall give considerable attention later to a very important case in which only **B** changes, the circuit (and hence the integration limits) remaining fixed.

The left side of Equation 1-2 can be given a physical interpretation in terms of work done by the field, **E**. Suppose the time rate of change of flux is constant so that the line integral of **E** is also constant in time. Then, from the definition of **E** as the force per unit positive charge, the line integral is the work per unit positive charge performed on a test charge as it traverses the closed path of integration. When this line integral is not zero, a charge free to move around the path increases (or in some cases decreases) its energy. It is because of this energy conversion that we use the term emf (Chapter 4, Section 6). Although the experimental evidence for the existence of the emf is the observation that a charge or charges receive energy, it is common practice to speak of the emf existing, even though no charges are present, such as on an imaginary curve in empty space over the area of which there is a time-varying flux.

2. Lenz's Law

In Figure 1 the vector associated with the area of the closed path has its positive sense out from the page (right-hand rule), and the magnetic flux density also points out from the page. Consequently, the product, $\mathbf{B} \cdot d\mathscr{A}$ is positive and Φ is positive. The negative sign in Faraday's law says that, for a positive change in Φ, the induced emf will be in a negative (that is, clockwise) sense around the closed path. This negative sign is a concise expression of a law enunciated by the Russian physicist, Heinrich Lenz, in 1834, which states that the effects of a change in an electrical system are always such as to oppose the change. In a sense Lenz's law is the electrical counterpart of Newton's third law in mechanics relating action and reaction.

There have been various mnemonic devices suggested to assist in determining the direction of the emf induced by a changing magnetic flux. The author of this book favors the use of the simple right-hand rule which has already been introduced for determining the direction of a vector cross product, for determining the direction of the flux density in relation to a current, and for determining the direction of a vector area relative to its perimeter. The rule is as follows: the sense of the emf induced by a changing flux is given by the fingers of the right hand when the thumb points in the direction *opposite* to the *change* of flux.

Let us consider two elementary examples to see how Lenz's law works. Two coils whose planes are parallel are separated from one another by a short distance. The left-hand coil is connected to a source of emf such that there is a constant current in it, and the right-hand one is simply a closed circuit. The first coil thus produces a flux density in the surrounding space, resulting in a net flux over the area of the second; but since this flux is not changing, no emf is induced. However, if the coil on the left is pushed toward the other coil, the flux over it increases. Using the right-hand rule, we see that the emf and the resulting current in the second coil is in the opposite sense to the current in the first. These oppositely directed currents repel each other, so that there is a force on the first coil opposing the motion. If there were no negative sign in Faraday's law, the currents would be in the same sense, and there would be a force of attraction between the coils. Experiment supports the former and not the latter prediction.

Figure 2. *Current is induced in coil 2 when coil 1 is moved.*

As another example, consider a U-shaped wire on which rests a crossbar free to slide. Normal to the plane of the U is a **B**-field of uniform intensity. By means of a string, say, the crossbar is pulled toward the right (Figure 3). The closed path formed by the U and the crossbar is thus increasing in area and the total flux over its area is increasing, resulting in an induced emf in a counterclockwise sense. Because the crossbar now carries a current, there is a force on it which is directed opposite to the force that originally produced the motion. Again, the absence of the negative sign in Faraday's law would predict a force aiding the original force.

Let us examine the energy balance in this example. Assume frictionless conditions, a constant velocity, and all the resistance of the conducting loop concentrated in the crossbar (to avoid changes of resistance with the motion). Take the length of the bar as a vector, **L**, in the positive y-direction, **B in the negative** z-direction, and **v** in the positive x-direction. The interaction force between the current and the field is

$$\mathbf{F} = i(\mathbf{L} \times \mathbf{B}) \tag{2-1}$$

The force of the string is opposite to this interaction, and the work done by this external force per unit time is

$$-\mathbf{F} \cdot \mathbf{v} = -i(\mathbf{L} \times \mathbf{B} \cdot \mathbf{v}) \tag{2-2}$$

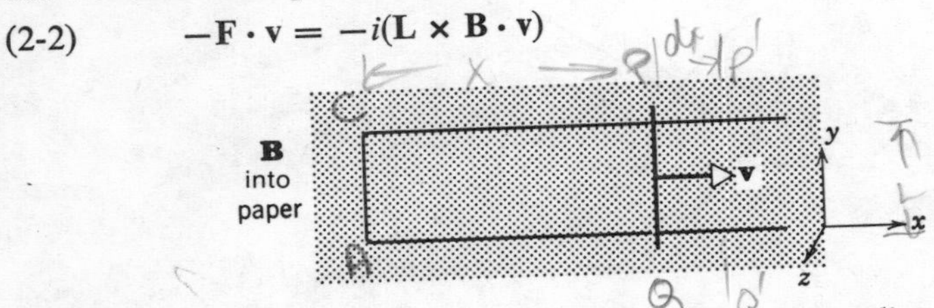

Figure 3. *Current is induced as crossbar is moved.*

By the rule for triple scalar products, this may be changed to

$$-\mathbf{F}\cdot\mathbf{v} = -i(\mathbf{v}\times\mathbf{L}\cdot\mathbf{B}) \tag{2-3}$$

$\mathbf{v}\times\mathbf{L}$ is the rate of change of the area of the loop. The quantity in the parentheses is thus the rate of change of the flux over the loop, which in turn is $-\mathscr{E}$, where $\mathscr{E}$ is the induced emf. We thus find that

$$-\mathbf{F}\cdot\mathbf{v} = i\mathscr{E} \tag{2-4}$$

where the right side is the electrical power, which for this case of constant current is converted into heat in the resistance. The reader should examine whether, without the negative sign in Faraday's law, this system would be a self-acting machine, generating its own power spontaneously.

3. A Simple Generator

A single-turn coil rotating in a uniform magnetic flux density is an example of an elementary electric generator. Figure 4 shows a view of such a coil. Consider the coil to be rotating with an angular velocity ω and to be at the moment in a position such that its area vector, $\mathscr{A}$, makes an angle θ with the field $\mathbf{B}$. If we count time starting when the area vector is parallel to the field, then $\theta = \omega t$. For a uniform B-field, Faraday's law can be written as

$$\mathscr{E} = -\frac{d}{dt}(B\mathscr{A}\cos\omega t) = \omega B\mathscr{A}\sin\omega t \tag{3-1}$$

Let the resistance of the coil be R. The current at the time t is, from Ohm's law

$$i = \frac{(\omega B\mathscr{A})}{R}\sin\omega t \tag{3-2}$$

From Joule's law the instantaneous rate of production of heat energy is

$$P = \frac{(\omega B\mathscr{A})^2}{R}\sin^2\omega t \tag{3-3}$$

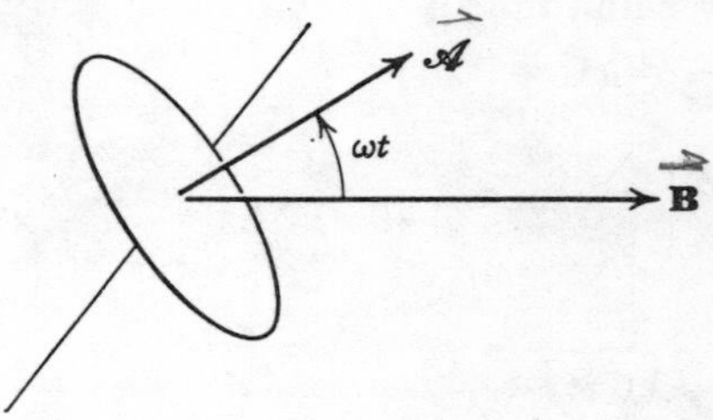

Figure 4. A coil rotating in a **B***-field.*

In Section 8 of Chapter 5 it was shown that a loop which carries a current i in a uniform field B experiences a torque given by

$$\mathbf{T} = i\mathscr{A} \times \mathbf{B} \tag{3-4}$$

so that the torque on the loop of Figure 4 has a magnitude

$$T = i\mathscr{A}B \sin \omega t \tag{3-5}$$

The mechanical power expended in turning the coil is the product of torque and angular velocity:

$$P = T\omega = i\mathscr{A}B\omega \sin \omega t = \frac{(\omega\mathscr{A}B)^2}{R} \sin^2 \omega t \tag{3-6}$$

which is to be compared to Equation 3-3. The mechanical energy required to rotate the coil reappears in the form of heat energy (energy dissipated against frictional mechanical forces is, of course, not included in this computation), and it is clear that the greater the resistance, the less is the effort required to turn the loop. In the limit, for an open circuited loop, no torque is required other than that to overcome friction.

Since the emf developed in such a loop is a function of the sine of the angle of rotation, it reverses itself every half-cycle. Consequently, in an external load resistance in series with the loop, the current is sinusoidally alternating and may be converted to a direct current by means of a commutator and brushes (see Figure 5).

Our analysis of the simple generator has assumed a constant B-field, and a closer examination shows that, when the generator is supplying current to a load, such an assumption is no longer valid. For the current, which has been assumed to have a maximum magnitude when $\sin \omega t = \pm 1$, itself produces a flux density which at this instant is normal to the external flux. Hence the net flux density at the coil is the vector sum of external and current-produced flux densities, and the rate of change of this sum is not likely to be the same as that used in Equation 3-1. The deviation from the assumed conditions is a function of the current. Particularly when a generator consists of several coils oriented symmetrically about the axis and with correspondingly greater number of commutator segments, does this coil-developed flux introduce difficulties with sparking at the commutator. The brushes are so positioned that they move from one segment to the next at the instant when the voltage differences between these segments is zero. Otherwise there is loss of efficiency of the generator and burning at the commutator.

To restore the conditions as assumed in the simple generator, electromagnetic poles are mounted above and below the rotating coil in such a position that their field is normal to the external flux density. The windings

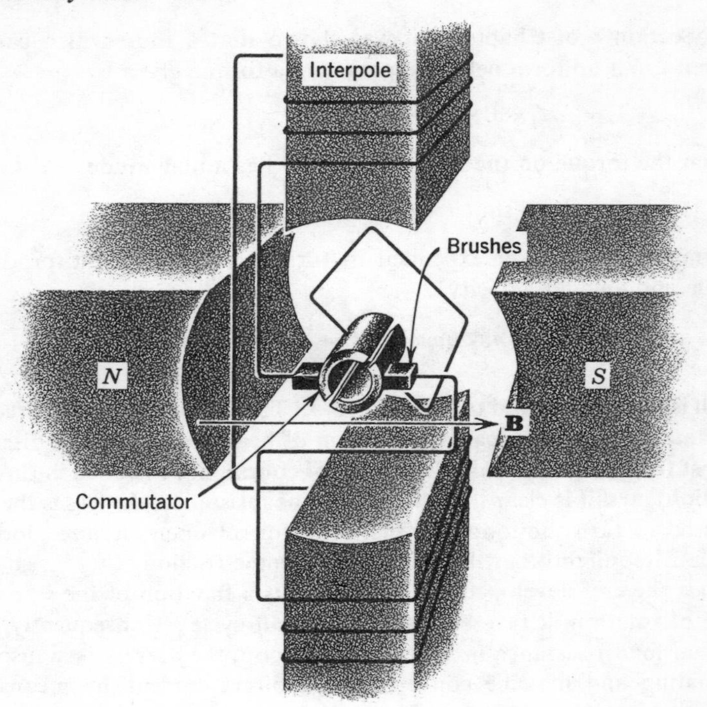

Figure 5. A simple generator with interpoles.

of these poles are connected in series with the rotating coil through the commutator so that their field is a maximum when the current in the coil is a maximum and is directed opposite to the field of the rotating coil itself. By properly constructing these poles (called interpoles), their field may be made closely to cancel the field of the coil, and hence closely to restore the conditions assumed in the original analysis.*

4. Inductance

In order to induce an emf into a circuit there must be a changing magnetic flux, but the origin of that flux is quite immaterial. It may originate in some external system, say, a current in another coil, or it may be the field which a current in the circuit itself sets up.

Let us picture two circuits (Figure 6) separated in some arbitrary fashion. A current in circuit 1 produces a magnetic flux which links

* See Harnwell, p. 389–90, for a somewhat fuller analysis.

circuit 1 and in part, at least, also circuit 2. In Section 6 of Chapter 5 we saw that at any given place in space the density of this flux is directly proportional to the associated current producing it. The integral of this flux density over the area of circuit 1 is then proportional to the current, I_1, as is also the integral over the area of circuit 2. The actual computation of these fluxes may often be complicated, but we are here concerned only with the fact of proportionality, which we express as follows:

$$\Phi_1 = L_1 I_1 \quad \text{and} \quad \Phi_2 = M_{12} I_1 \tag{4-1}$$

where the proportionality constants L_1 and M_{12} are called the self-inductance of circuit 1 and the mutual inductance between circuits 1 and 2, respectively. These constants, like capacitance, are determined by the geometrical relationships of the circuits and the magnetic properties of the medium in the space around the circuits.

Using Equation 4-1 and assuming that the circuits remain fixed in position, Faraday's law becomes

$$\mathscr{E}_1 = -L_1 \frac{dI_1}{dt} \quad \text{and} \quad \mathscr{E}_2 = -M_{12} \frac{dI_1}{dt} \tag{4-2}$$

For a current I_2 in circuit 2, the corresponding emf's are

$$\mathscr{E}_2 = -L_2 \frac{dI_2}{dt} \quad \text{and} \quad \mathscr{E}_1 = -M_{21} \frac{dI_2}{dt} \tag{4-3}$$

Both the self-inductance and the mutual inductance are measured in henrys or volt-seconds per ampere, as can be seen from Equations 4-2 and 4-3.

The negative sign in the self-inductance equations, from Lenz's law, implies that, given a changing current in a loop, the resulting emf is in such a direction as to oppose the change. That is, an increasing current generates an emf in a direction against the current; whereas a decreasing current generates an emf in the same sense as the current to tend to prevent it from decreasing.

The negative sign in the mutual inductance equations represents a somewhat more involved situation. A changing current in circuit 1 induces

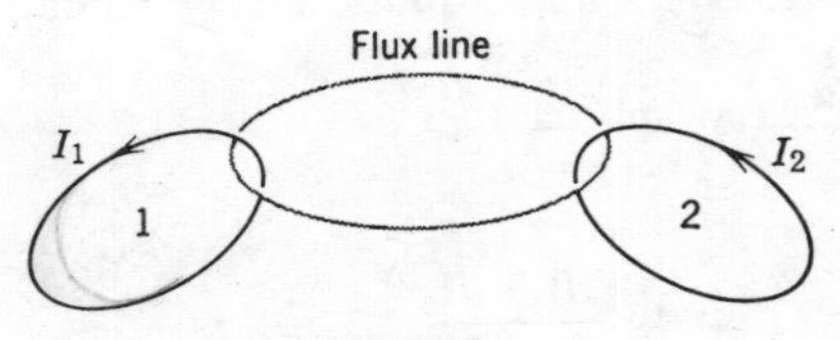

Figure 6

an emf in circuit 2. If circuit 2 is a conductor, the resulting current produces a magnetic flux at circuit 1 that opposes the original flux change. Circuit 1 may be thought of as sensing the presence of circuit 2 through this "return" magnetic flux.

5. Mutual Inductance

With the help of the vector potential introduced in Section 17 of Chapter 5, we now show that the mutual inductance between a pair of circuits is the same, in whichever sense the induction takes place. That is:

$$M_{12} = M_{21} \tag{5-1}$$

First consider the case in which an emf is induced in circuit 2 by a changing current in circuit 1. The flux at any instant at circuit 2 is

$$\Phi_2 = \int_2 \mathbf{B} \cdot d\mathscr{A}_2 \tag{5-2}$$

where **B** is the flux density associated with the current in circuit 1 and the integration is over an area of which circuit 2 is the perimeter. The magnetic flux density is replaced by the curl of the vector potential:

$$\Phi_2 = \int_2 \nabla \times \mathbf{A}_1 \cdot d\mathscr{A}_2 \tag{5-3}$$

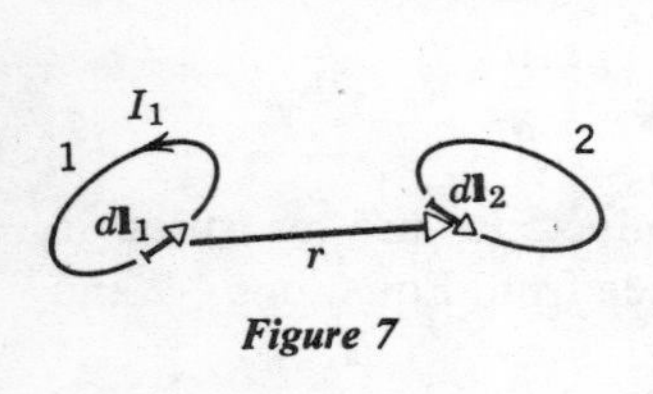

Figure 7

and Stokes's theorem then converts the area integral into a line integral:

$$\Phi_2 = \oint_2 \mathbf{A}_1 \cdot d\mathbf{l}_2 \tag{5-4}$$

where $d\mathbf{l}_2$ is an element of circuit 2. The situation is illustrated in Figure 7. The equation for the vector potential associated with a current such as in loop 1 is (see Section 17, Chapter 5):

$$\mathbf{A}_1 = \frac{\mu_0 I_1}{4\pi} \oint_1 \frac{d\mathbf{l}_1}{r} \tag{5-5}$$

where the integration is over the circuit giving rise to the vector potential, in this case circuit 1. With 5-5, Equation 5-4 becomes

$$\Phi_2 = \frac{\mu_0 I_1}{4\pi} \oint_2 \oint_1 \frac{d\mathbf{l}_1 \cdot d\mathbf{l}_2}{r} \tag{5-6}$$

and thus

$$M_{12} = \frac{\mu_0}{4\pi} \oint_2 \oint_1 \frac{d\mathbf{l}_1 \cdot d\mathbf{l}_2}{r} \tag{5-7}$$

A repetition of this reasoning, starting with a current in loop 2, leads again to Equation 5-7 but with subscripts reversed. The equation is not affected by such a reversal. Thus Equation 5-1 is verified and the subscripts on the mutual inductance are superfluous.

6. Self-Inductance

The nature of self-inductance is brought out by a study of the series circuit of Figure 8, where a coil of self-inductance L, a resistance R, and a cell of emf $\mathscr{E}$ are in series with a switch S. Although all current-carrying parts of a circuit produce magnetic flux and contribute to the self-inductance, this effect is most highly concentrated where the conductor is wound into a coil, and it is common practice schematically to show the self-inductance as a single coil. In like manner, although all conductors have some resistance, the total resistance is shown as a single element. The symbol for resistance in Figure 8 includes the resistance of the coil, of the connectors, and of any internal resistance there may be in the cell.

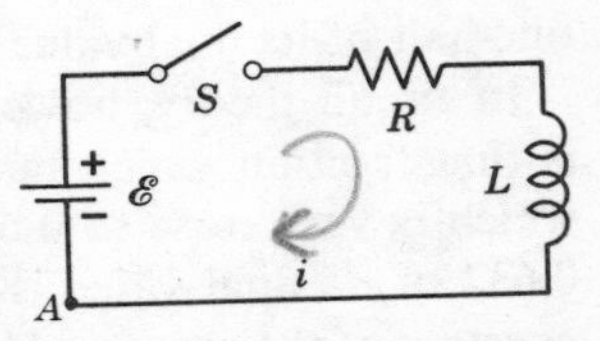

Figure 8. The current approaches a steady value asymptotically in an L-R circuit.

Let i be the current at the time t, where t is measured from the instant the switch is closed. When currents are changing with time, Kirchhoff's loop rule requires that the sum of the potential changes be zero at each instant of time. Thus, if we start at the point A and move clockwise around the circuit, we record, relative to A, a rise in potential through the cell of $\mathscr{E}$, a drop in passing through the resistance of iR, and a drop (check by Lenz's law) in passing through the coil of $L(di/dt)$. The sum of these changes must be zero:

$$\mathscr{E} - iR - L\frac{di}{dt} = 0 \tag{6-1}$$

Separation of variables gives:

$$\frac{di}{(i - \mathscr{E}/R)} = -\frac{R}{L}\,dt \tag{6-2}$$

and integration of the left side from $i = 0$ to $i = i$ and the right side from $t = 0$ to $t = t$ gives

$$i = \frac{\mathscr{E}}{R}(1 - e^{-(R/L)t}) \tag{6-3}$$

As Figure 9 shows, the current does not rise instantaneously, but approaches asymptotically to the value $\mathscr{E}/R$. The time required for the current to reach its final value is infinitely long, for any nonzero value of inductance. A graph of i versus t for several values of L/R yields distinct curves, each of which approaches asymptotically the same final value, $\mathscr{E}/R$. Although clearly the behavior of the three cases is different, they are not distinct in the time required to reach a steady state. But they can be labeled by using the time each requires to reach some agreed upon fraction of the steady-state value of current. A simple fraction (which is used in connection with equations that appear in radioactivity studies) is one-half, and Equation 6-3 shows that the time required for the current to rise to one-half of its final value is $0.693L/R$.

In circuit theory, however, the fraction that is commonly agreed upon is that fraction which makes the exponent of e equal to -1, a fraction which is very close to 0.632. The time required for the current to reach 0.632 of its final value is the ratio L/R, and this ratio is called the *time constant* of the circuit. The larger the time constant the slower the rate of rise of the current. In many common circuits encountered in the laboratory, this constant is of the order of a small fraction of a second, best observed with a cathode-ray oscillograph. In large electromagnets, however, where L may be quite large and R small, the time constant may be readily measured by timing the rate of rise of current with a watch.

What happens when the emf is shorted out? Equation 6-1 still applies with the difference that now $\mathscr{E} = 0$:

$$L\frac{di}{dt} + iR = 0 \tag{6-4}$$

whose solution may be found by a procedure similar to that used to solve

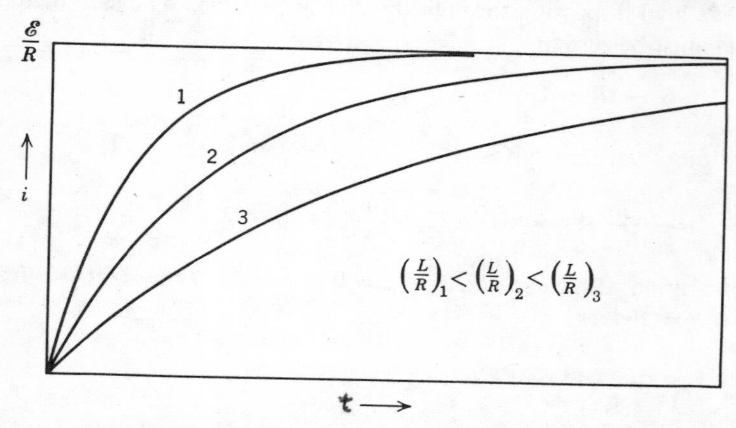

Figure 9. *i versus t in an L-R circuit for different values of L/R.*

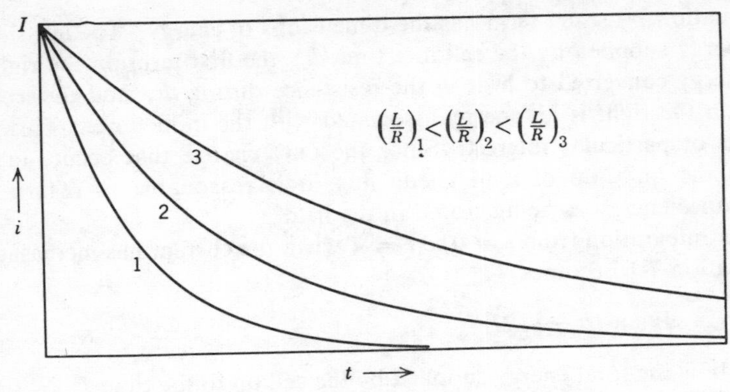

Figure 10. Decay of current in an L-R circuit.

Equation 6-2. The result is

$$i = Ie^{-Rt/L} \tag{6-5}$$

where I is the current when $t = 0$, the instant when the emf is removed. Curves showing the variation of current versus time for several values of L/R appear in Figure 10. Here again we see that although each curve approaches zero asymptotically, the shape of the curves depends on the ratio L/R. These curves are characterized by the time required for the current to fall very closely to 0.368 of the original current, at which time the exponent of e is -1 and the time constant is L/R. Not surprisingly, it is the same as for the rising current in the same circuit.

7. Energy Considerations in an *L-R* Circuit

The growth and decay of current in an *L-R* circuit has been introduced in this chapter, because it is instructive at this point to consider the energy relationships involved. We shall find that just as the static electric field represented stored energy, so does the magnetic flux density. The static electric field was associated with the potential energy of the configuration of charges that gives rise to the field; and the energy in the magnetic flux field is the energy associated with the moving charges (analogous to kinetic energy) in the currents that give rise to the magnetic field.

Multiply Equation 6-1 by dq, or its equivalent, $i\,dt$, the charge passing any cross section of the circuit conductor in a time dt to get

$$\mathscr{E}\,dq = Ri^2\,dt + Li\,di \tag{7-1}$$

In Equation 7-1, each term has the dimensions of energy. The left side is the energy supplied by the cell in a time dt; the first term on the right is the energy converted to heat in the resistance during dt; and the second term on the right is an energy associated with the inductance. This last term is of particular interest. Since the only change that occurs in the coil is the build-up of a magnetic flux, it is reasonable to regard the inductance energy as being stored in the field.

After integration from $t = 0$ to $t = T$ when the current has increased to I, Equation 7-1 becomes

$$W = H + \tfrac{1}{2}LI^2 \qquad \text{(7-2)}$$

where W is the total energy supplied by the cell up to the time T, H is the total heat developed in the resistance up to this time, and $\frac{1}{2}LI^2$ is the energy in the inductance. This last energy grows with the current I and, after I becomes steady, remains unchanged while the cell continues to supply energy only to develop heat in the resistance.

The concept of energy being stored in the magnetic flux field is further strengthened by seeing what happens when we remove the cell from the circuit and allow the current to die out. We now have no chemical source of energy, yet heat continues to be produced in the resistance. By counting time from the instant the cell is removed, we compute the total heat developed in the resistance from

$$H' = \int_0^\infty Ri^2\,dt \qquad \text{(7-3)}$$

The current i is known from Equation 6-5. Substituting for i and integrating from $t = 0$ to $t = \infty$, we get:

$$H' = I^2R\int e^{-(2Rt/L)}\,dt = I^2R\left(-\frac{L}{2R}\right)e^{-(2Rt/L)}\Big|_0^\infty = \tfrac{1}{2}I^2L \qquad \text{(7-4)}$$

which is just the energy originally stored in the magnetic flux of the coil when the emf was removed. This energy is transferred into heat in the resistance as the magnetic flux field dies out.

8. Magnetic Flux Energy in a Toroid

Certain interesting results are obtained from computing the energy stored in the magnetic flux of a geometrically simple device such as a large toroid. Suppose a toroid has a radius a, large compared to the dimensions of the cross section of its winding, the cross-sectional area of the winding is $\mathscr{A}$, the total number of turns is N, and the current is I. In

Chapter 5 the flux density B within such a toroid was found to be

$$B = \frac{\mu_0 NI}{2\pi a} \tag{8-1}$$

The total flux over the cross-section of this toroid is thus

$$\Phi = B\mathscr{A} = \frac{\mu_0 NI\mathscr{A}}{2\pi a} \tag{8-2}$$

For a changing current, the emf induced in each turn is $-d\Phi/dt$, and the total emf in the whole coil is

$$\mathscr{E} = -N\frac{d\Phi}{dt} = -\frac{\mu_0 N^2\mathscr{A}}{2\pi a}\frac{dI}{dt} \tag{8-3}$$

By comparing Equation 8-3 with the first of Equations 4-2, we find the inductance of the toroid:

$$L = \frac{\mu_0 N^2\mathscr{A}}{2\pi a} \tag{8-4}$$

and from Equation 7-2 we obtain the energy in the magnetic flux of this toroid:

$$\tfrac{1}{2}LI^2 = \tfrac{1}{2}\frac{\mu_0 N^2\mathscr{A}I^2}{2\pi a} \tag{8-5}$$

Since the flux density is essentially uniform throughout the volume $2\pi a\mathscr{A}$, it is meaningful to compute the energy per unit volume

$$w_e = \tfrac{1}{2}\mu_0\left(\frac{NI}{2\pi a}\right)^2 = \frac{B^2}{2\mu_0} \tag{8-6}$$

Although this expression for the energy density in a B-field has been obtained for an especially simple case, it is an expression of quite general application as we shall have occasion to see. It should be kept in mind, however, that we have in no sense proved that energy actually is stored in the space occupied by a magnetic flux, merely that it may be so regarded. Physically, all that is observed is that it requires energy to build up a magnetic flux and that this energy may be recovered when the flux disappears.

9. The Betatron

Limitations of space make it inadvisable to discuss in detail many experimental applications of electromagnetic theory, interesting and important though they may be. The design and construction of certain devices, however, constitute good illustrations of the theory and will be briefly discussed for that reason. The betatron, an especially instructive example of the operation of Faraday's law, is a device for the production of high-energy electrons, by means of which high-energy X-rays are produced. The device is basically an evacuated, doughnut-shaped, glass or ceramic tube equipped with an electron gun to inject electrons into the evacuated space. This doughnut tube is located between the poles of a large electromagnet whose field is alternating, usually with a 60-cycle per second frequency, and has a distribution in intensity that is circularly symmetric but decreasing outward along a radius.

Moving electrons have their path bent by the field. In particular, when the field is in the proper sense and the momentum of the electrons is right, the path is a circle of radius R (Figure 11). However, as the flux over the area of the path increases, the electrons gain energy and, unless the B-field at the path is right, they will not remain in the path of radius R. It is, therefore, necessary to determine the conditions that are necessary for the electron path to remain stable.

Let Φ be the flux over the area of the circular path of the electrons. As Φ changes, the emf around this path, from Faraday's law, is

$$\oint \mathbf{E} \cdot d\mathbf{l} = E \cdot 2\pi r = -\frac{d\Phi}{dt} \tag{9-1}$$

For a B-field directed upward (Figure 11) and increasing in magnitude, the emf is in a clockwise sense; hence electrons with a charge $-e$ will be accelerated counterclockwise. At any instant the force on the electron will be:

$$F = E(-e) = \frac{e}{2\pi r}\frac{d\Phi}{dt} = \frac{d}{dt}(mv) \tag{9-2}$$

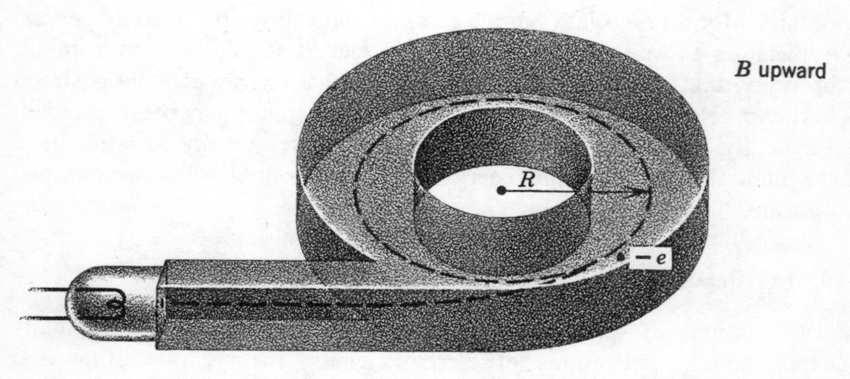

Figure 11. The doughnut-shaped chamber of a betatron.

where Newton's 2nd law in terms of the time rate of change of momentum is used because the electrons very quickly reach relativistic speeds where the mass is no longer constant. In fact, in a rather small betatron which accelerates the electrons to an energy of only 20 mev, the mass of the electron reaches approximately 40 times its rest mass. Roughly the electron mass increases by one rest mass for each half-million electron volts additional kinetic energy.

At the same time that the electron is being accelerated along its path, it is being deflected by the magnetic flux density. So that its path may be a circle of radius R, the force exerted by virtue of the flux density $\mathbf{B}$ at the path must equal the centripetal reaction:

$$\frac{mv^2}{r} = Bev \tag{9-3}$$

where B is the flux density at radius r. After dividing Equation 9-3 by v and differentiating with respect to the time, we have:

$$\frac{d}{dt}(mv) = er\frac{dB}{dt} \tag{9-4}$$

which relates the changes in B at the path to the changes in the momentum of the electron. Equation 9-2, on the other hand, relates the changes in momentum to the changes in total flux. Combining these two equations by eliminating $d(mv)/dt$, we get

$$\frac{dB}{dt} = \frac{1}{2\pi r^2}\frac{d\Phi}{dt} \tag{9-5}$$

Since the flux and the flux density each start from zero at the same time, integration of Equation 9-5

$$\int_0^B dB = \frac{1}{2\pi r^2}\int_0^\Phi d\Phi \tag{9-6}$$

gives

$$B = \frac{\Phi}{2\pi r^2} \tag{9-7}$$

a relation between the total flux and the necessary value of B at the orbit at any instant of time so that the electrons shall travel in a circular path of radius r. This expression has a rather simple physical interpretation, since πr^2 is the area of the electron orbit. Call this area $\mathscr{A}$. Then Equation 9-7 takes the form:

$$\mathscr{A}B = \Phi/2 \tag{9-8}$$

That is, for the electrons to continue to circulate in an orbit of constant radius as they gain energy and as the field increases, it is necessary that the total flux over the area of the orbit be twice as large as would exist if the flux density had a uniform value B (the value at the orbit) over the whole area. This added flux is obtained by making the flux density considerably stronger near the center of the field by proper shaping of the pole pieces of the electromagnet or by adding higher permeability magnetic material near the center.

The electron is injected into the orbit as the alternating magnetic field passes through zero and continues to acquire energy as the field increases. As long as the proper relation between total flux and orbital flux density is maintained, the way in which the field increases—linearly or sinusoidally, for instance—is not important. As the electron reaches its maximum energy, it is deflected either inwardly or outwardly toward a target by upsetting the balance between total flux and orbital field. This may be done by providing a central core of magnetic material which saturates (see Chapter 7) near the peak of the cycle, or by sending a pulse of current through an auxiliary winding on the central part of the magnet pole.

There are many interesting features and problems in the design of a betatron which are rewarding to study. They involve the design of the pole pieces of the magnet, not only in order to maintain the flux to orbital flux density relationship, but also to provide the proper shape of the magnetic field at the orbit to keep the electron from wandering above or below the plane of the orbit, and to force back into orbit an electron that has strayed to a larger or a smaller radius; the electronic control circuits to time the injection and ejection of the electrons at the proper points on the cycle; and the design of a multi-ton electromagnet subject to stresses alternating 120 times a second, this latter being an interesting problem in the boundary area between electrical engineering and mechanical engineering. But these problems are beyond the scope of this book.

10. Differential Form of Faraday's Law

Equation 1-2 is the integral form of Faraday's law. We shall now restrict our attention to those cases in which only B varies with time, the geometrical relations being fixed. In this case, the time derivative may be taken inside the integral and written as the partial derivative of B with respect to time. It has become common practice in electromagnetic theory notation to indicate a partial time derivative by means of a dot placed over the function thus:

$$\frac{\partial}{\partial t}\mathbf{B} = \dot{\mathbf{B}} \tag{10-1}$$

but a word of warning is in order here. There is a difference in convention between the use of such a dot in electromagnetism and in mechanics where a dot is used for the *total* derivative with respect to time. There is little danger of confusion, but should the possibility of confusion occur, the derivative symbol should always be written out in full.

Faraday's law, with the notation and the limitation just noted, is written:

$$\oint \mathbf{E} \cdot d\mathbf{l} = -\int \dot{\mathbf{B}} \cdot d\mathcal{A} \tag{10-2}$$

where the line integral on the left establishes the limits of the area integral on the right. By means of Stokes's theorem, the line integral of $\mathbf{E}$ can be converted to an area integral:

$$\int \nabla \times \mathbf{E} \cdot d\mathcal{A} = -\int \dot{\mathbf{B}} \cdot d\mathcal{A} \tag{10-3}$$

A relation which is valid for lines and areas of arbitrary size and shape. It therefore follows that the integrand functions themselves must be equal:

$$\nabla \times \mathbf{E} = -\dot{\mathbf{B}} \tag{10-4}$$

which is the differential form of Faraday's law. It is the generalization referred to at the beginning of this chapter, and is one of the relationships used by Maxwell to develop the equations of the electromagnetic field.

A further interesting consequence of this relation is found by introducing the vector potential. Since we can always make the substitution: $\mathbf{B} = \nabla \times \mathbf{A}$ we may rewrite Equation 10-4 as:

$$\nabla \times \mathbf{E} = -\frac{\partial}{\partial t} \nabla \times \mathbf{A} = -\nabla \times \dot{\mathbf{A}} \tag{10-5}$$

which becomes, by transposing terms to the left,

$$\nabla \times (\mathbf{E} + \dot{\mathbf{A}}) = 0 \tag{10-6}$$

A vector whose curl is zero may always be written as the gradient of a scalar function. Thus,

$$\mathbf{E} + \dot{\mathbf{A}} = -\nabla\phi \quad \text{or} \quad \mathbf{E} = -\dot{\mathbf{A}} - \nabla\phi \tag{10-7}$$

where ϕ is a scalar function, the negative sign being introduced for convenience. Here we have a generalization to time-varying conditions of the electrostatic relation: $\mathbf{E} = -\nabla V$. Under static conditions ϕ in Equations 10-7 becomes identical with the electrostatic scalar potential. It is a general scalar potential which, in combination with the vector potential, determines the electric (not necessarily electrostatic) field. The electric field is a function of the rate of change of the scalar potential in space and the rate of change of the vector potential in time.

Thus all the electromagnetic field relationships may be expressed in terms of these potentials. The flexibility of our being able to express the electrostatic field in terms of the static scalar potential was a considerable convenience; so also does the vector and general scalar potentials often ease computation and simplify the forms of theoretical relations.

Discussion Questions

1. One of Michael Faraday's experiments, which led to the law bearing his name, consisted of causing a wire carrying a current to approach another, parallel wire. A current was induced in the second wire in the direction opposite to that in the first during the approach. Upon withdrawing the wire, he found that the induced current was in the same direction as the inducing current. Can this be explained in terms of the rate of change of magnetic flux in a closed circuit?

2. In Problem 17 of Chapter 5 there is described the Faraday disk motor. If, instead of a current, I, in the disk, the rim and axle are connected to an ammeter and a torque is applied to the disk to turn it, the device becomes a generator, producing a current. Analyze this generator and in particular identify a closed loop in which there is a changing magnetic flux associated with the induced emf and current.

3. A conducting loop is mounted on pivots between the poles of a magnet. The magnetic poles are so mounted that they can be rotated about the same axis on which the loop pivots, but there is no mechanical connection between the poles and the loop. Reasoning from Lenz's law, predict the behavior of the loop when the magnet (and hence the magnetic field) is rotated. The discussion can be continued by developing the principles of the induction motor. [Ref: Harnwell, *Prin. of Elec. and Electromagnetism*, Chapter 12, Sec. 7.]

4. What property of the magnetic flux density, $\mathbf{B}$, is responsible for the fact that the area integral of $\mathbf{B}$ has the same value over all surfaces having the same curve as a perimeter?

5. In the discussion of the simple generator in Section 3 nothing was said about the effect of the inductance of the rotating loop on the performance. Qualitatively, what would you expect to be the effect of such inductance?

Problems

1. A wire is bent as in Figure P1 to form two square loops, each of side a, the planes of the loops being parallel to the yz and the xz coordinate planes, respectively. A magnetic flux density exists in this region given by

$$\mathbf{B} = (bt + c)(\mathbf{i} \cos 30° + \mathbf{j} \sin 30°)$$

Determine the magnitude and the polarity of a voltage applied at the points p and q which will prevent the flow of charges in the wire.

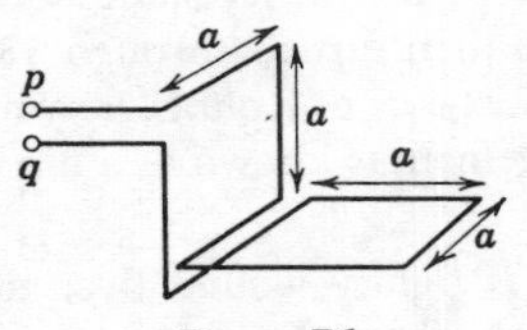

Figure P1

2. A crossbar rests across two parallel rails and is perpendicular to them as in Figure P2. A battery is connected to the ends of the rails so that a current of 5 amperes exists in the crossbar. Measurements show that there is a force acting on the crossbar in a direction parallel to the rails of 0.0002 newton. The battery is now removed and replaced by a conductor. The conductor and the rails have negligible resistance, but the resistance of the crossbar between the contacts with the rails is 0.1 ohm. Determine the current in the crossbar when it is moved parallel to the rails (and perpendicular to itself) with a speed of 100 meters per second.

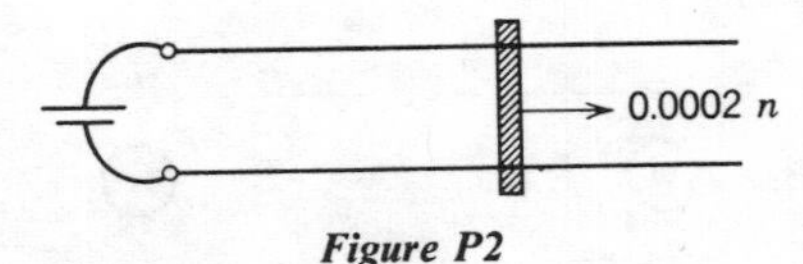

Figure P2

3. Two circular loops of wire are located near each other, as in Figure P3, in such a relation that a current in loop 1 produces half as much flux over loop 2 as it does over loop 1. The self-inductance of loop 1 is 10^{-3} henry. Loop 2 is connected to a cathode ray oscilloscope which measures the emf but permits negligible current. The current in loop 1 is given by the relation:

$$i_1 = 2 \sin 377t \text{ amperes}$$

What is the voltage as a function of time at the terminals of loop 2?

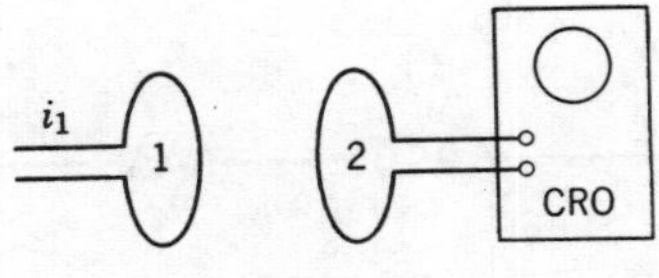

Figure P3

4. A pair of coils have a mutual inductance of 0.1 henry and self-inductance of 0.2 henry each. The resistance of each coil is 1 ohm. A 2-volt battery is suddenly connected in series with one coil. If the second coil is a closed circuit, what total charge will move through any cross section of its wire?

5. A flat coil of N turns with a mean radius R, is connected to the terminals of a ballistic galvanometer (see Chapter 8) whose deflection measures total quantity of charge passing through the instrument (rather than current). The coil is placed in a magnetic field with its plane normal to the direction of the magnetic flux density. It is then turned through 180°. If the coil and galvanometer circuit has a net resistance of r ohms, how much charge passes through the circuit? Such an arrangement is known as a flip coil and is used to measure magnetic flux density.

6. A coil of 500 turns is tightly wound over the center region of a long solenoid of 25 turns per cm. The radius of the solenoid is 1 cm. What is the mutual inductance between the solenoid and the 500-turn coil?

7. A toroid of 1000 turns is wound on a wooden core. The mean diameter of the toroid is 30 cm and the cross-sectional diameter is 1 cm. What is the self-inductance of this toroid? If another winding of 500 turns is placed uniformly over the 1000-turn winding on this toroid, what will be the mutual inductance between them?

8. Two parallel wires, each of radius a, are separated by a distance d between centers, as in Figure P8. The currents in the wires are equal and oppositely directed. Determine the inductance per unit length of the pair, with the assumption that a is small compared to d.

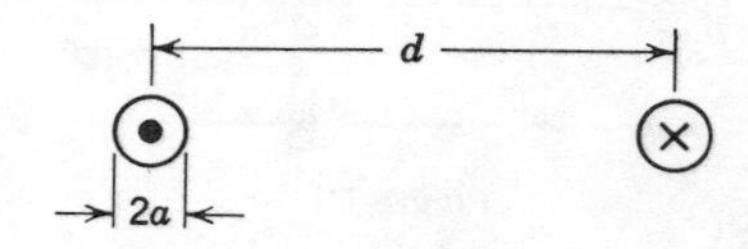

Figure P8

9. Two coils are connected in series with a battery and a switch. One coil has an inductance L_1 and resistance R_1, while the other has an inductance L_2 and a resistance R_2. Between them is a mutual inductance M. Find the time constant of the circuit. There are two possible answers.

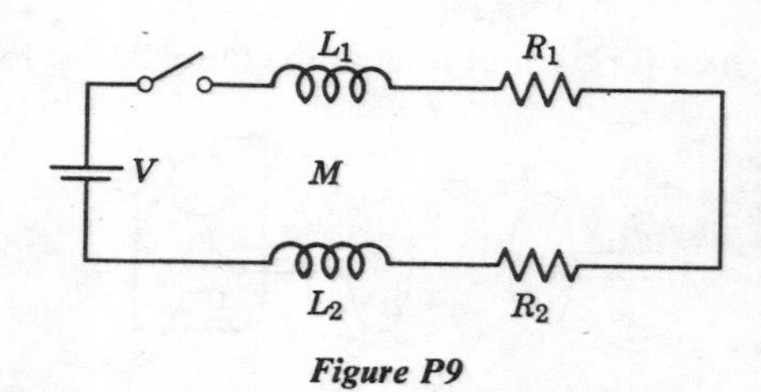

Figure P9

10. A certain electromagnet used on a cyclotron has a total resistance of 0.25 ohm and a time constant of 20 minutes. What is its inductance? What is the elapsed time between the closing of a switch to a constant voltage generator and the time when the current reaches 90% of its asymptotic value? By the time the current has reached 90% of its final value, what fraction of the total energy supplied by the generator has been stored in the magnetic field?

11. A wooden ring of circular cross section has an outside diameter of 21 cm and a cross-sectional diameter of 1 cm. Copper wire is wrapped closely on this ring to produce a tightly wound toroid. The wire has an over-all diameter, including insulation, of 1.2 mm; the diameter of the conductor is 1 mm, and the copper has a resistivity of 1.77×10^{-6} ohm-cm. What is the time constant of this toroid? At an applied potential difference of 0.1 volt, what energy is stored in the magnetic field? What must be the capacitance of a capacitor so that, with the same potential difference, it will store the same energy in its electric field? Identify any approximations used in the computations.

12. An electron, injected into a betatron orbit with an energy that is small compared to its final energy, very rapidly reaches a speed practically equal to the speed of light. It is a good approximation to say that it is traveling at the speed of light for the total time that it is in the orbit. Show that the final energy of the electron is given, under these assumptions, by

$$\mathscr{E} = \frac{ec\Phi_{max}}{2\pi r}$$

where r is the radius of its orbit in the betatron, e is the electronic charge, c is the speed of light, and Φ_{max} is the value of the total flux over the area of the orbit at the moment the electron is ejected toward the target. For 100 mev electrons in a betatron orbit of 0.5 meters radius, what is the value of Φ_{max} and of the corresponding magnetic flux density at the orbit? How much energy does the electron gain per revolution, and how many revolutions does it make before ejection? The field has a frequency of 60Hz.

REFERENCES

Most of the previously noted references constitute good collateral reading for the topics of this chapter. In particular the following should be helpful:

1. Scott, *op. cit.*, Chapter 7.
2. Reitz and Milford, *op. cit.*, Chapter 9.

7

The magnetic behavior of matter

Our study of the magnetic properties of matter will follow closely the pattern used in discussing the dielectric properties of matter. Inadequate information about the microscopic details of charge distributions and velocities led to the introduction of a new field vector, **D**; similar lack of detailed information about microscopic currents will lead us to introduce still another field vector, **H**, to account for the magnetic properties of matter.

1. "True" Currents and Electron Orbital Currents

In free space, the magnetic flux density, **B**, was found to be related (Chapter 5, Equation 13-9) to the current density by

$$\oint \mathbf{B} \cdot d\mathbf{l} = \mu_0 \int \mathbf{J} \cdot d\mathcal{A} \tag{1-1}$$

where we are now using the more conventional $d\mathbf{l}$ to indicate an element of path of a line integral. The integral on the right side represents the total current encircled by the path of the line integral on the left. This relation was developed by using filamentary conduction currents in wires of negligible cross section; but all kinds of current—an electron stream, an electron orbiting about a nucleus, or what in the classical sense may be looked upon as the circulation of charge due to electron spin—can be part of the integrand on the right of Equation 1-1 and are included in obtaining a line integral of **B** that passes through matter. Formally, a distinction is made between so-called "true" current densities, involving actual current densities in conductors or movements of charges on a macroscopic scale, and current densities associated with the medium, arising out of the summed effect of currents in the atoms. We indicate true current densities with **J**, without a subscript, and current densities associated with the presence of a medium as $\mathbf{J}_m$, so that Ampère's circuital law becomes

$$\oint \mathbf{B} \cdot d\mathbf{l} = \mu_0 \int \mathbf{J} \cdot d\mathscr{A} + \mu_0 \int \mathbf{J}_m \cdot d\mathscr{A} \tag{1-2}$$

The integral over $\mathbf{J}_m$ is analogous to the integral over the polarization charge densities required in the computing of the electric field in the presence of a dielectric. Further, just as in dielectrics the induced charge density was expressed in terms of a macroscopic dipole moment per unit volume, representing an average over many molecules, so in the magnetic case we express $\mathbf{J}_m$ in terms of a macroscopic property **M**, the magnetic dipole moment per unit volume or the intensity of magnetization.

2. The Magnetization and the Magnetic Field

The magnetic moment of a plane current loop is defined (Chapter 5, Section 8) as a vector whose magnitude is the product of the current and the area of the loop and whose direction is determined by the right-hand rule relative to the direction of the *conventional* current.

For an aggregate of such elementary current loops, each of moment **dm**, the magnetic moment per unit volume is

$$\mathbf{M} = \frac{\sum d\mathbf{m}}{\Delta \mathscr{V}} \tag{2-1}$$

where the summation is a vector summation over the volume $\Delta\mathscr{V}$. This volume is to be small compared to macroscopic dimensions but nevertheless large compared to atomic dimensions.

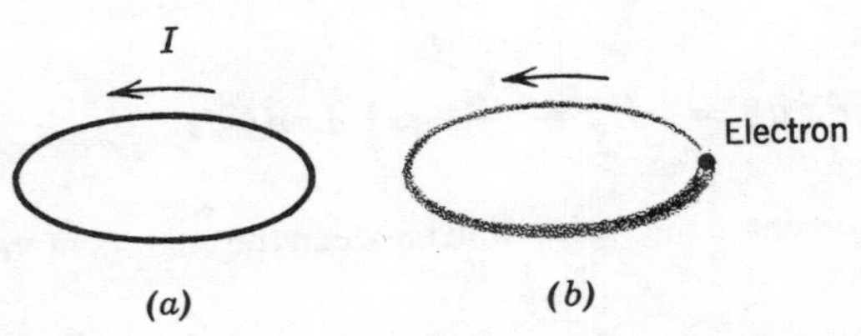

Figure 1. A conventional current loop and an electron orbit. Magnetic moment of (a) is out from page; (b) is into page.

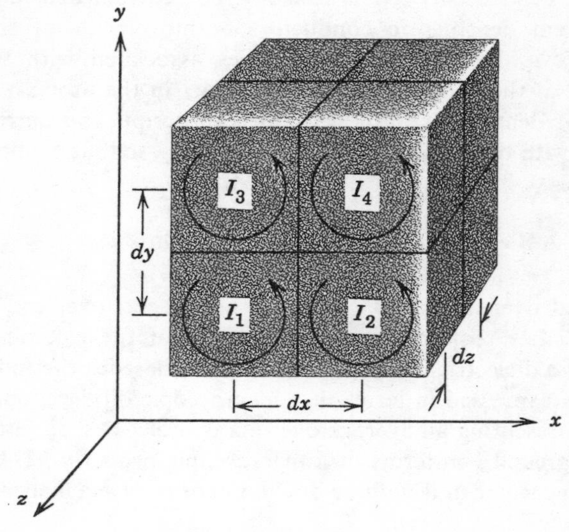

Figure 2

In a group of volume elements, each of dimensions $dx\,dy\,dz$ (Figure 2), let us consider only the z component of the magnetic moments. If M_z is the z component of the magnetic moment per unit volume, then the z component of the magnetic moment of volume element 1 is $M_z\,dx\,dy\,dz$ and that of volume element 2 is:

$$\left[M_z + \left(\frac{\partial M_z}{\partial x}\right) dx\right] dx\,dy\,dz$$

Let these magnetic moments be associated with effective currents, I_1 and I_2, respectively, circulating in each block, so that, from the definition of magnetic moment:

$$I_1\,dx\,dy = M_z\,dx\,dy\,dz \tag{2-2}$$

and

$$I_2\,dx\,dy = \left(M_z + \frac{\partial M_z}{\partial x}\,dx\right) dx\,dy\,dz$$

From the net current in the y-direction between the centers of these two volume elements we get

$$(I_1 - I_2)\,dx\,dy = -\left(\frac{\partial M_z}{\partial x}\,dx\right) dx\,dy\,dz \tag{2-3}$$

or

$$I_y = -\frac{\partial M_z}{\partial x}\,dx\,dz$$

from which the current density in the y-direction is

$$J_{my}' = -\frac{\partial M_z}{\partial x} \tag{2-4}$$

Similarly, associated with the x component of magnetic moment, there is an additional y component of current density

$$J_{my}'' = \frac{\partial M_x}{\partial z} \tag{2-5}$$

and the total y component of current density is thus

$$J_{my} = \frac{\partial M_x}{\partial z} - \frac{\partial M_z}{\partial x} \tag{2-6}$$

By similar arguments, the x and z components of current density are

$$J_{mx} = \frac{\partial M_z}{\partial y} - \frac{\partial M_y}{\partial z} \tag{2-7}$$

$$J_{mz} = \frac{\partial M_y}{\partial x} - \frac{\partial M_x}{\partial y}$$

and a comparison with the definition of the curl of a vector given in Equation 14-1 of Chapter 5 shows that

$$\mathbf{J}_m = \nabla \times \mathbf{M} \tag{2-8}$$

Thus, only if the magnetic dipole moment per unit volume is not constant in space can there be a net magnetization current density arising from the microscopic current loops, analogous to the situation in a dielectric where polarization charge density requires a space rate of change of polarization.

Combining Equation 2-8 and 1-2, we obtain

$$\oint \mathbf{B}\cdot d\mathbf{l} = \mu_0\int \mathbf{J}\cdot d\mathscr{A} + \mu_0\int \nabla\times\mathbf{M}\cdot d\mathscr{A} \tag{2-9}$$

or by applying Stoke's theorem:

$$\oint \mathbf{B}\cdot d\mathbf{l} = \mu_0\int \mathbf{J}\cdot d\mathscr{A} + \mu_0\oint \mathbf{M}\cdot d\mathbf{l} \tag{2-10}$$

Transposing terms in Equation 2-10 gives

$$\oint\left(\frac{\mathbf{B}}{\mu_0} - \mathbf{M}\right) \cdot d\mathbf{l} = \oint \mathbf{H} \cdot d\mathbf{l} = \int \mathbf{J} \cdot d\mathscr{A} \tag{2-11}$$

where

$$\mathbf{H} = \frac{\mathbf{B}}{\mu_0} - \mathbf{M} \tag{2-12}$$

H is a new vector called the magnetic field. Its line integral around a closed path is a function only of the true current densities, analogous to the way the surface integral of **D** was related to only the free charge densities.

With the exception of certain substances such as those containing compounds and alloys of iron, cobalt, and nickel, the intensity of magnetization, **M**, is proportional to **H**. For such materials Equation 2-12 is written, after regrouping (cf: Equation 6-2 of Chapter 3):

$$\mathbf{B} = \mu_0\mathbf{H} + \mu_0\chi_m\mathbf{H} = \mu_0\kappa_m\mathbf{H} = \mu\mathbf{H} \qquad \mu = \mu_0\kappa_m \tag{2-13}$$

where $\kappa_m = 1 + \chi_m$. κ_m is called the relative permeability of the medium, and χ_m the magnetic susceptibility. χ_m for free space is, of course, zero; but, different from the electric susceptibility discussed in Chapter 3, χ_m can be positive or negative and the relative permeability can be either greater than or less than unity.

The field **H** is a variable introduced to allow us to deal with the influence of a material medium in spite of our lack of information about the microscopic details of current distributions. But it should be clearly recognized that **H** is not a field vector that is confined to a material medium. **H** exists in all space, according to the definition in Equation 2-12; in a vacuum, $\mathbf{M} = 0$, and $\mathbf{H} = \mathbf{B}/\mu_0$. Although the *line integral* of **H** around a closed path depends only on the true currents linked by that path and is independent of the medium, the vector **H** itself does depend on the medium. For example, the line integral of **H** over a circle through whose center passes a long straight wire carrying a current I, is

$$\oint \mathbf{H} \cdot d\mathbf{l} = \frac{I}{2\pi a} \tag{2-14}$$

where a is the radius of the circular path. When the space around the wire is occupied by a vacuum or symmetrically by a homogeneous magnetic medium, the value of **H** is the same at all points on the path of integration and this value is independent of the nature of the medium. But if the conditions are not symmetrical—if, for example, only part of the path is in a medium of relative permeability κ_m while the rest of the path is in a vacuum—then the value of **H** is in general different at all points of the path from the value in the symmetrical situation. The line integral of **H**, however, is given by Equation 2-14 in all these cases.

3. Vector Potential of Magnetic Dipoles

When the microscopic dipoles of a medium are not randomly oriented, they produce a net magnetic flux density in space, which is most readily found by means of the vector potential function. We begin by finding the vector potential in the space around a single, circular current loop of radius a, starting with Equation 17-10 of Chapter 5. Let the current loop be in the yz-plane with center at the origin. Let the sense of the current be such that when the fingers of the right hand curve with the conventional current, the thumb points along the positive x-axis. Without loss of generality we can take the point at which we shall find the vector potential as being in the xy-plane, with coordinates x, y, 0. An element of length on the loop is located by the coordinates 0, y_0, z_0, and is at a distance r from the point P, whereas the center of the loop is distant R from P. From Figure 3 it is seen that

$$d\mathbf{l} = -\mathbf{j}\,dy_0 + \mathbf{k}\,dz_0 \tag{3-1}$$

$$r^2 = x^2 + (y - y_0)^2 + z_0{}^2 = R^2 + a^2 - 2yy_0$$

and the contribution of this current element to the vector potential at the point P is

$$d\mathbf{A} = \frac{\mu_0 I}{4\pi}\frac{(-\mathbf{j}\,dy_0 + \mathbf{k}\,dz_0)}{[R^2 + a^2 - 2yy_0]^{1/2}} \tag{3-2}$$

so that

$$\mathbf{A} = \frac{\mu_0 I}{4\pi}\left\{-\mathbf{j}\oint\frac{dy_0}{[R^2 + a^2 - 2yy_0]^{1/2}} + \mathbf{k}\oint\frac{dz_0}{[R^2 + a^2 - 2yy_0]^{1/2}}\right\} \tag{3-3}$$

Inspection shows that the first integral is zero, since it involves only y_0 and is integrated from a to a. Evaluation of the second requires elliptic

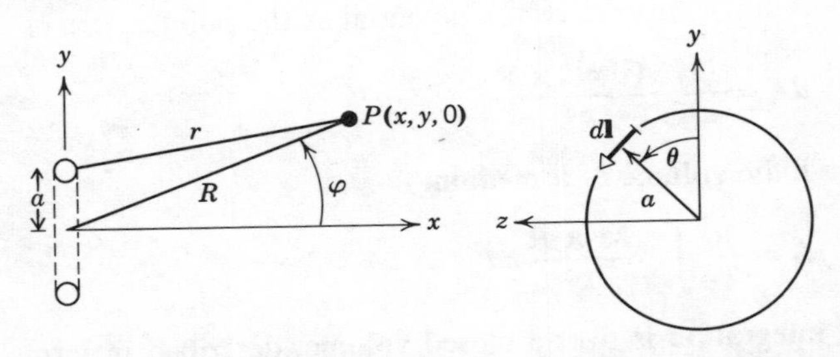

Figure 3. A current loop and associated axes, viewed edge-on and face-on.

integrals, but as we are chiefly concerned with the vector potential at large distances from the loop, the integral may be simplified by the use of the binomial expansion. For $R \gg a$, we obtain

(3-4) $$\mathbf{A} = \mathbf{k}\frac{\mu_0 I}{4\pi}\int \frac{dz_0}{R[1 + (a^2/R^2) - (2yy_0/R^2)]^{1/2}}$$

$$= \mathbf{k}\frac{\mu_0 I}{4\pi R}\int \left(1 + \frac{yy_0}{R^2}\right) dz_0$$

From Figure 3 it is seen that

(3-5) $$y_0 = a\cos\theta \qquad z_0 = a\sin\theta \qquad dz_0 = a\cos\theta\, d\theta$$

so that

(3-6) $$\mathbf{A} = \mathbf{k}\frac{\mu_0 I}{4\pi R}\int_0^{2\pi}\left(1 + \frac{ya\cos\theta}{R^2}\right) a\cos\theta\, d\theta$$

$$= \mathbf{k}\frac{\mu_0 I a}{4\pi R}\cdot\frac{ya\pi}{R^2}$$

but

(3-7) $$\mathbf{i} I\pi a^2 = \mathbf{m} \qquad \text{and} \qquad \frac{y}{R^3} = \frac{R\sin\varphi}{R^3}$$

where $\mathbf{m}$ is the magnetic moment of the loop, so that finally

(3-8) $$\mathbf{A} = \mathbf{k}\frac{\mu_0 m R\sin\varphi}{4\pi R^3} = \frac{\mu_0}{4\pi}\frac{\mathbf{m}\times\mathbf{R}}{R^3}$$

The latter form, being in vector notation, is independent of the location of the origin. Within the approximation of $R \gg a$ it can be shown that Equation 3-8 is true for arbitrary shapes of loops.

Over a small region $d\mathscr{V}'$, let the magnetic moment per unit volume be $\mathbf{M}$ and thus the magnetic moment be $\mathbf{M}d\mathscr{V}'$. Relative to an arbitrary origin, let the coordinates of this volume element be x', y', z', so that the contribution of $d\mathscr{V}'$ to the vector potential at the point x, y, z is

(3-9) $$d\mathbf{A} = \frac{\mu_0}{4\pi}\frac{\mathbf{M}\,d\mathscr{V}'\times\mathbf{R}}{R^3}$$

Then for a finite volume of a medium

(3-10) $$\mathbf{A} = \frac{\mu_0}{4\pi}\int_{\mathscr{V}'}\frac{\mathbf{M}\times\mathbf{R}}{R^3}\,d\mathscr{V}'$$

where the integration is over a closed volume, described in terms of the variables x', y', and z'. Some care must be used if the point xyz is within or close to the volume of integration because of the approximations used in obtaining Equation 3-10, but so long as the dimensions of the volume of integrations are large compared to molecular distances, the error is negligible. Equation 3-10 is not valid, however, for integrating over a microscopic volume around the point xyz.

Equation 3-10 can be transformed into the sum of a volume and a surface integral which lends itself to instructive physical interpretation. In what follows it is necessary to distinguish carefully the integrations and the differentiation operations that are with respect to the primed and with respect to the unprimed variables. Noting that

$$\mathbf{R} = \mathbf{i}(x - x') + \mathbf{j}(y - y') + \mathbf{k}(z - z') \tag{3-11}$$

it follows that

$$\nabla'\left(\frac{1}{R}\right) = \frac{\mathbf{R}}{R^3} \tag{3-12}$$

where

$$\nabla' = \mathbf{i}\frac{\partial}{\partial x'} + \mathbf{j}\frac{\partial}{\partial y'} + \mathbf{k}\frac{\partial}{\partial z'} \tag{3-13}$$

in terms of which Equation 3-10 becomes

$$\mathbf{A} = \frac{\mu_0}{4\pi}\int \mathbf{M} \times \nabla'\left(\frac{1}{R}\right) d\mathscr{V}' \tag{3-14}$$

With the aid of the vector relation

$$\nabla \times (u\mathbf{S}) = (\nabla u) \times \mathbf{S} + u\nabla \times \mathbf{S} \tag{3-15}$$

where u is a scalar function and $\mathbf{S}$ is a vector function, Equation 3-14 takes the form

$$\mathbf{A} = \frac{\mu_0}{4\pi}\left\{\int \frac{\nabla' \times \mathbf{M}}{R} d\mathscr{V}' - \int \nabla' \times \left(\frac{\mathbf{M}}{R}\right) d\mathscr{V}'\right\} \tag{3-16}$$

By applying Gauss's theorem to the divergence of the vector $\mathbf{S} \times \mathbf{T}$, where $\mathbf{T}$ is an arbitrary *constant* vector, it follows that

$$\int \nabla \times \mathbf{S}\, d\mathscr{V} = -\int \mathbf{S} \times d\mathscr{A} \tag{3-17}$$

which can be used to put Equation 3-16 into the form

$$\mathbf{A} = \frac{\mu_0}{4\pi}\left\{\int \frac{\nabla' \times \mathbf{M}}{R} d\mathscr{V}' + \int \frac{\mathbf{M}}{R} \times d\mathscr{A}\right\} \tag{3-18}$$

A comparison of the first integral with the results of Problem 25 of Chapter 5 shows that $\nabla' \times \mathbf{M}$ is equivalent to a current density $\mathbf{J}_m$ within the medium (cf: Equation 2-8) and **M** in the second integral has the magnitude of a surface current density. The vector surface current density on the surface of the volume of integration is $\mathbf{M} \times \mathbf{n}$, where $\mathbf{n}$ is a unit vector normal to the surface of integration. If the surface of the volume of integration is placed just outside the medium, where $\mathbf{M} = 0$, then the surface integral in Equation 3-18 drops out. But, for reasons similar to those given in Section 4, Chapter 3, it is sometimes advantageous to take the volume limits just at the surface of the medium and use both terms in 3-18. In a body, for example, in which the magnetic moment per unit volume is constant, the volume integral drops out, and the abrupt change in **M** at the surface leads to a surface current density in terms of which the vector potential is obtained.

It should be understood that this vector potential is only that due to the magnetization of the medium; any true currents make an additional contribution. As it stands Equation 3-18 is directly applicable to obtaining the vector potential of permanent magnets. The **B**-field is obtained from the curl of the vector potential, which is a differentiation operation with respect to the unprimed variables x, y, z.

4. Induced Magnetic Dipoles

When orbital magnetic moments of the atoms are so aligned as to produce a net magnetic moment per unit volume in a medium, we saw that these magnetic moments must be taken into account in finding the average **B**-field within and outside the medium. (This **B**-field within the medium is not necessarily the local microscopic field, but is an average determined in much the same way as the average **E** field within a dielectric was determined.) We shall now examine what the effect of an imposed **B**-field is on the orbital electrons, showing that such an imposed field induces a change in the magnetic moment of the orbit, the increment of magnetic moment being in a direction *opposite* to the imposed field.

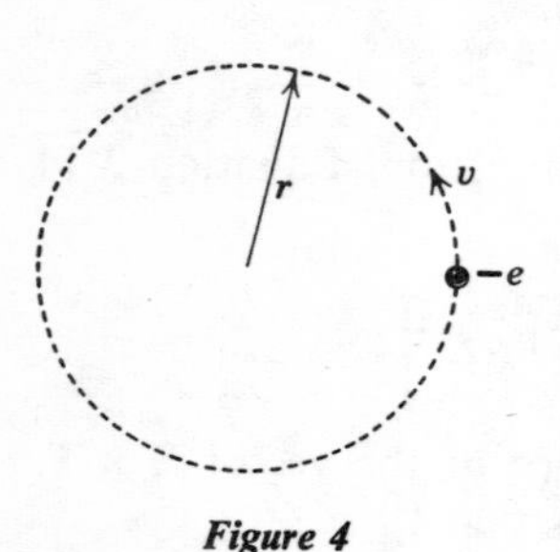

Figure 4

The simplest picture to consider is a circular electron orbit whose plane is perpendicular to the imposed field (Figure 4).

The motion of the electron with no imposed **B** is obtained by equating the centripetal acceleration to the inverse square centripetal force

$$\frac{K}{r^2} = \frac{m_0 v^2}{r} \tag{4-1}$$

where K is a constant including the effective central charge and the charge on the electron, v is the speed of the electron in the orbit, m_0 is the mass of the electron, and r is the radius of the orbit. Suppose now that a magnetic flux density, **B**, is established uniformly over the area of the orbit and directed outward from the diagram of Figure 4. There results an additional radial force on the electron given by the Lorentz force, $-e\mathbf{v}' \times \mathbf{B}$, where v' is the new electron speed. This force is directed toward the center when v' is in the sense indicated in the figure. Thus the new equilibrium is given by

$$ev'B + \frac{K}{r'^2} = \frac{m_0 v'^2}{r'} \tag{4-2}$$

where r' is the new radius and e is the charge on the electron. Arguments can be given to show that the change in the radius from r to r' is negligibly small, even without resort to quantum-mechanical restrictions.*

Pending a qualitative justification to be given later, we assume at this point that $r \simeq r'$. Introducing the angular velocity $\omega = v/r$ into Equations 4-1 and 4-2 and then eliminating K from 4-2, we get

$$eB\omega' = m_0(\omega'^2 - \omega^2) = m_0(\omega' - \omega)(\omega' + \omega) \tag{4-3}$$

With even the largest fields available in the laboratory, the fractional change in the angular velocity is small. Very closely, then, we may replace $(\omega' + \omega)$ with $2\omega'$. Writing $(\omega' - \omega)$ as $\Delta\omega$, we obtain from Equation 4-3

$$\Delta\omega = \frac{eB}{2m_0} \tag{4-4}$$

An electron making f revolutions per second passes a given point in the orbit f times in a second, so that the effective current in the orbit is $-fe$. Thus the magnetic moment of the orbit is

$$dm = -fe(\pi r^2) = -\frac{\omega e r^2}{2} \tag{4-5}$$

and the change in the magnetic moment as a result of the imposed B-field is

$$\Delta(dm) = -\frac{e^2 B r^2}{4m_0} \tag{4-6}$$

showing that the change in the magnetic moment is opposite to the direction of **B** and independent of the direction of rotation of the electron in the

* The reader is referred to R. W. Wood, *Physical Optics*, 3rd edition (Macmillan), p. 677, for an interesting qualitative as well as quantitative discussion; and to Page and Adams, *Electrodynamics* (Van Nostrand), p. 237, for a more elegant treatment.

orbit. A symmetrical atom, in which the sum of the individual orbital magnetic moments of the several electrons is zero, will thus have induced in it a magnetic moment directed opposite to the field **B**. From the arguments of Section 2 it follows that the magnetic susceptibility will be negative and the relative permeability will be less than unity. Such materials showing a negative susceptibility are called diamagnetic.

It can be shown that a similar result follows in the case of an orbit whose plane is inclined to the direction of the magnetic flux density. The orbit will then precess about the field direction with the angular velocity given by Equation 4-4, with a corresponding effect on the orbital magnetic moment.*

5. Energy Considerations for Induced Magnetic Moments

Some interesting features are brought out by examining more closely the physics of induced magnetic moments. Clearly the electron has a different energy after the application of the field than before. A moving charge, however, experiences only a force normal to its direction of motion in a magnetic field. Where, then, does the energy come from?

During the build-up of the field there exists a Faraday emf around the orbit which is determined by the negative rate of change of the total magnetic flux over the plane of the orbit. In Figure 4, for example, if **B** is increasing in value in a direction out from the plane of the page, the emf will be directed in a clockwise sense around the orbit, accelerating the electron in a counterclockwise sense. The electron thus continues to acquire energy from the field until the field becomes steady. There being no dissipative forces in the atom, it retains this energy until the field is removed, at which time the energy is returned to the field.

Why were we justified in assuming that the radius of the orbit remains constant? Because of the increase in angular velocity, the electron tends to move to a larger radius. But with B increasing at the same time, the centripetal force on the electron increases as well. (The reverse arguments naturally apply for an oppositely directed or an oppositely changing field.) There is no guarantee that these two influences will exactly cancel each other, but at least it is qualitatively plausible that the fractional change in r can be expected to be small. It is in fact a smaller quantity by at least another order of magnitude from the fractional change in the angular frequency.

---

* For a more detailed development of the general case see Goble and Baker, *Elements of Modern Physics* (Ronald), p. 71ff. R. A. Becker, *Introduction to Theoretical Mechanics* (McGraw-Hill), p. 304; or Bitter, *Currents, Fields, and Particles* (Wiley), p. 232ff.

These induced magnetic moments may be expected to be a property of all matter and the magnitude of the magnetic moment per unit volume resulting from them should be independent of temperature, provided that the number of molecules per unit volume does not change with temperature. Much of the analysis performed in connection with the theory of dielectrics may be carried over essentially intact to the theory of induced magnetic dipoles. An important difference, however, lies in the negative sign of the induced orbital dipole moment.

6. Permanent Magnetic Dipoles

Whereas the induced negative magnetic dipole moment is a general property of all molecular matter, some molecules have an electronic structure possessing a net permanent magnetic moment for the atom as a whole, analogous to a permanent electric dipole. A permanent magnetic dipole experiences a torque in an applied **B**-field (Chapter 5, Section 8), and this torque tends to align the dipole with the field.

A positive intensity of magnetization—in the same direction as the applied field—appears in the medium, often more than overcoming the negative magnetization from the induced dipoles. Just as with permanent electric dipoles, this effect is temperature-dependent because thermal agitation tends to oppose alignment; it also may show saturation effects, although laboratory fields are unlikely to be strong enough to saturate nonferro-magnetic materials.

Materials showing a positive magnetization in an applied field are called paramagnetic. Their susceptibility is positive and their relative permeability is greater than unity.

Ferromagnetic materials constitute a special class of paramagnetic materials in which the positive magnetization is greater by several orders of magnitude. The evidence points to the electron spins in the ferromagnetics as being responsible for this enhanced effect, but the orderly arrangement of atoms in a crystalline form is also necessary, the localized forces in the ferromagnetic crystal structure being such that large groups of atoms behave as a single magnetic unit. Such units, or domains, appear to have certain directions of easy magnetization, requiring relatively large applied fields to shift the direction of the magnetization.

A domain may shift its magnetization direction as a unit, a domain already magnetized in the direction of the applied field may grow at the expense of neighboring domains, or domains may subdivide. Experimental evidence shows that such changes do not take place smoothly or continuously, but go by jumps, especially when there are crystal imperfections or impurities present. This discontinuous growth in the magnetization can be readily demonstrated by winding two coils around a magnetically hard iron rod (magnetically hard iron is iron that will retain permanent magnetization) with a smoothly increasing current in one of the coils. The other coil is connected to some device for indicating the induced emf developed by the changing flux. An audio amplifier with loudspeaker gives a good qualitative indication. The small but sharp changes in the flux resulting from the stepwise changes in magnetization of the sample produce a roaring noise in the loudspeaker. On the other hand, a smoothly changing flux produces no sound, as can be demonstrated by performing the experiment without the iron in the coils.

At high enough temperature the special crystalline forces that cause groups of atoms to act magnetically in unison disappear, and a ferromagnetic material reverts to paramagnetic behavior. This temperature, at which ferromagnetism disappears, is characteristic of the material and is called the Curie temperature, because Pierre Curie did much of the pioneer work in investigating this effect.

7. Hysteresis

The discontinuous nature of the magnetizing process in a ferromagnetic sample indicates a reluctance of the domains to change their state of magnetization, and it is not surprising to find that the magnetization **M** and, hence, the flux density **B** do not pass through the same sequence of values with a decreasing applied field **H** as with an increasing field. Both **M** and **B** within such a sample are double-valued functions of **H**, a property which is called hysteresis. The simple relation $\mathbf{B} = \mu\mathbf{H}$, where μ is a scalar constant, is not valid for a ferromagnetic material, and the relationship between **B** and **H** is ordinarily shown graphically. A typical curve of this sort is shown in Figure 5. A flux density B_r exists in the sample even when the applied H-field has been reduced to zero. The magnitude of this flux density is a measure of the *retentivity* of the material. It requires a reversed field, H_c, to reduce the flux density to zero. The strength of this reversed field is called the *coercive force.*

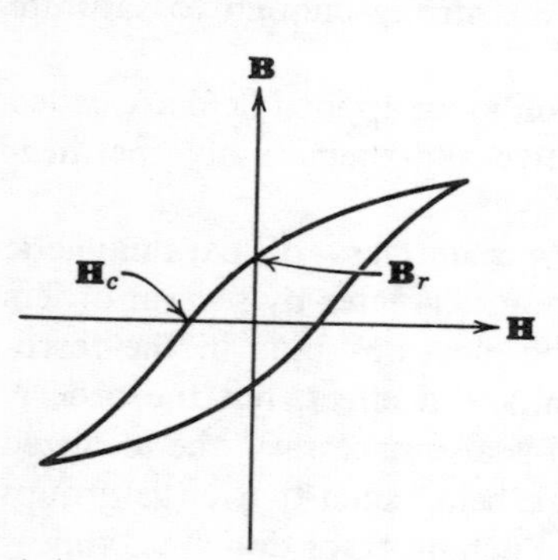

Figure 5. A hysteresis curve of a ferromagnetic material.

Ferromagnetic materials show considerable differences in the shapes and ranges of their hysteresis loops. Some so-called magnetically soft materials have very thin loops, almost but not quite a single line, whereas, at the other extreme, there are materials whose hysteresis loops resemble rectangles. These latter materials often reach saturation magnetization for very low applied H, and a large fraction of the magnetization is retained with the removal of the applied field. Such materials find use in permanent magnets.

8. Hysteresis Energy Losses

A doughnut-shaped ring of ferromagnetic material has a convenient geomentry for measuring the energy required to put the material through a hysteresis cycle. Such a ring is wound with insulated wire, the wire forming a toroid. We shall compute the energy required to carry the iron through a complete hysteresis cycle.

For a toroid of N turns, Equation 2-11 of this chapter becomes

$$\oint \mathbf{H} \cdot d\mathbf{l} = Ni \tag{8-1}$$

where i is the current in the toroid at any instant. The line integral of $\mathbf{H}$ about a closed path is a function only of the total current encircled and not of the material medium in the path. Symmetry requires that $\mathbf{H}$ have the same value at all points equidistant from the center of the toroid, so that, following the reasoning in Section 12 of Chapter 5, we write

$$\mathbf{H} = \frac{Ni}{2\pi r} \tag{8-2}$$

where r is the radius of a circular path within the toroid.

The instantaneous applied voltage at the terminals of the conductor must be equal to the resistance drop in the wire plus the back emf generated by the changing flux within the coil. It is only the energy associated with the latter that is of immediate interest. From Faraday's law the emf developed in each turn of the toroid is the negative of the rate of change of $\mathbf{B}$ multiplied by the cross-sectional area $\mathscr{A}$, provided that the toroid is large enough so that $\mathbf{B}$ may be considered effectively constant over the cross section of the ring. Thus the emf is

$$e = -N\mathscr{A}\frac{dB}{dt} \tag{8-3}$$

The energy that the source of power must supply in one cycle is the integral of the instantaneous power over a cycle

$$W = \int ei\,dt = \int N\mathscr{A}\frac{dB}{dt}\frac{2\pi rH}{N}\,dt = \mathscr{V}\oint H\,dB \tag{8-4}$$

where $\mathscr{V}$ is the volume of the iron within the toroid ring. The integral $\oint H\,dB$, which is the area within the hysteresis loop of Figure 5, represents the hysteresis energy loss per unit volume per cycle in the material. Consequently, the hysteresis power loss is proportional to the number of times the hysteresis loop is traversed per second:

$$P = f\mathscr{V} \oint H\,dB \tag{8-5}$$

where f is the frequency of the current in the windings.

Electrical transformers employ cores for which the hysteresis loop encloses as little area as possible. Other properties are, of course, also important: magnetic saturation should not take place within the range of currents for which the transformer is designed (however, there are certain special cases where saturation is desired); and the steepness of the hysteresis curve should be great, to provide as large a flux density as possible within the core. Actual transformer cores represent compromises among these demands.

The energy loss given by Equation 8-4 represents only part of the conversion to heat that occurs when a ferromagnetic core is cycled through its hysteresis loop. Because the material itself is a conductor, circulating currents, called eddy currents, will be induced in it by the changing flux, and these currents also generate heat. They are reduced by increasing the effective resistivity of the material, either by using the material in a finely divided powder form, with the particles being bonded together by an insulating cement, or by fabricating the material in thin sheets bonded by an insulating adhesive. The planes of the sheets are oriented normal to the directions of the induced currents.

A class of ferromagnetic materials called ferrites has found considerable application in recent years. Ferrites combine ferromagnetic properties with the low conductivities of insulators, and are particularly useful at very high frequencies where the conductivity of ordinary ferromagnetics makes them useless.

9. The Toroid with Gap

We now return to the ferromagnetic ring within a toroid, but instead of a complete ring, let there be an air gap of width x and let the current be steady. Since Equations 2-11 and 8-1 do not involve the properties of the material medium, they remain unchanged for this case. But although this integral of **H** around the closed path depends only on the true current in the coil, **H** itself is not similarly independent of the medium except in cases of special symmetry (such as the closed ring). We need to write

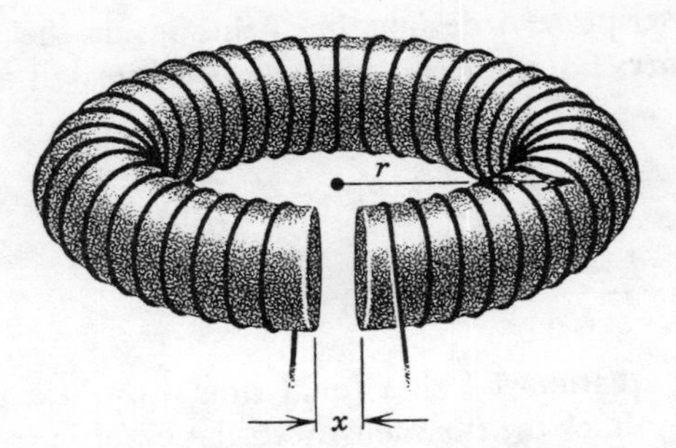

Figure 6. Iron core toroid with gap.

Equation 8-1 now as

$$\oint \mathbf{H} \cdot d\mathbf{l} = NI = \int_0^x \mathbf{H}_g \cdot d\mathbf{l} + \int_x^{2\pi r} \mathbf{H}_i \cdot d\mathbf{l} \tag{9-1}$$

where $\mathbf{H}_i$ is the value of $\mathbf{H}$ within the ferromagnetic material and $\mathbf{H}_g$ is the value of $\mathbf{H}$ in the gap.

Since $\nabla \cdot \mathbf{B} = 0$, the **B**-field lines, are continuous around the ring. There is, however, some fringing of the **B**-field at the edge of the gap, but for a gap width, x, small compared to the cross-sectional dimensions of the ring, this fringing is small, and we make the approximation of neglecting it, so that the flux density B is constant along the line of integration of 9-1.

At the current I, let the ratio $B/H = \mu_0 \kappa_m$, where κ_m is a relative permeability for this particular magnetic condition of the core. (κ_m will change with $\mathbf{H}_i$.) It follows that

$$H_i = \frac{B}{\mu_0 \kappa_m} \qquad \text{and} \qquad H_g = \frac{B}{\mu_0} \tag{9-2}$$

the relative permeability of the gap being unity, and Equation 9-1 becomes

$$NI = \int_0^x \frac{B}{\mu_0} \, dl + \int_x^{2\pi r} \frac{B}{\mu_0 \kappa_m} \, dl \tag{9-3}$$

$$= \frac{B}{\mu_0}\left[\left(\frac{2\pi r - x}{\kappa_m}\right) + x\right]$$

or

$$NI = \frac{B}{\mu_0}\left(\frac{L}{\kappa_m} + x\right) \tag{9-4}$$

where L is the path length of integration *within* the material.

With the subscript zero designating fields inside the toroid *without* a ferromagnetic core, Equation 9-4 can also be expressed as

$$2\pi r H_0 = \frac{2\pi r B_0}{\mu_0} = \frac{(2\pi r - x)B}{\mu_0 \kappa_m} + \frac{xB}{\mu_0} \tag{9-5}$$

or

$$\frac{B_0}{B} = \frac{L + \kappa_m x}{\kappa_m (L + x)} \tag{9-6}$$

It is seen from Equation 9-6 that for a ring without a gap ($x = 0$), the flux density is κ_m times the flux density for an empty toroid, and that for even a small value of gap width the flux density is considerably decreased. As a simple illustration, if κ_m is 1000 and x is 1/1000 of the circumference of the ring, the flux density is very close to one-half of the value without a gap.

Using Equations 9-6 in 9-2, we find that

$$H_g = \frac{\kappa_m H_0 (L + x)}{L + \kappa_m x} \tag{9-7}$$

and
$$H_i = \frac{H_0 (L + x)}{L + \kappa_m x}$$

so that

$$\frac{H_g}{H_i} = \kappa_m \tag{9-8}$$

The H-field within the gap is many times the H-field in the material.

For design purposes it is often useful to show this information in another way. Consider first the toroid and ring without gap ($x = 0$). Multiply Equation 9-4 by the cross-sectional area of the ring, $\mathscr{A}$, and then rearrange into

$$\mathscr{A} B = \Phi = \frac{NI}{L/\mu_0 \kappa_m \mathscr{A}} \tag{9-9}$$

where Φ is the total flux in the ring. This equation has a form reminiscent of Ohm's law, with Φ analogous to the current and NI, called the magnetomotive force, analogous to the voltage. The denominator, called the reluctance, has the same dependence on dimensions that resistance does, the permeability being analogous to the conductivity.

For a ring with a gap x, 9-4 becomes

$$\Phi = \frac{NI}{[(L/\mu_0 \kappa_m \mathscr{A}) + (x/\mu_0 \mathscr{A})]} \tag{9-10}$$

showing that the reluctances of different parts of a series magnetic circuit add as series resistances do.

The foregoing, of course, depends on approximations, within the validity of which the performance of transformers, electromagnets, generators, etc., where air gaps, if present, are small and where κ_m is large, can be predicted by manipulation of equations very similar to Ohm's law for d-c circuits.

10. Magnetic Poles

The magnetic pole is a familiar concept to anyone who has studied permanent magnets in elementary physics. Such poles were looked upon as the magnetic counterparts of electric charges, with the limitation that they cannot have independent existence on separated bodies, as can electric charges. But throughout the foregoing discussion, no mention of magnetic poles has been made and the question properly arises whether such a concept can fit into the theory of magnetic behavior as here presented.

Equation 2-12 is the definition of the magnetic field vector, **H**, whose divergence is

$$\nabla \cdot \mathbf{H} = -\nabla \cdot \mathbf{M} \tag{10-1}$$

since $\nabla \cdot \mathbf{B} = 0$. Thus, the divergence of the magnetic dipole moment per unit volume plays the same role here as the divergence of the polarization played in dielectric theory in the absence of free charges. The negative of the divergence of **M** may be formally replaced by a magnetic charge density, ρ_m, which acts as a polar source function for the magnetic field, **H**. The expression

$$\nabla \cdot \mathbf{H} = \rho_m \tag{10-2}$$

is then the magnetic equivalent of Gauss's law for the electric field. In Chapter 1 Gauss's law was shown to be essentially an expression of the inverse square law, and we can infer from 10-2 that the field **H** drops off with the inverse square of the distance from a point magnetic charge.

Equation 10-1 shows that a magnetic pole density exists where the divergence of **M** is nonzero. As an example, consider a permanent magnet in the form of a cylindrical rod lying along the x-axis and assume for the sake of simplicity that the magnetization has a uniform value over the

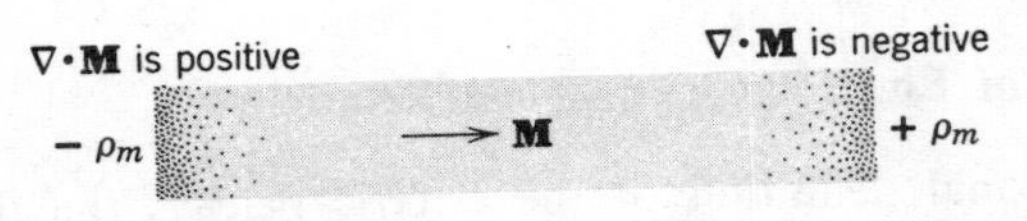

Figure 7. A bar magnet with poles at the ends where $\nabla \cdot \mathbf{M} \neq 0$.

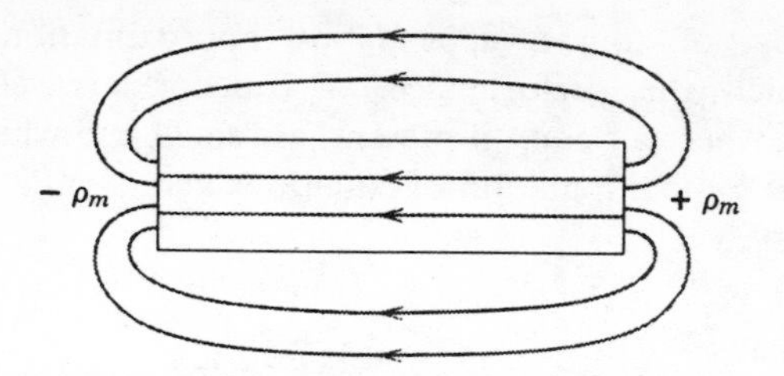

Figure 8. H-*field inside and outside a bar magnet.*

volume of the rod and is directed parallel to the axis. Then

$$\nabla \cdot \mathbf{M} = \frac{\partial M_x}{\partial x} \tag{10-3}$$

and this divergence is zero everywhere except at the ends, being negative at the right-hand end and positive at the left. There is thus a positive magnetic charge density at the right-hand end and a negative charge density at the left end. Following the reasoning used with dielectrics, the magnetic charges in this idealized case may also be treated as surface charge densities given by the normal components of **M**. In an actual bar magnet, however, **M** in the vicinity of the ends will not be strictly parallel to the axis and the poles will be somewhat distributed over the end regions.

According to the conclusions drawn from Equation 3-18, we see that there is no effective magnetization current density within the volume of this idealized magnet, but there is an effective surface current density circulating on the cylindrical surface. The magnet acts as if it were a solenoid whose **B**-field arises from these surface currents and whose **H**-field arises from the poles at the ends. The latter is a dipole field, being directed from the positive toward the negative charge. Figure 8 shows qualitatively the pattern of the **H**-field. From the relation

$$\mathbf{B} = \mu_0(\mathbf{H} + \mathbf{M}) \tag{10-4}$$

it is seen that $\mathbf{B} = \mu_0\mathbf{H}$ in space outside the body of the magnet, but that inside, **B** depends on the difference between **M** and **H**, these two vectors being oppositely directed. Should the bar be within a solenoid carrying true current, the **H**-field is the vector sum of that due to the solenoid and that due to the magnetic poles.

11. Summary of Electromagnetic Field Relations

The operational definition of the electric field is the force per unit positive charge acting on a test charge:

$$\mathbf{F} = \mathbf{E}q \tag{11-1}$$

The magnetic flux field is defined by the expressions:

(11-2) $$d\mathbf{F} = I(d\mathbf{l} \times \mathbf{B})$$

or

(11-3) $$\mathbf{F} = q(\mathbf{v} \times \mathbf{B})$$

where, in 11-2, $I\,d\mathbf{l}$ is a current element, and in 11-3, q is a point charge moving with a velocity $\mathbf{v}$. Operationally, a single measurement of the force on a current element, $I\,d\mathbf{l}$, gives only the magnitude and the direction of the component of $\mathbf{B}$ which is normal to the plane of $\mathbf{F}$ and $d\mathbf{l}$; and similarly, a single measurement of the force on the moving charge q fixes only the component of $\mathbf{B}$ normal to the plane of $\mathbf{F}$ and $\mathbf{v}$. At least two measurements in different directions are necessary to obtain both the magnitude and direction of the field $\mathbf{B}$.

Based on an experimentally observed inverse square law of force between point charges, the electric field due to a charge distribution is

(11-4) $$\mathbf{E} = \int \frac{\rho \mathbf{r}\, d\mathscr{V}}{4\pi\epsilon_0 r^3}$$

and based on an assumed inverse square law of force between current elements, the magnetic flux density due to a current density distribution is

(11-5) $$\mathbf{B} = \frac{\mu_0}{4\pi} \int \frac{\mathbf{J} \times \mathbf{r}}{r^3}\, d\mathscr{V}$$

where $\mathbf{J}$ is the current density.

Gauss's law for the electric field is another expression for the inverse square law, 11-4;

(11-6) $$\int \mathbf{E} \cdot d\mathscr{A} = \int \frac{\rho}{\epsilon_0}\, d\mathscr{V}$$

where ρ is the charge density and the left-hand integral is over the surface of the volume of the integral on the right. Equation 11-6, by the use of Gauss's theorem, is put into the point or differential form:

(11-7) $$\nabla \cdot \mathbf{E} = \frac{\rho}{\epsilon_0}$$

Similarly, Gauss's law for the magnetic flux density is the expression of the magnetic inverse square law, 11-5:

(11-8) $$\int \mathbf{B} \cdot d\mathscr{A} = 0$$

where again the integral is over a closed area. The corresponding differential relation is:

(11-9) $\nabla \cdot \mathbf{B} = 0$

In order to be able to describe the effect of a material medium on the electric field, a new field vector, the displacement, is defined by

(11-10) $\mathbf{D} = \epsilon_0 \mathbf{E} + \mathbf{P}$

where $\mathbf{P}$ is the dipole moment per unit volume. The inverse square law—Gauss's law—for the displacement is

(11-11) $\int \mathbf{D} \cdot d\mathcal{A} = \int \rho \, d\mathcal{V}$

where ρ is now only the free charge density, and the corresponding point form of 11-11 is

(11-12) $\nabla \cdot \mathbf{D} = \rho$

To describe the magnetic effects of a material medium, a vector, the magnetic field, is defined by

(11-13) $\mathbf{H} = \dfrac{\mathbf{B}}{\mu_0} - \mathbf{M}$

Relations for $\mathbf{H}$ comparable to 11-11 and 11-12 for $\mathbf{D}$ are

(11-14) $\int \mathbf{H} \cdot d\mathcal{A} = -\int \mathbf{M} \cdot d\mathcal{A} \qquad \nabla \cdot \mathbf{H} = -\nabla \cdot \mathbf{M} = \rho_m$

where ρ_m is formally a magnetic charge density.

In the special case of a linear, isotropic medium, $\mathbf{D}$ is proportional to $\mathbf{E}$ according to

(11-15) $\mathbf{D} = \epsilon_0 \kappa \mathbf{E} = \epsilon \mathbf{E}$

where κ is the relative dielectric constant. And correspondingly, $\mathbf{H}$ is proportional to $\mathbf{B}$, according to

(11-16) $\mathbf{B} = \mu_0 \kappa_m \mathbf{H} = \mu \mathbf{H}$

where κ_m is the relative magnetic permeability.

The line integral of the electric field around a closed path is

(11-17) $\oint \mathbf{E} \cdot d\mathbf{l} = -\dfrac{d}{dt} \int \mathbf{B} \cdot d\mathcal{A}$

for which the corresponding point or differential relation is

(11-18) $\nabla \times \mathbf{E} = -\dot{\mathbf{B}}$

which is a consequence of Equation 11-17 when the path of integration is itself not a function of time. A deduction from 11-17 is that a charge, in traversing a closed path in a static electric field, neither gains nor loses energy, and a consequence of 11-18 is that a *static* electric field can always be expressed in terms of the gradient of a scalar potential function:

$$\text{(11-19)} \qquad \mathbf{E} = -\nabla V$$

The line integral of B around a closed path is

$$\text{(11-20)} \qquad \oint \mathbf{B} \cdot d\mathbf{l} = \mu_0 I$$

for filamentary currents, where I is the current encircled or linked by the path of integration; and for distributed currents,

$$\text{(11-21)} \qquad \oint \mathbf{B} \cdot d\mathbf{l} = \mu_0 \int \mathbf{J} \cdot d\mathscr{A}$$

where the path of integration of the line integral is the perimeter of the area of integration for the current density. In Equation 11-21 the currents in the integrand include all currents, microscopic and "true," whereas in the similar relation for **H**;

$$\text{(11-22)} \qquad \oint \mathbf{H} \cdot d\mathbf{l} = \int \mathbf{J} \cdot d\mathscr{A}$$

the current densities on the right are only "true" currents.

The point relation corresponding to Equation 11-21 is

$$\text{(11-23)} \qquad \nabla \times \mathbf{B} = \mu_0 \mathbf{J} \qquad \text{(all currents)}$$

and to 11-22, is

$$\text{(11-24)} \qquad \nabla \times \mathbf{H} = \mathbf{J} \qquad \text{(only ``true'' currents)}$$

Finally from 11-5 is derived the fact that **B** can always be written as

$$\text{(11-25)} \qquad \mathbf{B} = \nabla \times \mathbf{A} \qquad \text{where } \mathbf{A} = \frac{\mu_0}{4\pi} \int \frac{\mathbf{J}}{r}\, d\mathscr{V}$$

A being called the vector potential. The use of Equation 11-25 in 11-18 gives

$$\text{(11-26)} \qquad \mathbf{E} = -\dot{\mathbf{A}} - \nabla \phi$$

where ϕ is a general scalar potential. Evidently Equation 11-26 is a general relation which reduces to 11-19 under static conditions.

This summarizes the high points of the development of the electromagnetic field relations obtained up to this point. Equations 11-23 and 11-24 will require closer scrutiny in Chapter 10, where it will be found that they need modification for application to time-varying conditions, after which they, together with Equation 11-7, 11-9, and 11-18 and their integral counterparts, will be the basis for the theory of electromagnetic waves.

But meanwhile, before we come to electromagnetic waves, we shall study the behavior of electric circuits under the influence of time-varying emf's and time-varying currents.

Discussion Questions

1. Carry out the argument of Section 4 for the case where the vector area of the electron orbit is not parallel to **B**. (See references given in footnote.)

2. In Equation 2-13 μ is the ratio of B to H. Suppose we have two ferromagnetic ring samples of the same geometry. They are wound with identical toroids and each is initially in an unmagnetized state and each is then magnetized to the point a where the flux density is B_a, so that the ratio of B to H is the same for each. The magnetization curves for the two samples are different, as indicated in Figure Q2. In the reasoning of Chapter 6, Section 8, the energy stored in the

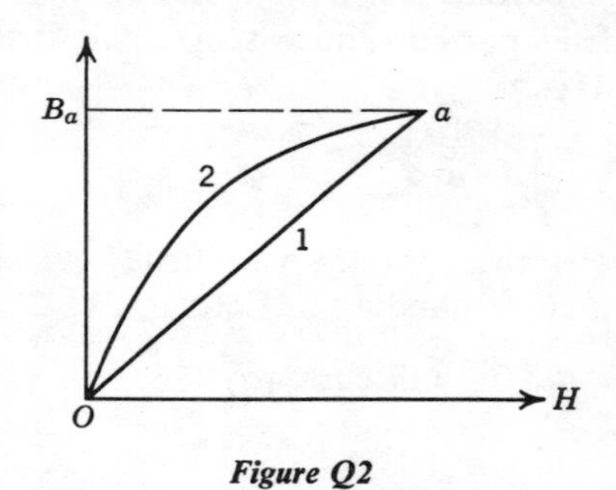

Figure Q2

magnetic flux density field is $B_a^2/2\mu$ joules per unit volume. But by the argument of Section 8 of this chapter, the energy per unit volume required to establish this state is $\int_0^a HdB$, which is not the same for the two samples. A casual reading of the above suggests that different energy inputs result in the same stored energy in the fields in the two systems. Analyze carefully what is wrong with this conclusion.

3. A region of space is in the vicinity of some true currents, but is so delimited that it is impossible within this region to describe a closed path that encircles a current. Show that, within this region, the magnetic field **H** can be described in terms of a magnetic scalar potential. Consider some of the computational advantages of this fact. Describe such a region in the vicinity of a long straight wire carrying a current; in the vicinity of a solenoid of length L; of a permanent magnet.

4. Two possible definitions of the magnetic flux density **B** have been given. Still a third may be formulated in terms of the torque on a magnetic dipole of moment **m**. Formulate and express such a definition analytically. Then discuss the relative merits of these definitions. The latter is in effect the one used when flux densities **B** are measured by the proton magnetic resonance method.

5. Suggest a method of experimentally checking the value of the exponent in the force law for magnetic poles using bar magnets. The results of Problem 11 may contain a suggestion.

6. What is wrong with the following reasoning? From

$$\mathbf{B} = \nabla \times \mathbf{A} \quad \text{and} \quad \mathbf{B} = \mu_0(\mathbf{H} + \mathbf{M}) \tag{1}$$

we can write

$$\oint \mathbf{B} \cdot d\mathbf{l} = \mu_0 \oint \mathbf{H} \cdot d\mathbf{l} + \mu_0 \oint \mathbf{M} \cdot d\mathbf{l} = \oint \nabla \times \mathbf{A} \cdot d\mathbf{l} \tag{2}$$

In a region where there are no true currents,

$$\oint \mathbf{H} \cdot d\mathbf{l} = 0 \tag{3}$$

Therefore, in such a region

$$\mu_0 \oint \mathbf{M} \cdot d\mathbf{l} = \oint \nabla \times \mathbf{A} \cdot d\mathbf{l} \tag{4}$$

and since this result does not depend on any particular path of integration within this region, then

$$\mu_0 \mathbf{M} = \nabla \times \mathbf{A} \tag{5}$$

(But Equations 1 and 5 are not consistent.)

Problems

1. Show that $\nabla \times \mathbf{H} = J_{\text{true}}$

2. A region in space is empty except for a permanent magnet. The magnetization within the volume of the magnet is **M**. Show that at every point in the region the relation

$$\mathbf{B} = \mu_0 \mathbf{M} - \nabla \psi$$

holds, where ψ is a scalar function of position.

3. Using the pattern of reasoning employed with dielectrics, find a relation by means of which the magnetic polarizability of the molecules of a gas can be obtained from experimental measurements.

4. The molecules of a gas have a permanent magnetic moment **m**. There are n molecules per unit volume, the gas is at an absolute temperature T, and is in a uniform flux density **B**. Show that, for small fields and high temperatures, that part of the magnetization of the gas due only to the permanent dipoles is given very closely by

$$\mathbf{M} = \frac{nm^2\mathbf{B}}{3kT}$$

5. A needle-like magnet has a length $2d$, a cross-sectional area $\mathscr{A}(d \gg \sqrt{\mathscr{A}})$, and a uniform magnetization **M** along the axis. A point P has the polar coordinates r, θ, where the polar axis is the axis of the magnet and the origin of coordinates is the middle of the magnet, and r is large compared to $2d$. Show that the magnitude of the vector potential at P is given by

$$A = \frac{\mu_0 m \sin\theta}{4\pi r^2}$$

where m is the magnitude of the magnetic moment of the magnet.

6. An idealized bar magnet has a circular cross section of area $\mathscr{A}$ and a length **L**. It is uniformly magnetized with a magnetic moment per unit volume **M**. Find the total magnetic moment of the bar. Show that associated with **M** is a current $I_m = \mathbf{M} \cdot \mathbf{L}$ on the cylindrical surface directed in the positive sense around the cylinder relative to **M**.

7. Show that the surface density of magnetic charge is equal to the normal component of **M** at the surface.

8. What are the units of magnetic charge density?

9. What are the units of magnetomotive force? Of magnetic flux? Of reluctance? Test their consistency with Equation 9-10.

10. Show, from Equation 3-10, that

$$\mathbf{B} = \frac{\mu_0}{4\pi} \int \left[\frac{3\mathbf{R}(\mathbf{M} \cdot \mathbf{R})}{R^5} - \frac{\mathbf{M}}{R^3} \right] d\mathscr{V}'$$

11. A needle-like magnet has a length $2d$, a cross-sectional area $\mathscr{A}(d \gg \sqrt{\mathscr{A}})$, and a uniform magnetization **M** along the axis. A point P is distant r from the center of the magnet on the axis; and a point Q is distant r on the perpendicular bisector. For both points $r \gg d$. Using the expression for **B** from the previous problem, find **B** at the points P and Q, and show that the result is μ_0 times the **H**-field obtained from the effective magnetic charges on the ends of the magnet. What is the ratio B_P/B_Q?

12. Show that where alternative flux paths are available, the flux divides between these paths as current divides between parallel resistances; hence, that parallel reluctances combine as parallel resistances.

13. A betatron magnet is constructed as in Figure P13, with a column of cross-sectional area $\mathscr{A}_1$ and relative permeability κ_{m1} at the center. The remaining region is air or vacuum. Assume that there is no fringing of the magnetic field at the edges of the gap and that the total area is $\mathscr{A}_t$. Find $\mathscr{A}_1$ in terms of

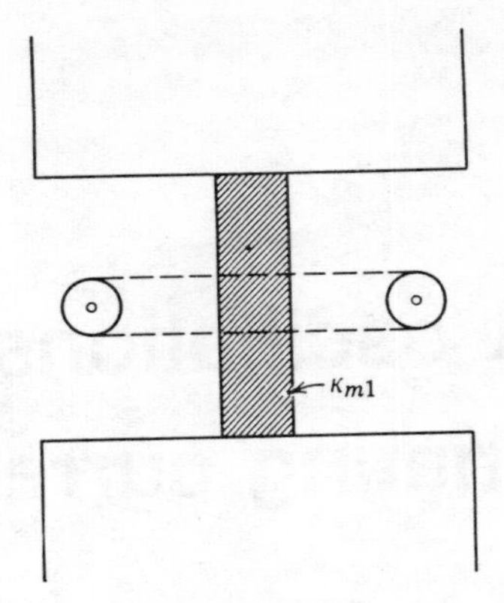

Figure P13

κ_{m1} and $\mathscr{A}_t$ in order that the condition for stable betatron orbits (Chapter 6, Section 9) is fulfilled. Discuss the various approximations involved in this calculation.

14. If, in the betatron magnet described in Problem 13, at a certain point in the cycle the value of κ_{m1} drops—that is, the center post approaches saturation—does the electron spiral outward or inward? Explain.

REFERENCES

1. Bitter, *Currents, Fields, and Particles* (Wiley), 1956. Chapter 5 and Chapter 6.
2. Scott, *The Physics of Electricity and Magnetism* (Wiley), 1959. Chapter 8.
3. Harnwell, *Principles of Electricity and Electromagnetism* (McGraw-Hill), 1938. Chapter XI.
4. Whitmer, *Electromagnetics*, 2nd edition (Prentice-Hall), 1962. Chapter 9.
5. Jackson, *Classical Electrodynamics* (Wiley), 1962. Chapter 5.
6. Panofsky and Phillips, *Classical Electricity and Magnetism*, 2nd edition (Addison-Wesley), 1962. Chapter 7 and Chapter 8.
7. Smythe, *Static and Dynamic Electricity* (McGraw-Hill), 1950. Chapter XII.

8

Transient oscillations and alternating currents

In mechanical systems transient phenomena, such as a struck bell, a plucked string, an automobile going over a bump, are familiar. A change in stimulus of relatively short duration produces a displacement which is followed by a return of the system to equilibrium. The return may, or may not be, oscillatory. Electrical systems respond to abrupt changes in stimuli in analogous fashions. So close is the analogy, in fact, that it is often possible to study the behavior of a mechanical system by observing its electrical analog. Such an electrical analog has the significant advantage that it is usually easier to introduce changes and study their effects than to produce the analogous changes in the mechanical system itself. But of at least comparable importance is the study of the way that electrical transients enter into the functioning of electrical systems themselves—power systems and communication systems—where their presence is sometimes undesirable but more often is part of the proper behavior of the system.

An example of an electrical transient has already been discussed in Chapter 6, where the current in an inductance-resistance circuit was analyzed after a voltage was suddenly inserted or removed from the circuit. This L-R circuit was introduced earlier than the general topic of transient behavior because it illustrated certain energy considerations associated with currents and magnetic fields.

We now examine a more general system containing inductance, capacitance, and resistance, of which the L-R circuit was a special case. We shall find that there are two general types of transient behavior, in one of which the system returns smoothly to equilibrium in a manner governed by exponential functions of time with real exponents; and in the other, the system returns in an oscillatory fashion, governed by exponential functions with complex exponents. The latter type is not only of intrinsic interest but also constitutes an introduction to the study of steady-state alternating currents.

1. The Series *R-L-C* Circuit

In the system of Figure 1, the switch is initially moved to the position x, thus charging the capacitor to a potential V. The switch is then moved to position y and the capacitor discharges through the resistor and inductor. We shall express both the charge on the capacitor and the current in the resistor and inductor as functions of the time from the instant of closing the switch to y. (Other means of initiating a transient response are available but this one lends itself readily to a simple specification of initial conditions.)

Since both the charge on the capacitor and the current may be oscillatory, it is necessary to specify to which plate we refer when we identify the charge on the capacitor and to specify a positive sense for the current. Our choice, which is of course arbitrary, shall be that the charge q in the subsequent relations is the charge on plate A (Figure 2) of the capacitor and that the positive sense of current is the clockwise sense. This convention is consistent with the relation between charge and current:

$$i = \frac{dq}{dt} \tag{1-1}$$

That is, an increase in charge q is associated with the positive sense of current. (Our convention also means that the initial charge applied to the capacitor by the method of Figure 1 is negative.)

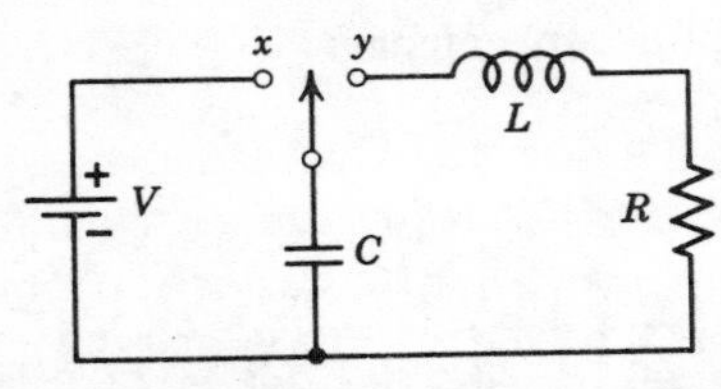

Figure 1. Circuit for studying the transient behavior of an R-L-C system.

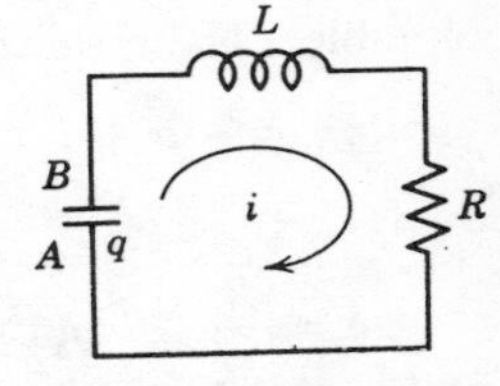

Figure 2. The description of the electrical behavior is in terms of the charges on plate A and current that is positive in the clockwise sense.

Where currents and potential differences are a function of time, Kirchhoff's loop rule expresses a relation among potential differences that occur at the same instant of time. In a clockwise sense around the circuit, there is a drop in potential across the inductance of $L(di/dt)$; a drop in potential across the resistance or Ri; and a drop in potential across the capacitor of q/C. Thus

$$-L\frac{di}{dt} - Ri - \frac{q}{C} = 0 \tag{1-2}$$

which, after introducing Equation 1-1, becomes

$$L\frac{d^2q}{dt^2} + R\frac{dq}{dt} + \frac{q}{C} = 0 \tag{1-3}$$

Such a linear differential equation of the second degree is encountered frequently in physics, and the reader should become thoroughly familiar with the method of solving it. To satisfy Equation 1-3, the derivatives of q with respect to t must have the same functional dependence on time as q itself. Therefore, we try an exponential function of the time:

$$q = Me^{mt} \tag{1-4}$$

where M and m are constants to be determined in such a way as to satisfy Equation 1-3 and to fit the initial conditions. With the introduction of 1-4, Equation 1-3 becomes:

$$Me^{mt}\left(Lm^2 + Rm + \frac{1}{C}\right) = 0 \tag{1-5}$$

and thus only the two values of m:

$$m = -\frac{R}{2L} \pm \left[\left(\frac{R}{2L}\right)^2 - \frac{1}{LC}\right]^{1/2} \tag{1-6}$$

are acceptable. These we shall write, for convenience:

$$m_1 = -a + b \quad \text{and} \quad m_2 = -a - b \tag{1-7}$$

where

$$a = \frac{R}{2L} \quad \text{and} \quad b = \left[\left(\frac{R}{2L}\right)^2 - \frac{1}{LC}\right]^{1/2} = \left(a^2 - \frac{1}{LC}\right)^{1/2}$$

With either m_1 or m_2, Equation 1-4 satisfies Equation 1-3, and since we are dealing with a linear equation, the sum of the two solutions is also a solution:

$$q = M_1e^{(-a+b)t} + M_2e^{(-a-b)t} \tag{1-8}$$

An examination of Equation 1-7 shows that, whereas a is always a real quantity, b may be real or imaginary, depending on whether a^2 is greater than or less than, $1/LC$. Of these two cases, we first take up the simpler one, where b is a real number.

2. The Overdamped Case

Where

$$\frac{R^2}{4L^2} > \frac{1}{LC} \tag{2-1}$$

a is always greater than b, and thus both exponents in Equation 1-8 are negative. After the switch is closed, the charge on the capacitor flows through the inductor and the resistor until the system is quiescent.

Equation 1-8 must be valid at the instant the switch is closed ($t = 0$), when the charge on plate A is $-Q(Q = CV)$ and the current is zero. At this instant

$$q = -Q = M_1 + M_2 \tag{2-2}$$

and

$$i = \frac{dq}{dt} = 0 = M_1(-a + b) + M_2(-a - b) \tag{2-3}$$

These two simultaneous equations determine M_1 and M_2 to be

$$M_1 = \frac{-Q(a + b)}{2b} \qquad M_2 = \frac{-Q(-a + b)}{2b} \tag{2-4}$$

so that the capacitor charge at any time, t, is

$$q = -\frac{Qe^{-at}}{2b}[(a + b)e^{bt} - (a - b)e^{-bt}] \tag{2-5}$$

and the current is

$$i = \frac{Q(a^2 - b^2)}{2b} e^{-at}[e^{bt} - e^{-bt}] \tag{2-6}$$

Equations 2-5 and 2-6 are plotted as functions of time in Figure 3.

3. The Underdamped Case

It is not unreasonable to expect a quite different dependence of charge on time when

$$\frac{R^2}{4L^2} < \frac{1}{LC}$$

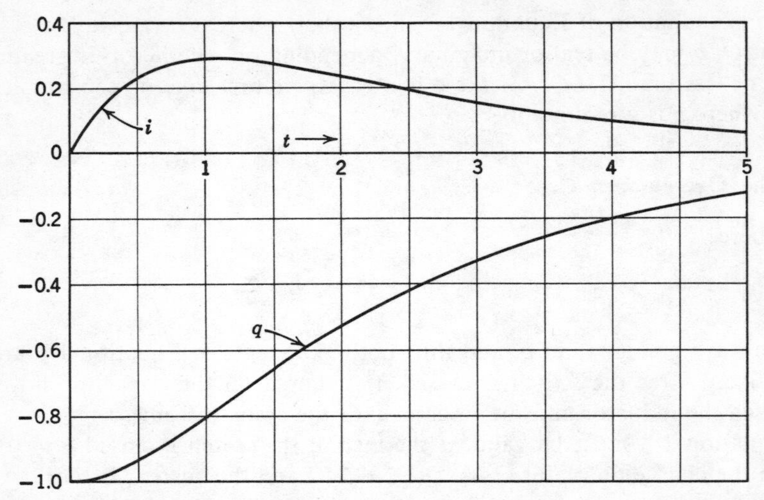

Figure 3. *Plot of Equation 2-5 (q versus t) and Equation 2-6 (i versus t). Here $Q = 1$ unit of charge; $a = 1$ per unit of time; $b = 0.5$ per unit of time.*

which makes b an imaginary quantity. Let us write

$$\text{(3-1)} \qquad b = j\beta \qquad \text{where } \beta = \left(\frac{1}{LC} - \frac{R^2}{4L^2}\right)^{1/2} \text{ and } j = \sqrt{-1}$$

(In the present text j, rather than i, is used for $\sqrt{-1}$ because i is already preempted for current.)

Equation 1-8 thus becomes

$$\text{(3-2)} \qquad \begin{aligned} q &= M_1 e^{(-a+j\beta)t} + M_2 e^{(-a-j\beta)t} \\ &= e^{-at}(M_1 e^{j\beta t} + M_2 e^{-j\beta t}) \end{aligned}$$

All the information concerning the dependence of charge on time is in this expression, but by changing the form we can better visualize what is happening in the circuit.

The constants M_1 and M_2 are arbitrary constants determined by initial conditions. We write these two constants in terms of two new constants, D and θ, such that

$$\text{(3-3)} \qquad M_1 = De^{j\theta} \qquad M_2 = De^{-j\theta}$$

(In a problem at the end of this chapter the reader is asked to show that such a substitution can always be made.)

Substituting Equation 3-3 into 3-2, we get

$$\text{(3-4)} \qquad \begin{aligned} q &= De^{-at}[e^{j(\beta t+\theta)} + e^{-j(\beta t+\theta)}] \\ &= 2De^{-at} \cos(\beta t + \theta) \end{aligned}$$

from which the behavior of the system is more readily interpreted. The variation of charge with time has the form of a sinusoidal oscillation of frequency $\beta/2\pi$, whose amplitude decreases with time in a manner governed by the exponential function, e^{-at}. A graph of q versus t is shown in Figure 4. The magnitude of a depends on the ratio R/L, and when this ratio becomes large enough, oscillations no longer occur and the over-damped condition of Section 2 exists. The transition condition between oscillation and simple exponential decay is called the critically damped condition and occurs for

$$R_{cr} = 2\left(\frac{L}{C}\right)^{1/2}$$

where R_{cr} is called the critical damping resistance. It is the smallest value of R for which the system returns to the quiescent state without oscillation.

Sometimes it is necessary to compare the rates at which oscillations die out in different systems. For this purpose a quantity called the logarithmic decrement is defined in the following way. Let A_1 and A_2 be two successive maxima of charges (or current), and t_1 and t_2 be the times at which these maxima occur. Then

$$\frac{A_1}{A_2} = \frac{e^{-at_1}}{e^{-at_2}} = e^{a(t_2-t_1)} = e^{aP} \tag{3-5}$$

where $P = t_2 - t_1$, the period of the oscillation. The logarithm of this amplitude ratio, commonly given the symbol $\delta(=aP)$, is the logarithmic

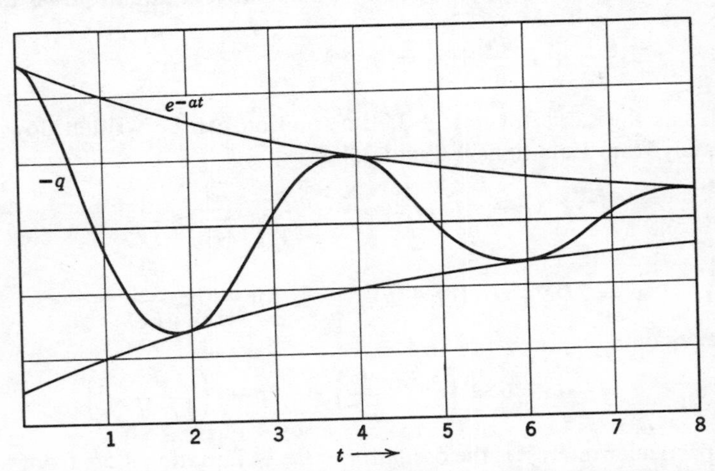

Figure 4. Damped oscillations. q versus t from Equation 3-4, showing exponential damping factor. Initial conditions the same as for graph in Figure 3. Here $Q = 1$; $a = 0.224$; $\beta = \pi/2$.

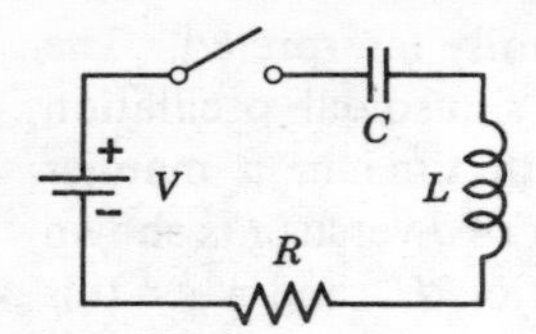

Figure 5. A transient voltage exists on the capacitor after the switch is closed.

decrement. It is independent of the choice of which successive peaks are compared.

4. Another Example: Capacitor Voltage Transient

Suppose the battery voltage remains in the *L-C-R* circuit after the switch is closed, as is the case in the circuit of Figure 5. Any internal resistance in the battery is included in R. The Kirchhoff loop rule gives an equation similar to Equation 1-2 except that the constant voltage V is also included:

$$L\frac{di}{dt} + Ri + \frac{q}{C} = V \tag{4-1}$$

We wish to examine the voltage on the capacitor instead of the charge, and we introduce it by the following substitutions:

$$q = Cv_c \qquad i = \frac{dq}{dt} = C\frac{dv_c}{dt} \qquad \frac{di}{dt} = C\frac{d^2v_c}{dt^2} \tag{4-2}$$

where v_c is the instantaneous capacitor voltage. Equation 4-1 becomes

$$LC\frac{d^2v_c}{dt^2} + RC\frac{dv_c}{dt} + v_c = V \tag{4-3}$$

Writing $u = v_c - V$ and dividing by C, we convert Equation 4-3 to

$$L\frac{d^2u}{dt^2} + R\frac{du}{dt} + \frac{u}{C} = 0 \tag{4-4}$$

which has the same form as 1-3. Its solution can be written down immediately from Equations 1-8 and 3-4 as

$$u = M_1e^{(-a+b)t} + M_2e^{(-a-b)t} \qquad \text{for } \frac{R^2}{4L^2} > \frac{1}{LC} \tag{4-5a}$$

$$u = 2De^{-at}\cos(\beta t + \theta) \qquad \text{for } \frac{R^2}{4L^2} < \frac{1}{LC} \tag{4-5b}$$

where, again,

$$a = \frac{R}{2L} \qquad b = \left(a^2 - \frac{1}{LC}\right)^{1/2} \qquad \beta = \left(\frac{1}{LC} - a^2\right)^{1/2}$$

Of particular interest is the oscillatory case of Equation 4-5*b*, from which we obtain the capacitor voltage, v_c:

$$v_c = V + 2De^{-at}\cos(\beta t + \theta) \tag{4-6}$$

If time is counted from the instant the switch is closed, the initial conditions are

$$\text{(4-7)} \qquad \text{at } t = 0, \qquad v_c = 0 \qquad i = C\frac{dv_c}{dt} = 0$$

from which the constants $2D$ and θ are found to be

$$\text{(4-8)} \qquad 2D = -\frac{V}{\cos\theta} = \frac{V}{\beta\sqrt{LC}}$$

$$\tan\theta = -\frac{a}{\beta} \qquad \cos\theta = \frac{-\beta}{\sqrt{a^2 + \beta^2}} = -\beta\sqrt{LC}$$

so that Equation 4-6 becomes

$$\text{(4-9)} \qquad v_c = V\left[1 + \frac{e^{-at}}{\beta\sqrt{LC}}\cos(\beta t + \theta)\right]$$

Thus, for values of resistance less than the critical damping resistance, the capacitor voltage is a damped oscillation centered on the battery voltage. It takes on values higher than the battery voltage during alternate half-cycles; in fact, for values of resistance approaching zero, the first maximum approaches twice the battery voltage. In some circuitry, unless this fact is taken into account in the design, electrical breakdown in the capacitor dielectric may occur during this initial surge.

Transient oscillations can appear in electrical circuits as the result of any sudden change in stimulus, such as an induced voltage pulse from a neighboring system, or, in power systems, an induced pulse from lightning in the vicinity. In sound reproducing systems, a sharp change in intensity, such as when a drum is struck in an orchestral piece, may cause the system to "ring," introducing spurious tones. On the other hand, this oscillatory property is used to advantage in certain types of electronic generators, amplifiers, and frequency converters.

5. A Ballistic Galvanometer System

A ballistic galvanometer is a moving coil galvanometer, particularly adapted to the measurement of a quantity of charge, rather than current. Its moving coil is constructed with a relatively large moment of inertia, causing it to respond rather slowly to an impulse and to move slowly thereafter. The details of its operation are discussed in books on electrical instruments* and we shall not consider them at length here; sufficient for

* Smith and Wiedenbeck, *Electrical Measurements* (McGraw-Hill) 5th edition, 1959. Chapter 6. Also Scott, *The Physics of Electricity and Magnetism* (Wiley), p. 393ff.

our purpose to note that if a pulse of current is applied to the galvanometer, with the pulse completed before the coil has moved appreciably from its rest position (hence the large moment of inertia), the coil will slowly deflect and then return to rest. The magnitude of the maximum deflection is proportional to the total charge which passed through the galvanometer during the current pulse. Whether the coil oscillates about the rest position after the initial swing, or returns directly to rest depends on the nature of the circuit of which the galvanometer is a part; what is significant to our discussion is that under any circumstances the magnitude of the initial swing is proportional to the charge in the original pulse.

We shall discuss in some detail one type of circuit that is used with such a galvanometer, because the analysis is a further illustration of the transient phenomena that we have been discussing in the previous sections. In Figure 6 there is a mutual inductance M between the two coils L_1 and L_2, a cell of emf V, a ballistic galvanometer, and a switch. R_1 and R_2 represent the total resistances in their respective loops, and as usual we count time from the instant that the switch is closed.

Kirchhoff's rule applied to each loop yields two simultaneous differential equations:

$$L_1 \frac{di_1}{dt} + M \frac{di_2}{dt} + R_1 i_1 = V \tag{5-1a}$$

$$L_2 \frac{di_2}{dt} + M \frac{di_1}{dt} + R_2 i_2 = 0 \tag{5-1b}$$

where the term involving M represents in each equation the emf induced from the other loop. (M itself may be either positive or negative.)

From these two differential equations we obtain expressions for each current as a function of the time. Physical intuition suggests that the current in the primary will rise to a final value determined by V and R_1 and that the current in the secondary will be a pulse induced while the primary current is changing. Our ultimate aim is to find the total charge associated with the current pulse, which is found from the integral of i_2, but it is instructive to find the dependence of i_1 on time and on the circuit constants as well.

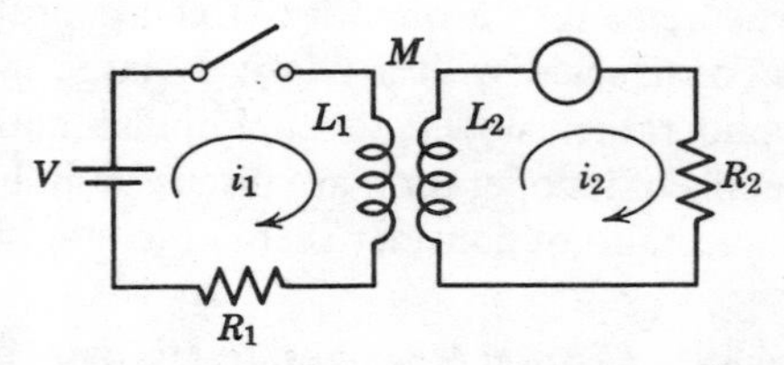

Figure 6. *A ballistic galvanometer-mutual inductance system.*

Equations 5-1 can be solved by differentiation and substitution to separate the variables. It is left to the reader to show that the primary and secondary currents are

$$i_1 = \frac{e^{-at}}{M}\left[-K_1\left(L_2 - \frac{R_2}{a-b}\right)e^{bt} - K_2\left(L_2 - \frac{R_2}{a+b}\right)e^{-bt}\right] + K_3 \tag{5-2a}$$

$$i_2 = e^{-at}(K_1 e^{bt} + K_2 e^{-bt}) \tag{5-2b}$$

where $$a = \frac{L_1R_2 + L_2R_1}{2(L_1L_2 - M^2)}$$

$$b = \left[\frac{(L_1R_2 + L_2R_1)^2 - 4R_1R_2(L_1L_2 - M^2)}{4(L_1L_2 - M^2)^2}\right]^{1/2}$$

These expressions contain the three constants of integration, K_1, K_2, and K_3, which are evaluated in terms of known initial and final conditions. Thus,

$$\text{at } t = 0, \quad i_1 = 0 \quad \text{and} \quad i_2 = 0 \tag{5-3}$$

$$\text{and at} \quad t = \infty, \quad i_1 = \frac{V}{R_1} = I \quad \text{and} \quad i_2 = 0$$

and Equations 5-2 become

$$i_1 = I\left\{1 + \frac{(a^2 - b^2)}{2bR_2}e^{-at}\left[\left(L_2 - \frac{R_2}{a-b}\right)e^{bt} - \left(L_2 - \frac{R_2}{a+b}\right)e^{-bt}\right]\right\} \tag{5-4a}$$

$$i_2 = -\frac{MI(a^2 - b^2)}{2bR_2}e^{-at}(e^{bt} - e^{-bt}) \tag{5-4b}$$

A non-oscillatory case has been graphed in Figure 7.

Finally the total charge which flows through the galvanometer during the current pulse i_2 is obtained by integrating i_2 over the time $t = 0$ to $t = \infty$:

$$\int_0^\infty i_2\,dt = -\frac{MI(a^2 - b^2)}{2bR_2}\left[\frac{e^{-(a-b)t}}{-(a-b)} - \frac{e^{-(a+b)t}}{-(a+b)}\right]_0^\infty = -\frac{MI}{R_2} = -\frac{MV}{R_1R_2} \tag{5-5}$$

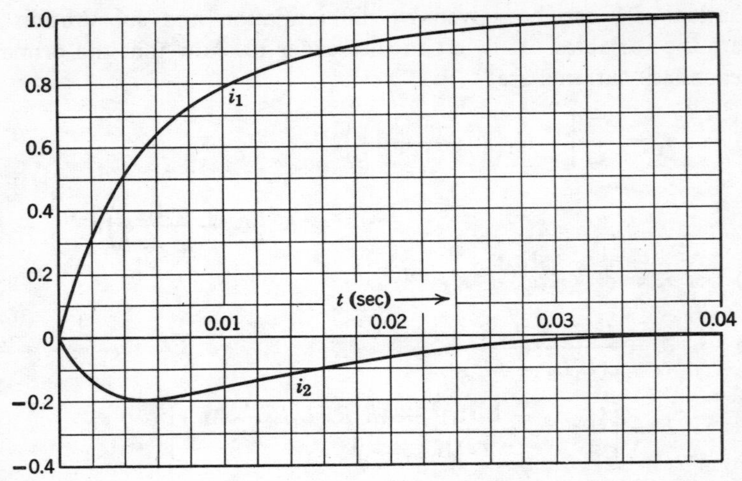

Figure 7. i_1 *and* i_2 *versus* t *from Equation 5-4.* $I = 1$ *amp*, $R_1 = R_2 = R, L_1 = L_2 = L$, $M = L/2$, $R/L = 150$.

It is interesting to note that this total charge is independent of the self-inductance in the system. Thus, if the galvanometer has been previously calibrated and if the emf and the resistances are known, the mutual inductance can be experimentally determined. Or, contrariwise, with a known mutual inductance, the galvanometer can be calibrated.

If there is no interest, theoretical or experimental, in the primary current, the result of Equation 5-5 can be obtained directly and very simply from Equation 5-1*b* by integrating from $t = 0$ to $t = \infty$:

$$L_2 \int_0^\infty \frac{di_2}{dt}\,dt + M \int_0^\infty \frac{di_1}{dt}\,dt + R_2 \int_0^\infty i_2\,dt = 0 \tag{5-6}$$

$$L_2 \int_0^0 di_2 + M \int_0^I di_1 + R_2 Q = 0$$

The first integral drops out because i_2 starts and ends with the value zero; the second integral is just the final current, I, in the primary; and the third integral is the total charge which flows in the secondary. Thus, from 5-6 we get:

$$MI + R_2 Q = 0 \tag{5-7}$$

in agreement with Equation 5-5. The detailed nature of the currents, however, is not brought out by this short cut.

Problems at the end of this chapter show other illustrations of transient phenomena, as well as other ways in which a ballistic galvanometer can be put to practical use.

The reader may sense that the handling of systems still more complicated than the one just considered can become very cumbersome. A method based on the use of Laplace transforms has been developed, which in effect converts the problem from one of solving differential equations to one of solving a group of algebraic simultaneous equations. It is a very powerful tool, but its elaboration is beyond the scope of this book.*

6. Alternating-Current Systems

The transient oscillations just discussed have a frequency fixed by the components of the circuit itself, and they die out because energy is lost from the system in various ways. At low frequencies this loss is principally through conversion to heat. If this lost energy is replaced by applying voltage pulses whose repetition rate is the same or a submultiple of the natural frequency of the system, the amplitude of the oscillations can be stabilized at a level determined by the rate at which energy is supplied, just as the oscillations in a child's swing can be maintained by supplying energy in the form of short pulses at the proper intervals.

It is also possible to drive an *R-L-C* system at an arbitrary frequency, f, by supplying a sinusoidal voltage of that frequency. The amplitude of the resulting sinusoidal current depends on the amplitude of the voltage and on the frequency, being greatest at or near the natural frequency of the system.

It may seem that to specify a sinusoidal alternating voltage unnecessarily limits the discussion to a special case. For should not any repetitive voltage form produce currents of the same form in an *R-L-C* system?

A function whose graph is a repeating pattern can be expressed mathematically in the form of a Fourier series, which in general is an infinite sum of terms, each of which is a sine or a cosine function of the independent variable. An example of such a series, corresponding to a function $V(t)$ is:

$$V(t) = A_0 + A_1 \cos 2\pi ft + A_2 \cos 4\pi ft + A_3 \cos 6\pi ft + \cdots \\ + B_1 \sin 2\pi ft + B_2 \sin 4\pi ft + B_3 \sin 6\pi ft + \cdots \tag{6-1}$$

where f is the repetition frequency of the pattern of V.

Because of the linearity of the differential equations describing the currents in an *R-L-C* circuit where the components are not functions of the current, simultaneously applied voltages produce a current which is the sum of the currents which each voltage would produce individually. Consequently, by studying the response of a linear system to sinusoidal voltages, we are able to predict the response to an arbitrary voltage pattern by

* For a brief discussion of Laplace Transforms, see Scott. *op. cit.*, Chap. 9, p. 472ff.

combining the system responses to all the sinusoidal voltages of which the voltage pattern is composed. In Appendix C there is a brief discussion of Fourier series, including the determination of the coefficients in Equation 6-1.

When core hysteresis and saturation in an inductance, or similar behavior in the dielectric of a capacitor, causes these elements to be functions of the current or the voltage, or where resistances are functions of the voltage or current, such as in semiconductors or where heating is appreciable, the simple application of Fourier series as described above is not valid. The differential equations are no longer linear and exact solutions become either very difficult or impossible. In some cases, however, an approximate solution may still be found.

The rest of this chapter is devoted to a consideration of linear R-L-C systems subjected to applied voltages of the form $V \cos \omega t$ or $V \sin \omega t$, where V is the amplitude of the alternating voltage and ω is the angular frequency—that is, $\omega = 2\pi f$, where f is the number of cycles per second.

7. The Series *R-L-C* Circuit with Alternating Voltage

A series R-L-C circuit is shown schematically in Figure 8. Kirchhoff's loop rule applies to this system just as it does with d-c and with transient currents, with the qualification again that the sum of the potential differences must be taken at one instant of time. Thus, the differential equation for the circuit of Figure 8 is

$$L\frac{di}{dt} + Ri + \frac{q}{C} = V \cos \omega t \tag{7-1}$$

where, instead of the fixed voltage V used in Equation 1-3, we now use the varying voltage $V \cos \omega t$. Any internal resistance in the voltage source is to be considered as included in R.

The charge q is eliminated by differentiating with respect to the time and by introducing the relation $i = dq/dt$.

$$L\frac{d^2 i}{dt^2} + R\frac{di}{dt} + \frac{i}{C} = -\omega V \sin \omega t \tag{7-2}$$

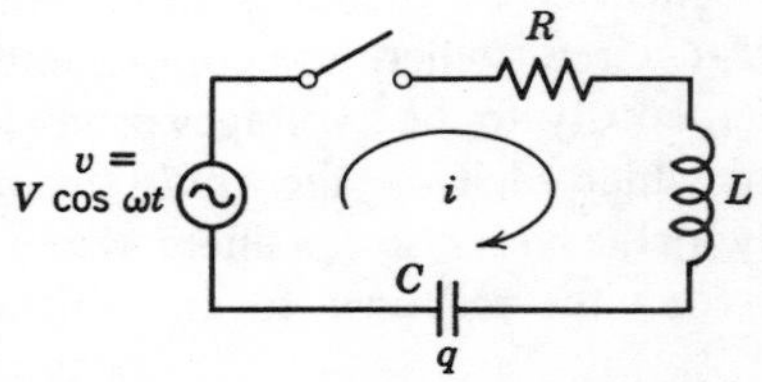

Figure 8. A series circuit with alternating applied voltage.

for which we try a solution in the form:

$$i = I \cos(\omega t - \theta) \tag{7-3}$$

where I is the amplitude of the current and θ is a phase constant introduced because we have no assurance that the voltage and the current will necessarily reach their maxima at the same instant of time. Because of the linearity of the differential equation, a driving voltage of frequency f produces a current of the same frequency, so long as the components R, L, and C are constants independent of the voltage and current. I and θ are to be determined in order that the solution 7-3 satisfies 7-2. Putting Equation 7-3 into 7-2, we get

$$\left(\frac{I}{C} - \omega^2 LI\right) \cos(\omega t - \theta) - \omega RI \sin(\omega t - \theta) = -\omega V \sin \omega t \tag{7-4}$$

The trigonometric identities for the sine and cosine of the difference of two angles enable us to write Equation 7-4 as

$$\left[\omega V + I\left(\frac{1}{C} - \omega^2 L\right) \sin\theta - \omega RI \cos\theta\right] \sin \omega t + \left[I\left(\frac{1}{C} - \omega^2 L\right) \cos\theta + \omega RI \sin\theta\right] \cos \omega t = 0 \tag{7-5}$$

For this relation to be true at any value of t, the coefficients of $\sin \omega t$ and $\cos \omega t$ must be separately zero, a requirement which leads to two simultaneous equations that yield I and θ:

$$I = \frac{V}{\sqrt{R^2 + X^2}} \qquad \tan\theta = \frac{X}{R} \tag{7-6}$$

where $X = \omega L - \dfrac{1}{\omega C}$. With 7-6, the current 7-3 becomes, in full:

$$i = \frac{V}{\sqrt{R^2 + X^2}} \cos\left(\omega t - \arctan \frac{X}{R}\right) \tag{7-7}$$

and we now proceed to examine the implications of this solution.

Before we do, however, one additional point should be mentioned. In form, the left side of Equation 7-2 is the same as the left side of Equation 1-3. Therefore, a function of the form

$$i = e^{-at}(Ae^{bt} + Be^{-bt}) \tag{7-8}$$

can be found which makes the left side of Equation 7-2 zero. The addition of 7-8 to 7-7 yields a function which also satisfies the differential equation 7-2:

$$i = e^{-at}(Ae^{bt} + Be^{-bt}) + \frac{V}{\sqrt{R^2 + X^2}} \cos(\omega t - \theta) \tag{7-9}$$

The total current existing in the circuit of Figure 8 is thus the sum of two currents, one of which may be considered the result of closing the switch and the other the result of the driving voltage. The first current dies out with time in an exponential manner, whereas the second continues as long as the driving voltage exists. The constants A and B are the two arbitrary constants of integration necessary in any second-order differential equation and are determined by the initial conditions. We shall not consider them further because our concern is with the steady-state current. We postulate that enough time has elapsed since the closing of the switch for the transient part of the current to have decayed to an insignificant magnitude.

8. The Phase Angle θ

As a first step in discussing the physical significance of Equation 7-7 we plot in Figure 9 both $v(= V \cos \omega t)$ and i against ωt. Although both v and i are cosine functions, the current is displaced with respect to the voltage along the time axis by an amount fixed by the phase angle θ. (For positive θ the voltage reaches its peak before the current.) Since $\tan \theta$ is determined by the ratio X/R and since the range of X is from ∞ to $-\infty$, depending on

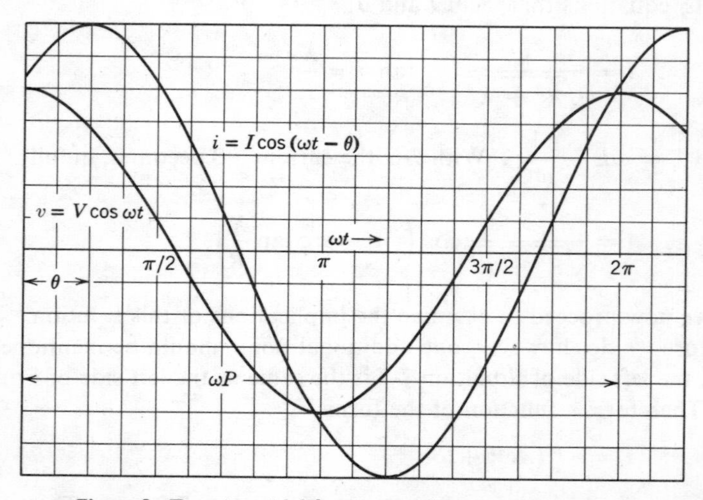

Figure 9. Two sinusoidal functions with a phase difference θ.

the relative magnitude of ωL and $1/\omega C$, the angle θ can take on values ranging from $\pi/2$ to $-\pi/2$, and the current can be as much as a quarter cycle ahead or behind the voltage.

Three limiting cases in the circuit of Figure 8 are important.

CASE 1. $L = 0$. In this case $\tan\theta$ is negative and θ itself lies between 0 and $-\pi/2$. Consequently, the current in an C-R series circuit reaches its peak before the voltage. Circuits of greater complexity, in which the current leads the applied voltage, are by analogy said to behave capacitatively.

CASE 2. $C = \infty$. This is the case of *no* capacitor in the circuit. (As the plates of a capacitor come closer together, the capacitance rises until, on touching, the capacitance is infinite.) Tan θ is positive and θ lies between 0 and $+\pi/2$. The current reaches its peak after the applied voltage in such an L-R circuit. More complex circuits, in which the voltage leads the current, are said to behave inductively.

CASE 3. $L = 0$ and $C = \infty$. In this case $\tan\theta$ is zero, independent of the frequency, the current and voltage are in phase, and many of the properties of the circuit are similar to those of a d-c circuit. Again, in more complex circuits, where current and voltage are in phase, the circuits are said to be resistive.

9. Power in an Alternating-Current Circuit

The phase angle plays an important role in computing the power absorbed by the circuit from the generator. It was shown in Chapter 4, Section 5, that the power supplied to a d-c circuit is determined by the product of generator voltage and current. In an a-c circuit the power so computed is the *instantaneous* rate at which energy is supplied. In the circuit of Figure 8 the power at the time t is:

$$p = vi = V(\cos\omega t)I\cos(\omega t - \theta) \tag{9-1}$$

There are, in general, intervals during which the voltage and the current have opposite signs, and the instantaneous power is negative. At these times the circuit is returning energy to the generator, whereas when the instantaneous power is positive the circuit receives energy from the generator. With an expansion of $\cos(\omega t - \theta)$, Equation 9-1 becomes

$$p = VI(\cos^2\omega t\cos\theta + \cos\omega t\sin\omega t\sin\theta) \tag{9-2}$$

Since $\cos^2\omega t$ can be converted into a function of $\cos 2\omega t$ and $\cos\omega t\sin\omega t$ into a function of $\sin 2\omega t$, the instantaneous power varies at twice the frequency of the voltage or the current.

Again, some special cases are instructive.

CASE 1. A simple circuit, consisting only of a generator and a capacitor with zero resistance and inductance. Tan θ is negative infinite so that $\cos\theta = 0$ and $\sin\theta = -1$. The instantaneous power is

$$p = -\frac{VI}{2}\sin 2\omega t \tag{9-3}$$

which is alternately negative and positive with a frequency of $2\omega t$. During times of positive power the generator supplies energy to build up the electric field in the capacitor, and during times of negative power this energy is discharged back to the generator.

CASE 2. A pure inductor in series with the generator, no capacitor and no resistor. Tan θ is positive infinite, $\cos\theta = 0$ and $\sin\theta = +1$. The instantaneous power is again given by Equation 9-3 except for a positive sign on the right side. During times of positive power the generator supplies energy to build up the magnetic field in the inductor (increasing current), and during times of negative power (decreasing current) the energy in the field returns to the generator. In both Case 1 and Case 2 the intervals of positive and negative power are of equal duration, with the result that an average of the power over a whole number of cycles is zero.

CASE 3. A pure resistance in series with the generator, no inductor and no capacitor. Tan $\theta = 0$, $\cos\theta = 1$, and $\sin\theta = 0$. The instantaneous power is:

$$p = \frac{VI}{2}(1 + \cos 2\omega t) \tag{9-4}$$

The power is always positive, varying between a maximum of VI and a minimum of zero with a mean value of $VI/2$.

Seldom is there much interest directly in the instantaneous power. Of greater interest is the average power which the generator supplies to the circuit or which is absorbed in a part of a circuit. The reader should satisfy himself that the time average of the right side of Equation 9-2, taken either over a whole number of periods of oscillation of the voltage or over a time which is long compared to one period, is:

$$\bar{p} = \left(\frac{VI}{2}\right)\cos\theta \tag{9-5}$$

With θ varying only over the range from $-\pi/2$ to $+\pi/2$, $\cos\theta$ cannot be negative, and the average power is either zero or positive, not surprising since we do not expect a group of passive circuit elements to supply, on the average, power back to the generator. An important inference from Equation 9-5, however, is that conditions can exist in which, with the phase angle near $\pi/2$ either the current or the voltage is, or both are, quite high and yet the average power absorbed by the circuit is low.

10. Effective Values of Voltage and Current

The quantities V and I appearing in Equation 9-5 are the peak values of the voltage and current, respectively. Specifying these peak values, together with the frequency, is one way of identifying sinusoidal a-c voltages and currents. But any other number, proportional to the peak value may be used as well.

Equation 9-5 can be expressed in the form

$$\bar{p} = P = \frac{V}{\sqrt{2}}\frac{I}{\sqrt{2}}\cos\theta \tag{10-1}$$

$V/\sqrt{2}$ and $I/\sqrt{2}$ are called the effective values of the voltage and current, respectively, or, alternatively, the root-mean-square values. The latter designation comes from the result of averaging the square of a sinusoidally varying function over a whole number of cycles. If $s = S\sin\omega t$, then the average value of s^2 is

$$\overline{s^2} = \frac{S^2\int_0^{nP}\sin^2\omega t\,dt}{\int_0^{nP}dt} = \frac{S^2}{2} \tag{10-2}$$

Consequently, $V/\sqrt{2}$ is the square root of the average square of the voltage or root-mean-square value; similarly for $I/\sqrt{2}$.

In a circuit consisting of a generator and a pure resistance, Equation 10-1 becomes simply

$$P = \frac{V}{\sqrt{2}}\frac{I}{\sqrt{2}} = V_{\text{eff}}I_{\text{eff}} \tag{10-3}$$

which has the same form as the power dissipated as heat in a d-c circuit. Thus the same heat is developed in a resistance R when R is connected to a d-c source of voltage $V_{\text{d-c}}$ supplying current $I_{\text{d-c}}$ as when R is connected to a sinusoidal a-c source whose effective voltage is V_{eff} supplying current I_{eff}, provided $V_{\text{d-c}} = V_{\text{eff}}$ and $I_{\text{d-c}} = I_{\text{eff}}$.

The effective values of nonsinusoidal voltages and currents are also found by an averaging process like Equation 10-2, where the integration is over a whole number of periods of the nonsinusoidal pattern. Equation 10-3 also applies to such nonsinusoidal quantities.

Several types of commonly used a-c meters give the same deflection for a d-c current (or voltage) as for an effective a-c (sinusoidal or nonsinusoidal) current (or voltage). These meters may thus be calibrated by using a d-c current (or voltage), it being understood that, in a-c circuits, they read effective values.

In what follows, effective values of voltage and current are assumed unless otherwise specified. There need be no confusion since circuit equations apply with equal validity for consistently used effective values or peak values.

11. Impedance and Reactance

In Equation 7-6:

$$I = \frac{V}{\sqrt{R^2 + X^2}}$$

the function $\sqrt{R^2 + X^2}$ acts in a way analogous to resistance in a d-c circuit. It is called the impedance and commonly designated by Z, and, in a sense, is a measure of the opposition to a-c current. It has the units of ohms, its two parts being the resistance and the reactance. In a simple series circuit the frequency appears only in the reactance, but in parallel and series-parallel combinations the resistive term may also depend on the frequency.

The reactance for a series circuit, $\omega L - (1/\omega C)$ contains one part which is a linear function of frequency, and one which is an inverse function of frequency. These two parts and their difference are plotted in Figure 10.

X is zero when ωL, the inductive reactance, equals $1/\omega C$, the capacitative reactance. At this frequency Z has its minimum value, R. The current amplitude or effective value, I, reaches its maximum value and $\tan \theta = 0$. In fact, the series R-L-C circuit behaves at this special frequency as if L and C were not present. Conditions of maximum current or minimum current or zero phase angle are in general called conditions of resonance (sometimes minimum current is called antiresonance) and it is peculiar to a simple series circuit that both maximum current and zero phase angle occur at the same frequency. In general, current resonance (maximum or minimum current) may occur at one frequency and phase resonance ($\tan \theta = 0$) at another, although very often these two frequencies are so close together that no distinction need be made.

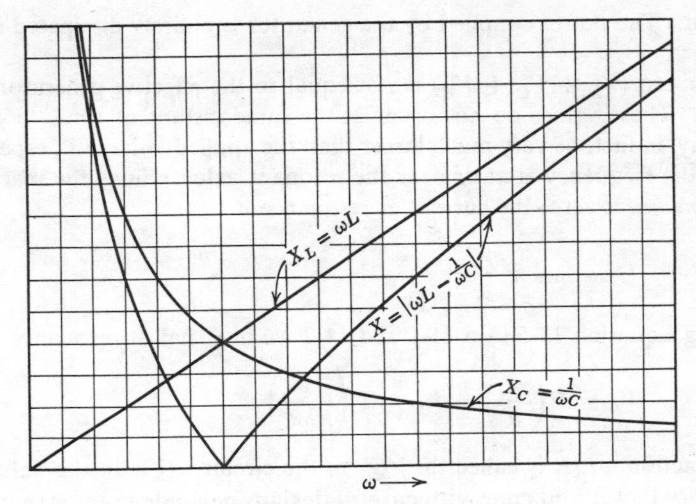

Figure 10. Reactance versus angular frequency.

Applying Equation 7-6 to the voltage and current in the inductor alone in a series circuit, we find for the voltage V_L:

$$(11\text{-}1) \qquad V_L = I\omega L \qquad \cos\theta_L = 0 \qquad \theta_L = \frac{\pi}{2}$$

where I is the same in each of the components. Similarly applying Equation 7-6 to the voltage and current in the capacitor alone, we find for the voltage V_C:

$$(11\text{-}2) \qquad V_C = \frac{I}{\omega C} \qquad \cos\theta_C = 0 \qquad \theta_C = -\frac{\pi}{2}$$

From the discussion of Section 9 we saw that neither the inductor nor the capacitor, on the average, dissipates power.

Again applying Equation 7-6 to the resistor alone, the voltage across the resistor V_R is

$$(11\text{-}3) \qquad V_R = IR \qquad \cos\theta_R = 1 \qquad \theta_R = 0$$

and on the average the power dissipated in the resistor is

$$(11\text{-}4) \qquad P = V_RI = (IR)I = \frac{VRI}{\sqrt{R^2 + X^2}} = VI\cos\theta$$

where V without a subscript is the effective voltage of the generator and θ without a subscript is the phase angle between the generator voltage and current. The power supplied by the generator is entirely dissipated in the resistor.

The sum $(V_L + V_C + V_R)$ is not equal to the effective generator voltage V. These values do not occur at the same instant of time. V_L and V_C may, in fact, be very much larger than the applied voltage V, especially when the frequency is at or near the resonant value. Since the reactance *at resonance* is zero, the current at resonance is

$$I_{\text{resonance}} = \frac{V}{R} \tag{11-5}$$

Putting Equation 11-5 into 11-1 and 11-2 we find that at resonance

$$V_L = V\left(\frac{\omega L}{R}\right) = V_C = V\left(\frac{1}{R\omega C}\right) \tag{11-6}$$

The fraction $\omega L/R$ is called the "Q" of the circuit. Q is in the nature of a *quality* factor and can, with careful design, be made greater than 100. The effective voltages across the inductor and the capacitor separately can thus be several orders of magnitude greater than the generator voltage. Not only is the current a maximum at the resonant frequency in a series circuit, but voltages across parts of the circuit may be greatly amplified over the driving voltage.

12. Series and Parallel Combinations of Impedances

We saw in Section 7 that the voltage and current for an element or a combination of elements in series are given by

$$\begin{aligned} v &= V_m \cos \omega t \\ i &= \frac{V_m}{Z} \cos (\omega t - \theta) \end{aligned} \tag{12-1}$$

where v and i are the instantaneous values of the voltage and the current, V_m is the voltage amplitude, Z is the impedance given by $\sqrt{R^2 + X^2}$, and $\tan \theta = X/R$. It should be self-evident that an alternative form to Equations 12-1 is

$$\begin{aligned} i &= I_m \cos \omega t' \\ v &= I_m Z \cos (\omega t' + \theta) \end{aligned} \tag{12-2}$$

where I_m is the amplitude of the current and $\omega t' = \omega t - \theta$. Because it is unlikely that both Equations 12-1 and 12-2 will be used in the same computation, the prime is often dropped in 12-2.

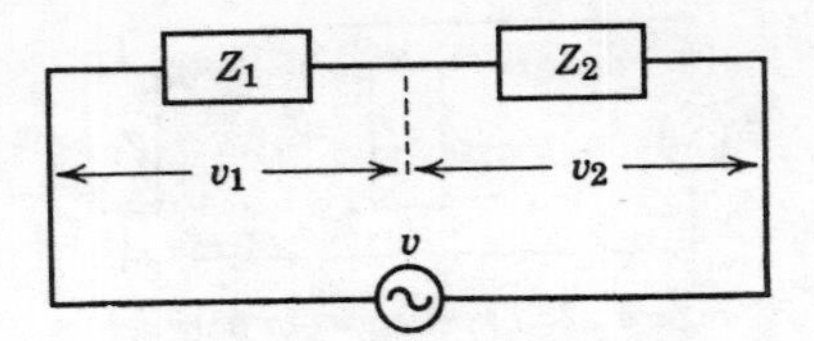

Figure 11. Impedances in series.

Suppose we have two impedances, Z_1 and Z_2, in series with a generator of instantaneous voltage v, as in Figure 11. The generator voltage must at every instant be the sum of the voltages across the individual impedances:

$$v = v_1 + v_2 \tag{12-3}$$

Since in a series circuit the current is the same in both impedances, we can, by using Equation 12-2, transform 12-3 into

$$\begin{aligned} v &= I_m[Z_1 \cos(\omega t + \theta_1) + Z_2 \cos(\omega t + \theta_2)] \\ &= I_m[Z_1(\cos \omega t \cos \theta_1 - \sin \omega t \sin \theta_1) \\ &\quad + Z_2(\cos \omega t \cos \theta_2 - \sin \omega t \sin \theta_2)] \\ &= I_m[\cos \omega t(Z_1 \cos \theta_1 + Z_2 \cos \theta_2) \\ &\quad - \sin \omega t(Z_1 \sin \theta_1 + Z_2 \sin \theta_2)] \end{aligned} \tag{12-4}$$

where the subscripts refer to the individual impedances and the primes have been dropped. With the substitutions:

$$\begin{aligned} Z_1 \cos \theta_1 + Z_2 \cos \theta_2 &= Z_T \cos \theta_T \\ Z_1 \sin \theta_1 + Z_2 \sin \theta_2 &= Z_T \sin \theta_T \end{aligned} \tag{12-5}$$

Equation 12-4 takes the form

$$\begin{aligned} v &= I_m Z_T(\cos \omega t \cos \theta_T - \sin \omega t \sin \theta_T) \\ &= I_m Z_T \cos(\omega t + \theta_T) \end{aligned} \tag{12-6}$$

Now, if

$$\begin{aligned} Z_1 &= \sqrt{R_1^2 + X_1^2} \qquad Z_2 = \sqrt{R_2^2 + X_2^2} \\ \tan \theta_1 &= \frac{X_1}{R_1} \qquad \tan \theta_2 = \frac{X_2}{R_2} \end{aligned} \tag{12-7}$$

then from 12-5 we find that

$$Z_T = \sqrt{(R_1 + R_2)^2 + (X_1 + X_2)^2} \qquad \tan \theta_T = \frac{X_1 + X_2}{R_1 + R_2} \tag{12-8}$$

That is, in a series circuit we combine the resistances to get the total

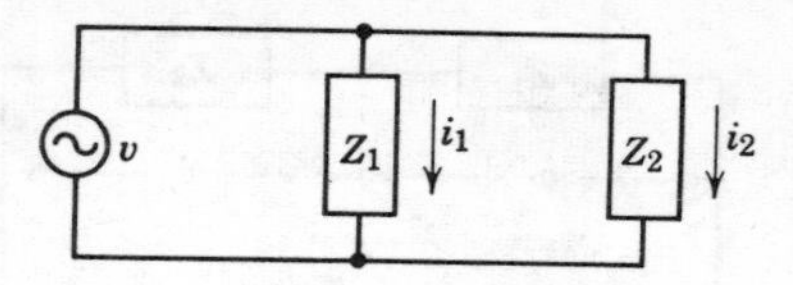

Figure 12. Impedances in parallel.

resistance, and combine the reactances (with due regard to sign) to get the total reactance. Then from these we obtain the total impedance of the combination. *Impedances* in series do not add algebraically as do resistance in a d-c series group.

Now, consider two impedances connected in parallel as in Figure 12. In this case the generator voltage is applied across each impedance and the generator current is the sum at every instant of the individual currents:

$$i_T = i_1 + i_2 \tag{12-9}$$

Using the forms of Equation 12-1, the total current can be written:

$$\begin{aligned} i &= V_m\left[\frac{\cos(\omega t - \theta_1)}{Z_1} + \frac{\cos(\omega t - \theta_2)}{Z_2}\right] \\ &= V_m\left[\cos\omega t\left(\frac{\cos\theta_1}{Z_1} + \frac{\cos\theta_2}{Z_2}\right) + \sin\omega t\left(\frac{\sin\theta_1}{Z_1} + \frac{\sin\theta_2}{Z_2}\right)\right] \\ &= V_m\left[\cos\omega t\left(\frac{R_1}{Z_1{}^2} + \frac{R_2}{Z_2{}^2}\right) + \sin\omega t\left(\frac{X_1}{Z_1{}^2} + \frac{X_2}{Z_2{}^2}\right)\right] \\ &= V_m[\cos\omega t(G_1 + G_2) + \sin\omega t(B_1 + B_2)] \end{aligned} \tag{12-10}$$

where $G_i = R_i/Z_i{}^2$ and $B_i = X_i/Z_i{}^2$. By making an analogous substitution to 12-5, we get

$$G_1 + G_2 = \frac{\cos\theta_T}{Z_T} \qquad B_1 + B_2 = \frac{\sin\theta_T}{Z_T} \tag{12-11}$$

Equation 12-10 takes the form

$$i_T = \frac{V_m}{Z_T}\cos(\omega t - \theta_T) \tag{12-12}$$

where

$$\frac{1}{Z_T} = \sqrt{(G_1 + G_2)^2 + (B_1 + B_2)^2} \qquad \tan\theta_T = \frac{B_1 + B_2}{G_1 + G_2} \tag{12-13}$$

The extension to more complex systems is straightforward in principle but tedious.

The reciprocal of the impedance used in Equation 12-13 is called the *admittance*, usually given the symbol Y. The G's are conductances and the B's susceptances. Equation 12-13 is the admittance of a pair of impedances in parallel, and it follows that the admittance of a single impedance is

$$Y = \sqrt{G^2 + B^2} \tag{12-14}$$

where the definitions of G and B are inherent in 12-10. Evidently impedances in parallel also do not combine in the simple fashion of resistances in parallel.

The concepts of admittance, conductance, and susceptance are in a sense reciprocal to impedance, resistance, and reactance, and any given circuit problem may be solved in terms of either group. Usually series circuits are easier to handle in terms of impedances, while, parallel circuits are easier to handle in terms of admittances.

At this point the reader will rightly contemplate without enthusiasm the solving of any but the simplest circuit problems by the above method. We now proceed to develop a more elegant method of solving circuit problems by the use of the algebra of complex numbers, a method which reduces the problem in many respects to the simplicity of a d-c problem.

13. Complex Representation of Alternating Current Quantities

For the series circuit of Figure 8, Kirchhoff's loop rule yielded Equation 7-1:

$$L\frac{di}{dt} + Ri + \frac{q}{C} = V\cos\omega t$$

Should this circuit, with the same components, have instead a generator whose voltage is represented by $V \sin \omega t$, Kirchhoff's loop rule yields

$$L\frac{di}{dt} + Ri + \frac{q}{C} = V\sin\omega t \tag{13-1}$$

The same method as used in Section 7 leads to the solution to Equation 13-1

$$i = I\sin(\omega t - \theta) \tag{13-2}$$

where

$$I = \frac{V}{\sqrt{R^2 + X^2}} \qquad \tan\theta = \frac{X}{R} \qquad X = \omega L - \frac{1}{\omega C}$$

The amplitude and the phase angle are the same in either circuit. By mathematically combining the solutions for these two circuits, we produce a method which leads to a simplification in the treatment of the steady-state behavior of more complex systems.

Although the amplitudes and the phase angles of the currents in the two circuits are the same, the instantaneous currents are not. Identify the instantaneous current and charge in Equation 7-1 as i' and q_1' and the instantaneous current and charge in Equation 13-1 as i_2' and q_2'. Now multiply Equation 13-1 by $j = \sqrt{-1}$ and add the resulting expression to 7-1:

$$L\frac{d}{dt}(i_1' + ji_2') + R(i_1' + ji_2') + \frac{q_1' + jq_2'}{C} = V(\cos\omega t + j\sin\omega t) \tag{13-3}$$

Letting

$$i' = i_1' + ji_2' \qquad q' = q_1' + jq_2' \quad \left(\text{where } i' = \frac{dq'}{dt}\right), \qquad \cos\omega t + j\sin\omega t = e^{j\omega t} \tag{13-4}$$

Equation 13-3 becomes:

$$L\frac{di'}{dt} + Ri' + \frac{q'}{C} = ve^{j\omega t} \tag{13-5}$$

We shall distinguish between real number magnitudes (capital letters), complex numbers (small letters), and complex variables which include the time function $e^{j\omega t}$ (small letters primed). The voltage amplitude in Equation 13-5 has been changed to v to allow for the possibility that it may be complex.

Differentiating Equation 13-5 once with respect to time to get rid of the charge q gives:

$$L\frac{d^2i'}{dt^2} + R\frac{di'}{dt} + \frac{i'}{C} = j\omega ve^{j\omega t} \tag{13-6}$$

where R, L, and C are real numbers. As a trial solution for 13-6 use:

$$i' = ie^{j\omega t} \tag{13-7}$$

where the amplitude i is a complex number. Putting 13-7 into 13-6 gives:

$$-\omega^2 Lie^{j\omega t} + j\omega Rie^{j\omega t} + \frac{i}{C}e^{j\omega t} = j\omega ve^{j\omega t} \tag{13-8}$$

which requires that the amplitude i have the form:

$$i = \frac{v}{R + jX} \tag{13-9}$$

where again,

$$X = \omega L - \frac{1}{\omega C}$$

The complex denominator of 13-9 can also be written:

$$R + jX = Ze^{j\theta} \tag{13-10}$$

where

$$Z = \sqrt{R^2 + X^2} \qquad \text{and} \qquad \tan\theta = \frac{X}{R}.$$

Z and θ thus have the same significance as in previous sections, and the solution 13-7 may be written

$$\begin{aligned} i' &= \frac{v}{Z} e^{j(\omega t - \theta)} = \frac{v}{Z} \cos(\omega t - \theta) + j\frac{v}{Z} \sin(\omega t - \theta) \\ &= i_1' + ji_2' \end{aligned} \tag{13-11}$$

Thus the complex number solution 13-7 contains within it the solutions to both of the original differential equations.

What have we gained by this procedure? First, the complex number form for the current amplitude, $v/(R + jX)$, contains all the information about the impedance magnitude and the phase angle. It is not necessary to write two expressions unless an explicit computation requires the phase angle and the impedance magnitude separately. Second, we shall see that the combining of complex impedances is a simple algebraic operation closely analogous to the combining of resistances in d-c circuits.

In the series circuit of Figure 11, for example, the currents in each impedance may be written separately:

$$i_1' = \frac{v_1 e^{j\omega t}}{z_1} \qquad i_2' = \frac{v_2 e^{j\omega t}}{z_2} \tag{13-12}$$

where z_1 and z_2 are complex impedances. But since the current is continuous in a series circuit, we have:

$$\frac{v_1 e^{j\omega t}}{z_1} = \frac{v_2 e^{j\omega t}}{z_2} \tag{13-13}$$

The instantaneous voltage of the generator is the sum of the voltages across the two parts:

$$v_T' = v_1 e^{j\omega t} + v_2 e^{j\omega t} = v_1 e^{j\omega t}\left(1 + \frac{z_2}{z_1}\right) \tag{13-14}$$

$$= \frac{v_1 e^{j\omega t}}{z_1}(z_1 + z_2)$$

$$= i'(z_1 + z_2)$$

and thus the total complex impedance, which is the total complex voltage divided by the complex current, is:

$$z_T = z_1 + z_2 \tag{13-15}$$

This expression contains all the information of Equation 12-8. The complex impedance of a series combination is the algebraic sum of the individual complex impedances.

In the parallel circuit of Figure 12, the generator voltage is common to the two branches and the instantaneous currents in the branches are summed to obtain the generator current:

$$i_T' = i_1' + i_2' \tag{13-16}$$

$$= \frac{ve^{j\omega t}}{z_1} + \frac{ve^{j\omega t}}{z_2}$$

$$= v'\left(\frac{1}{z_1} + \frac{1}{z_2}\right)$$

where

$$v' = ve^{j\omega t}$$

Thus the total impedance of a parallel combination is obtained from

$$\frac{1}{z_T} = \frac{1}{z_1} + \frac{1}{z_2} \tag{13-17}$$

The reciprocal of the total impedance is the sum of the reciprocals of the individual impedances, similar to a parallel combination of resistances in a d-c circuit. If the complex form of the total impedance has been found, its magnitude and phase angle is obtained through the use of Equation 13-10. It is common practice to speak of the "angle of the impedance" instead of the "phase angle between voltage and current."

Kirchhoff's rules may be generalized in terms of complex quantities. The *Loop rule* becomes: The algebraic sum of the complex voltages around a closed circuit loop is zero.

And the *Point rule* becomes: The algebraic sum of the complex currents at any junction is zero.

These statements of Kirchhoff's rules reduce to the forms for d-c currents used in Chapter 4 when the frequency is zero, since in a d-c system all quantities are pure real numbers.

The results of this section may be stated in still another way. A system with an applied voltage given by $V \cos \omega t$ (or $V \sin \omega t$) may be solved (for currents or voltages) by the use of complex numbers. The actual currents and voltages are the real parts (or the complex parts) of the complex currents and voltages in the solution.

14. Applications of the Complex Number Method

Applications of the complex method to the solution of a-c circuits appear in the problems at the end of this chapter and in the treatment of filters and transmission lines in Chapter 9. At this point the method will be illustrated by its application to the analysis of alternating current bridges (circuits analogous to the d-c Wheatstone's bridge). There is a rich variety of a-c bridges for the measurement of self-inductance, mutual inductance, capacitance, resistance, and frequency. They are important tools in the laboratory and analyzing them is an instructive exercise in the handling of a-c circuits.

Most bridge circuits can be reduced schematically to the general form shown in Figure 13. When each of the four impedances is a pure real quantity and the generator frequency is zero, the circuit of Figure 13 reduces to the Wheatstone's bridge used in d-c measurements. In general,

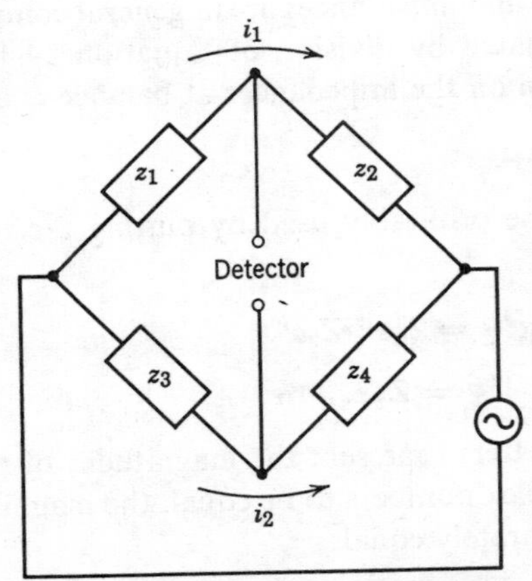

Figure 13. General form of an a-c bridge.

two or more of the four impedances are adjustable, and one of the four may be an unknown to be determined in terms of the other three by adjusting to a null reading on the detector. In some bridges the frequency must be known, in others it need not be. Where the frequency enters into the conditions for balance of the bridge, it is in principle, and in some cases in practice, possible to use the bridge for the measurement of frequency when all four impedances are known.

The detector is usually a voltage sensitive device which requires negligible current. As in the d-c bridge, the supply voltage need not be known, although its magnitude influences the bridge sensitivity. But, especially in such bridges in which the frequency is required in the computation, it is desirable that the output be as closely sinusoidal as possible. This is still convenient, although less vital, with those bridges whose operation is in theory frequency-independent. As in the d-c bridge, the values of the bridge arms are adjusted until the detector indicates a null, but whether such a null is achievable or not depends on the nature of the impedances.

We start with the assumption that such a null has been found. Thus, there is no potential difference across the detector and there is no current in the detector. Therefore, the current i_1 is common to z_1 and z_2, and the current i_2 is common to z_3 and z_4. Moreover, the complex voltage across z_1 equals that across z_3, with a like condition for z_2 and z_4. Analytically, this is expressed as

$$i_1 z_1 = i_2 z_3 \quad \text{(14-1}a\text{)}$$

$$i_1 z_2 = i_2 z_4 \quad \text{(14-1}b\text{)}$$

where the currents and impedances are in general complex quantities. The ratio i_1/i_2 is eliminated by division of Equation 14-1*a* by 14-1*b* and the following condition on the impedances at balance is obtained:

$$z_1 z_4 = z_2 z_3 \quad \text{(14-2)}$$

This relation can be profitably used by putting the impedances into the exponential form:

$$Z_1 e^{j\theta_1} Z_4 e^{j\theta_4} = Z_2 e^{j\theta_2} Z_3 e^{j\theta_3} \quad \text{(14-3)}$$

or

$$Z_1 Z_4 e^{j(\theta_1 + \theta_4)} = Z_2 Z_3 e^{j(\theta_2 + \theta_3)}$$

where the capital letters represent the magnitudes of the complex quantities. For two complex numbers to be equal, the magnitudes and the phase angles must be separately equal:

$$Z_1 Z_4 = Z_2 Z_3 \qquad \theta_1 + \theta_4 = \theta_2 + \theta_3 \quad \text{(14-4)}$$

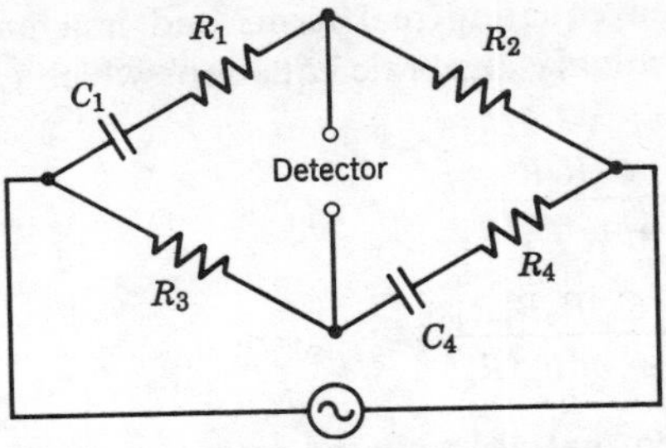

Figure 14. This bridge cannot be balanced.

If the conditions of Equations 14-4 cannot be fulfilled, it is not possible to find the balance originally assumed. We may therefore examine any arbitrary group of impedances in the form of a bridge and say whether a condition of balance is possible or not. As an example of a combination which cannot be balanced, consider a bridge in which z_1 and z_4 each consist of a capacitor and a resistor in series, while z_2 and z_3 are each pure resistors (Figure 14). Whatever the magnitudes of z_1 and z_4, their phase angles are negative, whereas the phase angles of z_2 and z_3 are zero. Hence the second condition in 14-4 cannot be satisfied.

On the other hand, the bridges shown in Figure 15 can be balanced by a proper adjustment of the impedance values.

Let us analyze in detail diagram (*c*) in Figure 15, a circuit which is referred to in the literature as a Hay bridge. Assume that the impedance in the upper left arm is unknown and is to be computed from those values in the other three arms which produce balance. For this analysis the balance conditions as expressed in Equation 14-2 are more convenient than 14-3. As applied to cirucit (*c*) in Figure 15, Equation 14-2 becomes:

$$(R_1 + j\omega L_1)\left(R_4 - \frac{j}{\omega C_4}\right) = R_2 R_3 \qquad (14\text{-}5)$$

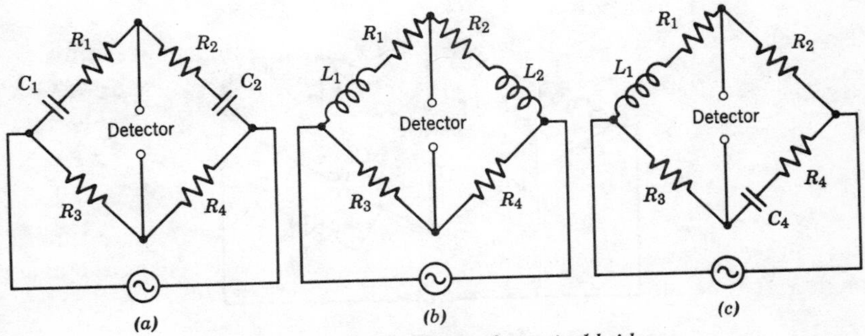

Figure 15. Three examples of practical bridges.

Expanding and then equating real terms and imaginary terms on both sides, we get two ordinary algebraic equations which can be solved for L_1 and R_1:

$$L_1 = \frac{C_4 R_2 R_3}{1 + \omega^2 C_4^2 R_4^2} \tag{14-6a}$$

$$R_1 = \frac{\omega^2 C_4^2 R_2 R_3 R_4}{1 + \omega^2 C_4^2 R_4^2} \tag{14-6b}$$

from which it is seen that not only the resistances and the capacitance in the other arms must be known, but also the frequency.

The fact that the frequency is required can be a disadvantage in cases where either the frequency is not known to sufficient accuracy or the supply voltage is not purely sinusoidal and thus contains more than one frequency. This disadvantage, however, is not always serious. The quality factor of an inductance is defined as the ratio of its reactance to its resistance, $\omega L/R$ (this quantity is a function of the frequency, but often only a slowly varying function because the resistance of a coil tends to rise with increasing frequency). From Equation 14-6 this quality factor for branch 1 is

$$Q = \frac{\omega L_1}{R_1} = \frac{\omega C_4 R_2 R_3}{\omega^2 C_4^2 R_2 R_3 R_4} = \frac{1}{\omega C_4 R_4} \tag{14-7}$$

With Q greater than, say, 10 (Q is dimensionless), the denominators in 14-6 differ from unity by virtue of 14-7 by less than 1%. Often L is not required to a greater accuracy than 1%, in which case the frequency term in the denominator of Equation 14-6*a* can be ignored. Although R_1 still depends on the frequency in the numerator, often the accuracy to which it needs to be known is much less than for L.

A variant of (*c*) in Figure 15 is shown in Figure 16. It is called a Maxwell bridge, and its balance is, interestingly enough, independent of frequency.

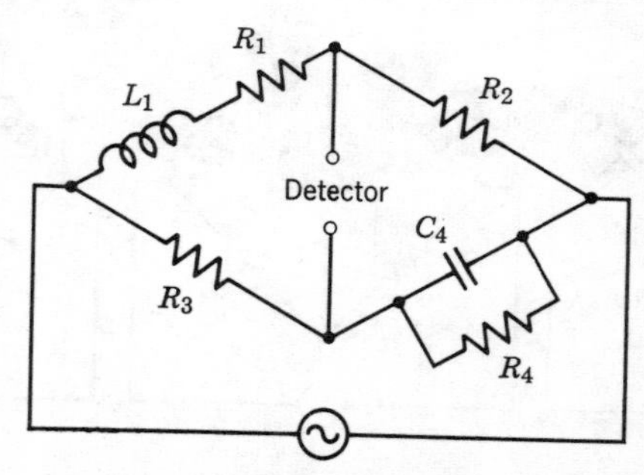

Figure 16. *The Maxwell bridge.*

The application of Equation 14-2 to the Maxwell bridge gives:

$$(R_1 + j\omega L_1)\left[\frac{-(jR_4/\omega C_4)}{R_4 - (j/\omega C_4)}\right] = R_2R_3 \tag{14-8}$$

leading to the following expressions for L_1 and R_1:

$$L_1 = C_4R_2R_3 \qquad R_1 = \frac{R_2R_3}{R_4}\,. \tag{14-9}$$

in which the frequency does not appear. The expression for Q turns out to be ωC_4R_4, instead of the reciprocal of this quantity as for the Hay bridge. In practice, the Maxwell bridge is best suited for the measurement of low-Q inductances. With high-Q inductances, the value of R_4 becomes uncomfortably large. In commercially available units a switch is provided, enabling the user to choose either a Hay bridge or a Maxwell bridge configuration, depending on the inductance to be measured.

Other examples of bridges are included in problems at the end of this chapter.

15. Complex Admittance, Conductance, and Susceptance

It has already been noted that in some problems, particularly where parallel circuits are involved, the admittance is a more useful quantity than the impedance. In its complex form the admittance is written in this text as:

$$y = G + jB = \frac{1}{R + jX} = \frac{R - jX}{R^2 + X^2} \tag{15-1}$$

although in some books the form $y = G - jB$ is used. The real part of the admittance is called the conductance and the imaginary part the susceptance. From Equation 15-1 the relationships between R, X, G, and B are:

$$G = \frac{R}{R^2 + X^2} \qquad B = \frac{-X}{R^2 + X^2} \tag{15-2}$$

$$R = \frac{G}{G^2 + B^2} \qquad X = \frac{-B}{G^2 + B^2}$$

The negative sign in the relation between the reactance and the susceptance should be noted.

16. Graphical Representation of Alternating Current Quantities

There are two ways of picturing a-c quantities graphically. In one, called a phasor diagram, a real instantaneous quantity s, where

$$s = S\cos\omega t \tag{16-1}$$

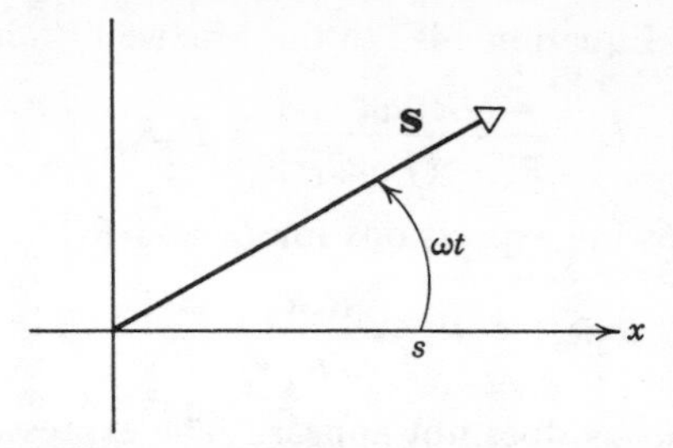

Figure 17. *S rotates counterclockwise, making an angle with the x-axis of ωt at time t.*

with amplitude S is shown as the projection on the x axis of a vector line **S** which is rotating about the origin with an angular velocity ω (Figure 17). The sum of s with another sinusoidally varying quantity

(16-2) $$t = T\cos(\omega t + \theta)$$

is

(16-3) $$s + t = S\cos\omega t + T\cos(\omega t + \theta)$$

and is represented as the projection of a vector **S** + **T** which is rotating at an angular velocity ω. The trigonometric aigument to show that $s + t$ is actually this projection is not difficult and is left to the reader.

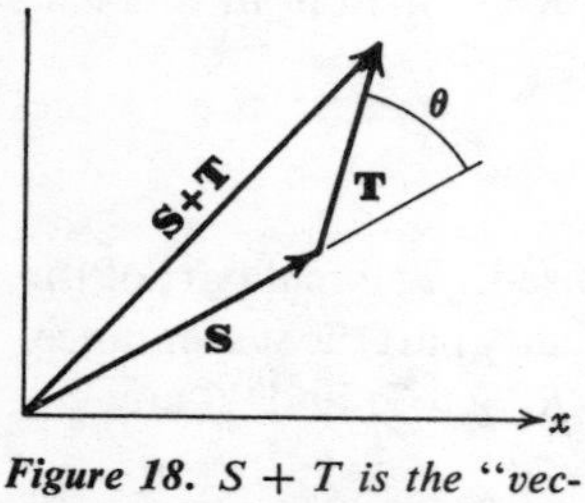

Figure 18. *S + T is the "vector" sum of S and T and rotates at an angular velocity ω.*

With a scaled ruler and compass the relations between the component voltages and the total voltage of a series circuit can thus be found graphically, as can the relations in a parallel circuit among the component currents and the total current. For example, let the current in a series R-L-C circuit be i' with amplitude I, so that

(16-4) $$i' = I\cos\omega t$$

Relative to the current, the voltage in the resistance has a phase angle zero, in the inductance, $\pi/2$, and in the capacitance, $-\pi/2$. The amplitudes of these voltages are shown in Figure 19 with phase angles relative to I. The vector sum of the three component voltages is the total voltage and it is seen that the phase angle is given by

(16-5) $$\tan\theta = \frac{V_L - V_C}{V_R} = \frac{I\omega L - (I/\omega C)}{IR} = \frac{\omega L - (1/\omega C)}{R} = \frac{X}{R}$$

These vectors are to be thought of as rotating as a group with the angular

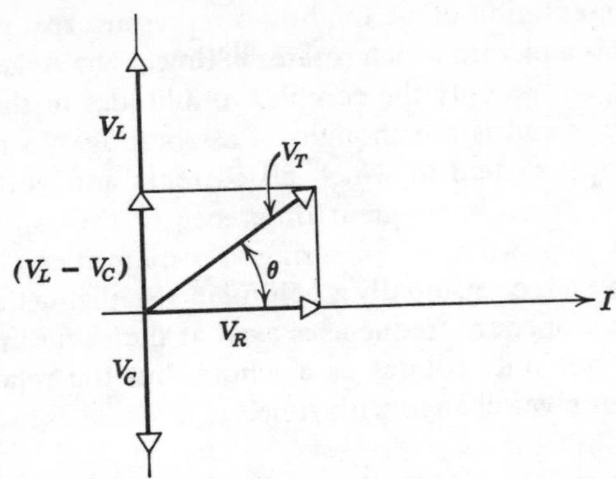

Figure 19. Component voltages and total voltage in a series R-L-C circuit shown on a phase diagram.

velocity ω. The vector current is common to each of the components and fixes the reference direction for the group.

As a second example we apply the vector representation to the circuit of Figure 12 (page 218). Here the voltage v is common to both elements, the current i_1 lags behind the voltage by θ_1 and i_2 by θ_2. The total current has a phase angle relative to the voltage of θ_T, as shown in Figure 20.

The second and closely related representation of a-c quantities is by the use of the Argand diagram for complex numbers. The real term of a complex number is plotted on the axis of reals, by convention taken as the horizontal axis, and the imaginary term on the axis of imaginaries, the vertical axis. Its magnitude is formally the vector sum of these two parts (Figure 21), and its phase angle is the angle between the magnitude vector and the axis of reals. Consistent with the rules of complex algebra the magnitudes of complex numbers are added graphically as vectors by the parallelogram rule (or by the adding of components).

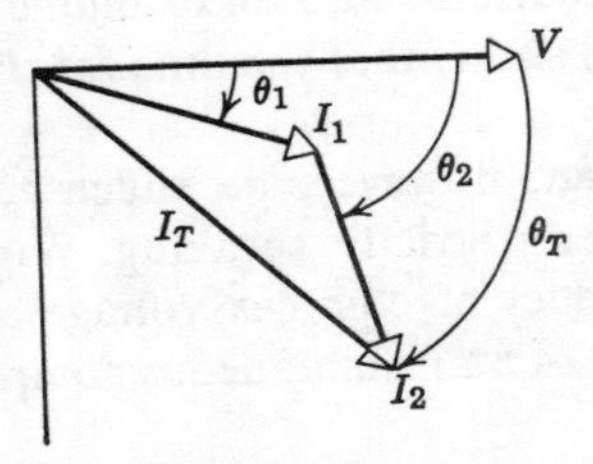

Figure 20. Phasor diagram of currents in a parallel circuit.

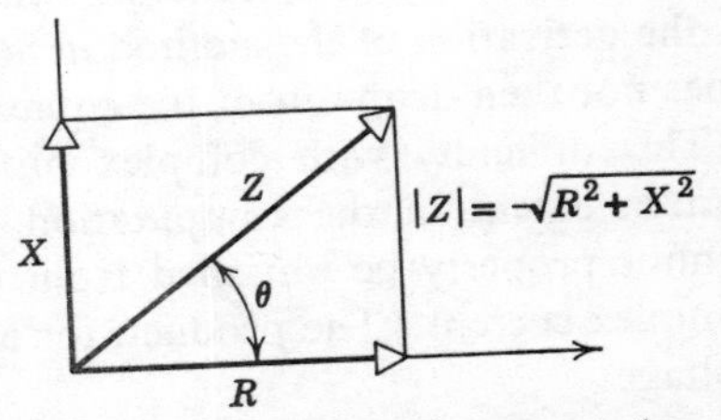

Figure 21. Graphical representation of a complex impedance.

The phasor representation of a-c quantities represents real numbers and, at least in thought, is a picture which rotates in time. The Argand diagram, on the other hand, shows only the complex amplitudes in the case of a-c voltages and currents and is not thought of as rotating. In the representation of an electrical system in which all currents and voltages have a common frequency, there is no great difference in the information obtained from these two ways of picturing a-c quantities. The phasor representation, however, occasionally is helpful in visualizing the conditions in circuits where two or more frequencies exist at the same time. In such a case the pattern not only rotates as a whole, but the relations of the vectors among themselves change with time.

17. Summary of the Complex Method

The analysis of a-c circuits is converted, by the use of complex quantities, from one expressed in terms of differential equations to one expressed in terms of algebraic equations. Formally, the differential equations of a-c circuits are altered by the device of using complex amplitudes instead of instantaneous voltages and currents and by replacing the operator d/dt by $j\omega$. Thus the complex voltage across a pure resistance is Ri, where i is the complex amplitude of the current; the complex voltage across a pure inductance is $j\omega Li$; and across a pure capacitance is $-ji/\omega C$. The solutions of the algebraic equations are complex quantities whose real parts are the amplitudes of the physical quantities if the applied voltages are cosine functions of the time, and whose imaginary parts are the physical amplitudes if the applied voltages are sine functions of the time.

The complex number method is a valuable tool, but certain limitations should be kept in mind. It is applicable only where there is a sinusoidal dependence on time and is thus limited to steady-state a-c problems. The treatment of transients requires a different approach. It is applicable only to linear circuits in which the impedance elements are not functions of the currents or voltages. Where impedance elements are nonlinear, there are product terms of instantaneous voltages or currents and an examination of the derivation of the method in Section 13 shows that the time function does not then drop out of the equations.

This difficulty with complex products can, however, be elided to a certain extent in the computation of power. Strictly speaking, power cannot properly be obtained from the product of complex voltage and complex current. The product, for example, of an instantaneous complex voltage

$$v' = Ve^{j\omega t} \tag{17-1}$$

and an instantaneous complex current

(17-2) $$i' = Ie^{j(\omega t - \theta)}$$

is

(17-3) $$v'i' = VIe^{j(2\omega t - \theta)}$$
$$= VI[\cos(2\omega t - \theta) + j\sin(2\omega t - \theta)]$$

which is to be compared with Equation 9-2. This complex product obviously has an average value of zero, both for the real and the imaginary terms. On the other hand, in the product of the instantaneous voltage and the complex *conjugate* of the instantaneous current,

(17-4) $$v'i'^* = Ve^{j\omega t}Ie^{-j(\omega t - \theta)}$$
$$= VIe^{j\theta} = VI(\cos\theta + j\sin\theta)$$

the real part is the average power delivered to the system, provided that effective values of voltage and current are used as the magnitudes. The imaginary term is the power returned from the system to the generator, sometimes called the "reactive power." The justification for the operation in Equation 17-4 is its agreement with the results of Section 9. It does not follow logically out of the application of complex numbers to a-c circuits.

Discussion Question

1. The angular frequency of a free *R-L-C* circuit was shown to be

(*a*) $$\beta = \left(\frac{1}{LC} - \frac{R^2}{4L^2}\right)^{1/2}$$

When the same elements are connected in series with an alternating emf of variable frequency, the resonant frequency is found to be

(*b*) $$\omega_{\text{res}} = \left(\frac{1}{LC}\right)^{1/2}$$

Discuss physical reasons why the second term on the right side of (*a*) does not appear in (*b*).

Problems

1. Justify that the replacements of Equation 3-3 can always be made.

2. Analyze the *C-R* circuit in the diagram as the *L-R* and the *L-C-R* circuits have been treated, obtaining the charge on the capacitor and the current in the resistor as functions of time.

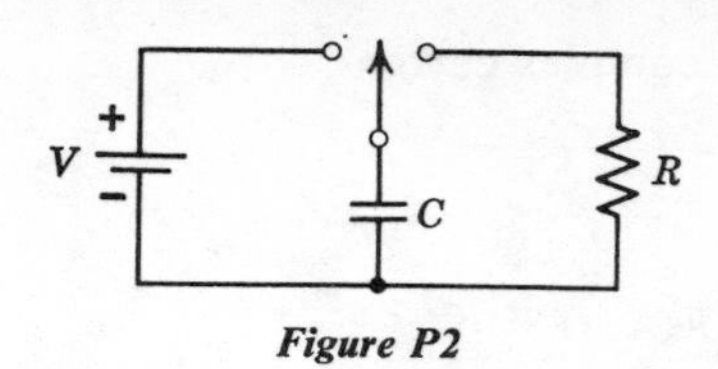

Figure P2

3. In the equation for the underdamped transient behavior of an *R-L-C* system (Equation 3-4)

$$q = 2De^{-at} \cos (\beta t + \theta)$$

evaluate the constants D and θ for the initial conditions: at $t = 0, q = -Q$, and $i = 0$.

4. In the circuit of Figure 2 (page 197), $(R/2L)^2 = 1/LC$ is a condition intermediate between the underdamped and the overdamped cases. Note that Equation 2-5 for the overdamped case and the results of the last problem for the underdamped case are both indeterminate under this condition. Show that for the overdamped case, as $b \rightarrow 0$, and for the underdamped case, as $\beta \rightarrow 0$, the charge as a function of time approaches:

$$q = -Q(1 + at)\, e^{-at} \qquad \text{where } a = 1/\sqrt{LC}$$

5. For transient oscillations in a series *R-L-C* circuit, show that the charge on the capacitor is a maximum when $\beta t = n\pi$, where $n = 0, 1, 2, 3, \cdots$.

6. For the conditions graphed in Figure 4, determine the difference in phase between the charge and the current oscillations.

7. In the circuit of Figure 5 (page 202), show that the ratio of the first voltage maximum on the capacitor to the applied voltage of the battery is given by

$$\frac{v_{C\max}}{V} = 1 + e^{-\pi a/\beta} = 1 + e^{-\pi/[(R_c/R)^2 - 1]^{1/2}}$$

where R_c is the critical damping resistance. For $L = 0.01$ henry and $C = 0.005$ microfarad, what is the value of R_c? What value of R results in $v_{C\max} = 1.9$ V?

8. Carry out the operations to obtain Equations 5-2*a* and 5-2*b*.

9. In Figure P9, A is a "flip" coil whose plane is perpendicular to a magnetic flux density **B**. The coil is to be turned quickly through 180°, thus producing a pulse of current in the circuit containing the ballistic galvanometer, G. R is the total resistance of the galvanometer, the flip coil, and the secondary of the mutual inductance. The galvanometer is calibrated in a separate experiment in which the switch is closed to the primary circuit of the mutual inductance, the resulting charge passing through the galvanometer being computed in terms of the final current read on the ammeter. Show that the magnetic flux density, **B** over the area of the flip coil is given by

$$B = \frac{MId}{2N\mathcal{A}d_c}$$

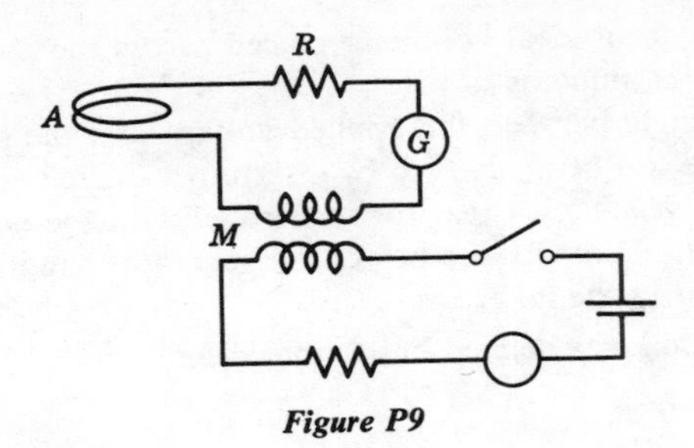

Figure P9

where the area of the flip coil is $\mathscr{A}$, the number of turns is N, d is the galvanometer deflection when the flip coil is rotated through 180°, d_c the deflection when the switch in the primary is closed.

10. A factory requires a power of 10 kilowatts from a power line, the energy being delivered at a frequency of 60 Hz and a potential of 880 volts. The phase angle between the voltage at, and the current in, the factory is 60°.

(*a*) What is the power factor at the factory?

(*b*) What current is required by the factory?

(*c*) The transmission line between factory and generator has a total resistance of 5 ohms. How much power is lost in heating the wires?

(*d*) What must the voltage be at the power station so that the voltage at the factory is 880 volts?

(*e*) If the power factor at the factory were unity and the potential at the factory still 880 volts, what would be the heat loss in the transmission line?

(*f*) What might be done at the input terminals to the factory to make the power factor unity? (Changes within the factory are not allowed.)

11. An inductance of 0.01 henry, whose resistance is 600 ohms, is in parallel with a capacitor C and an alternating-current source of 10 kHz frequency. Determine the value of the capacitance required so that the current in the source and the source voltage are in phase. Compare this to the capacitance required for zero phase difference when the capacitor and the inductor are in series.

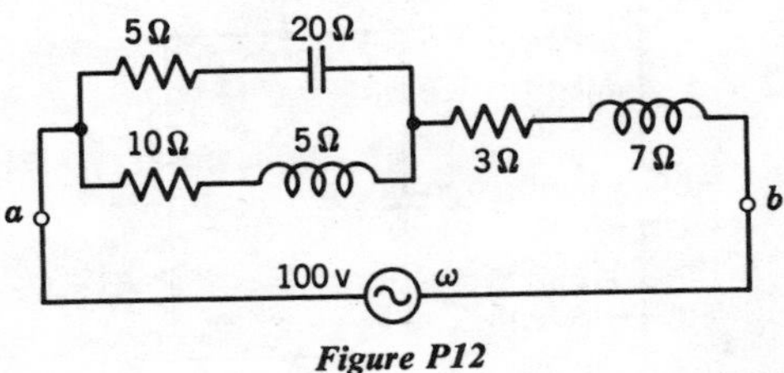

Figure P12

12. Values of resistances and reactances are indicated in Figure P12. Find:

(*a*) The complex currents in the capacitor and each inductor.

(*b*) The complex voltage across the parallel branch and across the series inductor.

(*c*) The reading of an ideal voltmeter placed across the parallel branch and also across the series inductor.

(*d*) The phase angle between the applied voltage and the generator current. This phase angle can be made zero by inserting a capacitor in series (at, say, the point *b*) or by placing a capacitor in parallel with the entire circuit (from point *a* to point *b*). Compare the power the generator supplies in the original case with the power in the latter two cases.

13. The Q of a coil was defined in Section 11 as

$$Q = \frac{\omega L}{R}$$

Show that this definition is equivalent to

$$Q = \frac{2\pi U_L}{W_L}$$

where U_L is the maximum energy stored in the magnetic field of the coil and W_L is the heat dissipated per cycle in the resistance.

14. An inductance L (including a resistance r) and a capacitor C are in series with a generator of terminal voltage V and frequency ω. The inductance L and the capacitance C are to be chosen to effect a condition of resonance at the generator frequency. Show that the ratio of the magnitudes of the voltage across the capacitor to the generator voltage is a function of the ratio, L/C. Suppose the inductor consists of a number of uniformly spaced turns on a ceramic core, the value of L being adjusted by moving a contact from one turn to another. L is then closely proportional to the square of the number of turns, whereas the resistance of the coil is closely proportional to the first power of the number of turns. Show that the capacitor to generator voltage ratio is then proportional to $\omega\sqrt{L}$.

15. In Figure P15, determine the frequency at which the current i_0 has a minimum magnitude, and the frequency at which the current i_0 and the applied voltage

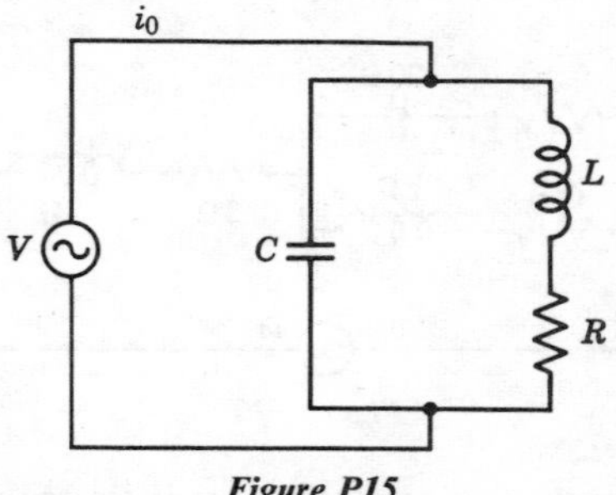

Figure P15

are in phase. Show that as $R \to 0$, these two frequencies approach the same value.

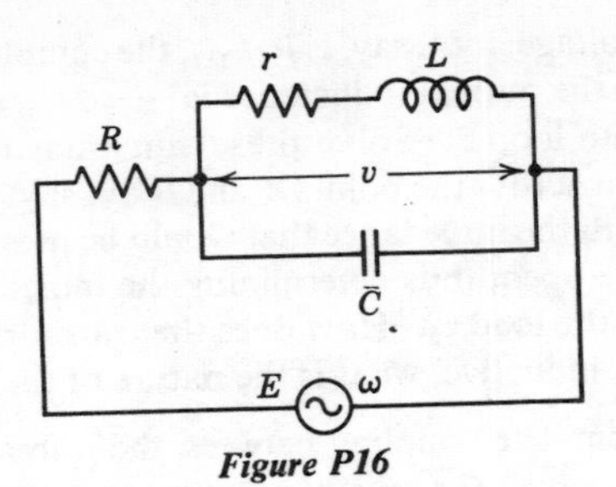

Figure P16

16. Determine the current in the inductance and the current in the capacitance in Figure P16. For a fixed ω, at what value of capacitance will the voltage v be a maximum?

17. Assuming that the components in Figure P17 are ideal (no resistances), show that there are two frequencies at which the impedance measured between

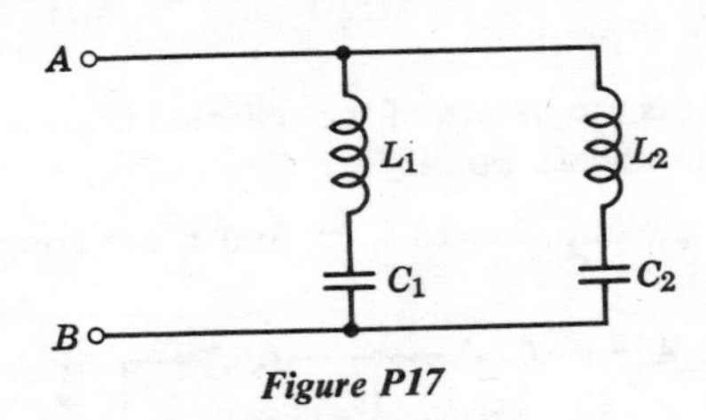

Figure P17

the points A and B is zero and one frequency at which the impedance is infinite. Show that the frequency for infinite impedance always falls between the two frequencies for zero impedance, unless the three frequencies coincide. (Under what circumstances does this latter situation occur?)

18. An a-c generator, operating at an angular frequency, ω, has an internal impedance $z_0 = r_0 + jx_0$. It is supplying power to an impedance, $z = r + jx$. Assume that z_0 cannot be changed but that the components of z are adjustable. How must the components of z be related to those of z_0 so that the power supplied to z is maximum? Compare to the analogous d-c case of maximum power transfer.

19. Apply the rules of complex algebra of a-c circuits to the transformer circuit shown in Figure P19. Write an equation for each loop. (Note that

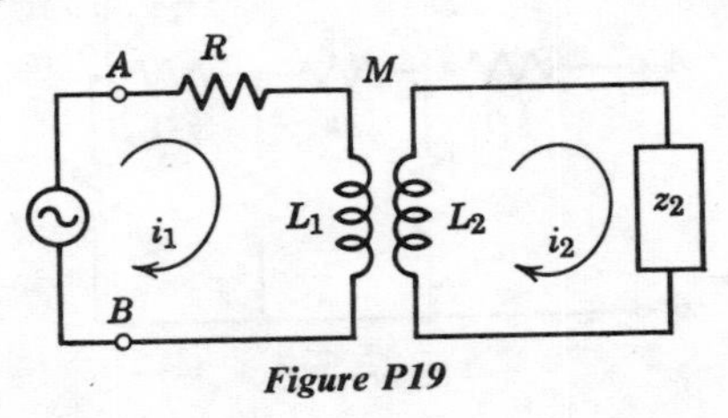

Figure P19

whereas the complex voltage in L_1, say, is $j\omega L_1 i_1$, the complex form of the voltage induced into loop 1 by the mutual inductance is $\pm j\omega M i_2$, and a similar form for the voltage induced into loop 2.) Solve these equations to find the impedance that would be measured across the points A and B by, say, an impedance bridge. Compare the results with the impedance that would be measured if the secondary of the transformer were open, thus determining the impedance which the transformer "reflects" from the load z_2. How does the transformer alter this load on reflection? If z_2 is, say, inductive, what is the nature of the reflected impedance?

20. When there is complete coupling between the primary and the secondary of a transformer (the flux over the area of any turn in one winding is the same as over any turn of the other winding), there is a particularly simple relation, $M^2 = L_1L_2$, between the mutual inductance and the self-inductances. Suppose this situation exists in the circuit of the previous problem, and suppose further that ωL_2 is very much larger than the load impedance z_2. Show that the impedance measured at the terminals A and B under these approximations, is given by

$$R + (L_1/L_2)z_2$$

If z_2 is inductive, what is the nature of the reflected impedance? Compare with the previous case. [Ref.: Scott, pp. 437ff.]

21. In Figure P21, the impedances z_1, z_2, and z_n are complex impedances.

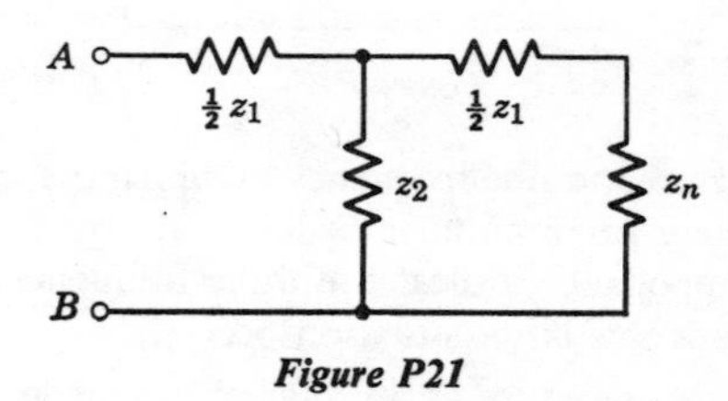

Figure P21

(*a*) Find the impedance between the terminals A and B.

(*b*) What must be the relation of z_n to the impedances z_1 and z_2 so that the input impedance between A and B is equal to z_n (in both magnitude and phase)?

22. In Figure P22, show that the relation of z_n to the impedances z_1 and z_2 required so that the input impedance between A and B shall be equal to z_n is the

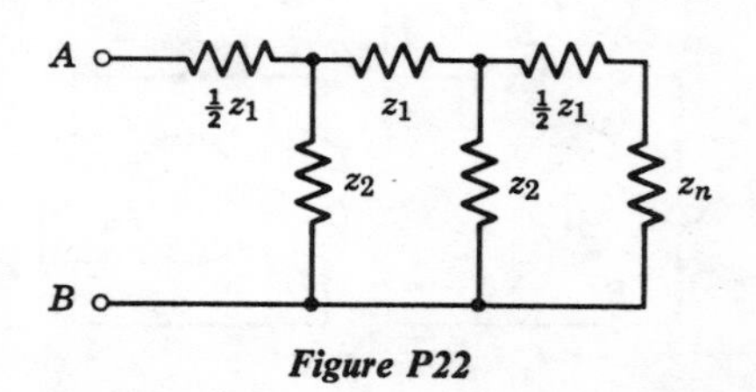

Figure P22

same relation as found in the previous problem, provided z_1 and z_2 are the same impedances as appeared before. Can you generalize from this?

23. Figure P23 shows a Schering bridge. Demonstrate that a bridge with this arrangement of elements can be balanced. Then consider the element in branch 2

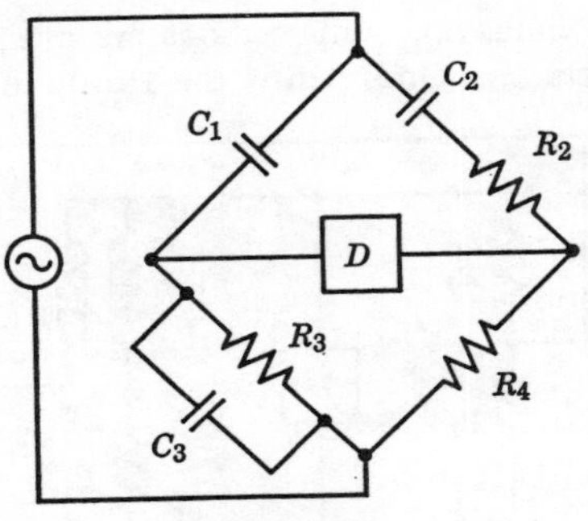

Figure P23

as the unknown, R_2 being the resistance associated with the capacitor, C_2, and show how R_2 and C_2 can be computed from the known values of the other elements at balance. Since two conditions must be fulfilled to produce balance in the bridge, necessarily two of the known components must be adjustable. Discuss which two of the four are best chosen.

24. Figure P24 is called a Wien bridge. Obtain the conditions of balance and note that the condition is not independent of the frequency. The bridge can

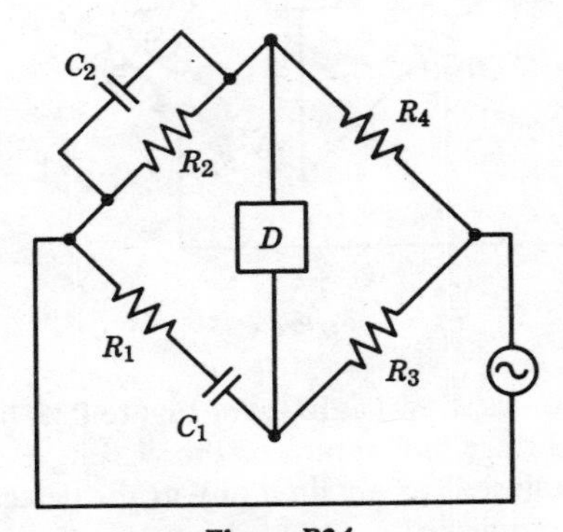

Figure P24

therefore be used to determine the frequency of the generator, if all the bridge elements are known. Show that the conditions for balance are

$$\omega^2 = \frac{1}{R_1 R_2 C_1 C_2} \qquad \text{and} \qquad \frac{R_3}{R_4} - \frac{R_1}{R_2} = \frac{C_2}{C_1}$$

Usually the bridge is used in such a way that the frequency is determined from

$$\omega = \frac{1}{R_1C_1}$$

How is this simplification accomplished?

25. Several circuits employing null methods are available for the measurement of mutual inductance. One such is the Heydweiller mutual inductance

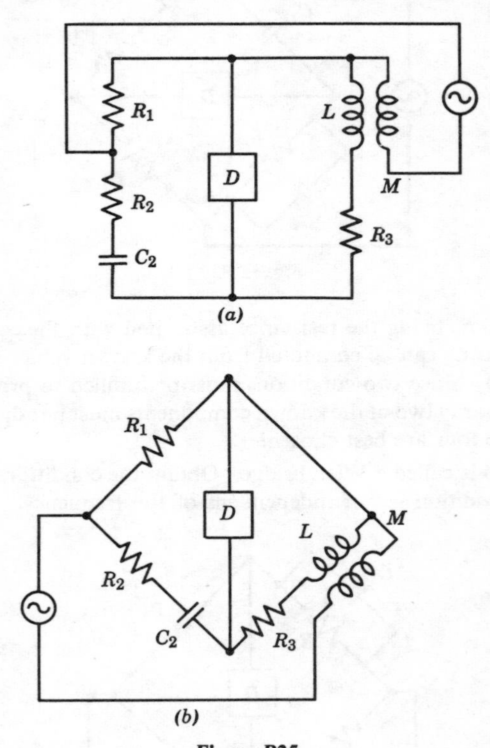

Figure P25

bridge, which is sometimes shown as in (*a*) of Figure P25, but may be redrawn as in (*b*) to suggest the bridge-like nature of the circuit. As with other bridges, the components are adjusted to obtain a null at the detector. Show that when this adjustment is accomplished,

$$M = C_2R_1R_3 \qquad L = C_2R_3(R_1 + R_2)$$

where, however, R_3 may not be completely known, since it includes both a known resistance and the resistance of the secondary of the mutual inductance. If C_2 and R_3 are the adjustable elements, consider how the resistance of L might also be found.

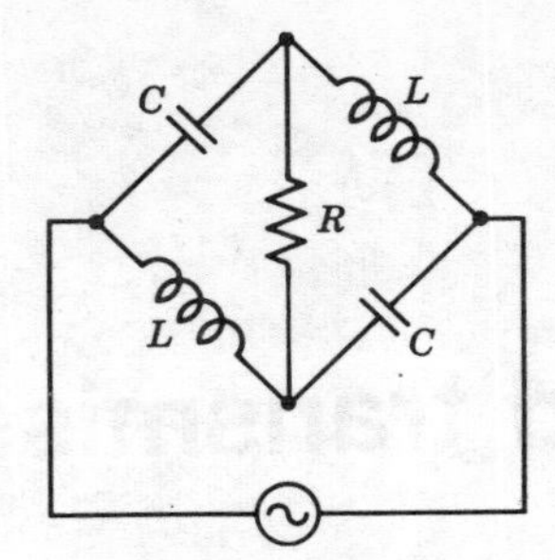

Figure P26

26. If the inductors and capacitors in Figure P26 are pure reactances, and if $LC = 1/\omega^2$, show that the current in R is independent of R and is 90° out of phase with the generator voltage, V.

REFERENCES

1. Page and Adams, *Principles of Electricity* (Van Nostrand), 2nd edition, 1949. Chapter XIII.
2. Scott, Chapter 9.
3. Reitz and Milford, Chapter 13.
4. Whitmer, Chapter 10.

9

Filters and transmission lines

The tools for handling both a-c and d-c circuits have now been developed. The fundamental laws—Coulomb's law, Ampère's law, and Faraday's law—together with the conservation laws of energy and charge have been used to formulate rules for expressing relations among potential differences and currents in electrical circuits. These laws are used in later chapters as a foundation for a theory of electromagnetic waves.

As a transition from ordinary a-c circuit theory to electromagnetic waves, this chapter discusses filters and high-frequency transmission lines. These may be profitably considered from the viewpoint of a-c circuits, but nevertheless represent examples of quasi-continuous and continuous media that have features in common with the space in which electromagnetic waves propagate. Some properties of filters and transmission lines are readily understandable in terms of familiar circuit concepts already studied, whereas others—propagation of waves, the concept of a characteristic impedance, the exponential absorption with distance, etc.—are common to the analysis of waves in space. Although filter circuits, consisting of lumped circuit elements of resistance, inductance, and capacitance, are best treated by the methods of complex circuit analysis, transmission lines are rigorously treated as a special case of guided waves. But it is a happy fact that many interesting properties of transmission lines can be developed as well, and from the point of view of the intermediate student, more simply, by the use of the techniques of complex circuit analysis.

Filters and transmission lines thus are a pedagogical bridge from circuit theory to electromagnetic waves.

1. Filters

The theory of filters in their full engineering detail is beyond the scope or purpose of this book. We shall, instead, deal with only the simplest T-section circuit, aiming to develop an understanding of the basic behavior of filters and an appreciation of those properties which filters have in common with transmission lines and radiation in space.

The name "filter" comes from the frequency selective property of these circuits. A filter is a four-terminal circuit. Two terminals, the input terminals, are connected to a source of a-c energy which will be symbolized as a generator. The remaining two terminals, the output terminals, are connected to a load which receives energy from the generator by way of the filter. The presence of the filter between the generator and the load favors the transmission of energy at certain frequencies and discriminates against the transmission of energy at other frequencies.

Many kinds of systems have filtering properties. One example, at least superficially familiar to the reader, is the tuning circuit in a radio receiver. The antenna plays the role of generator for the receiver, supplying voltages from many transmitters simultaneously. The tuning system filters out the undesired signals and passes the desired frequencies to the amplifiers. The systems we shall study, although different in detail, also have this property.

2. An Elementary Filter

A schematic diagram of a simple type of filter circuit is shown in Figure 1, where the filter proper is enclosed within the rectangle. The input terminals to the filter lead from a generator of effective emf V, angular

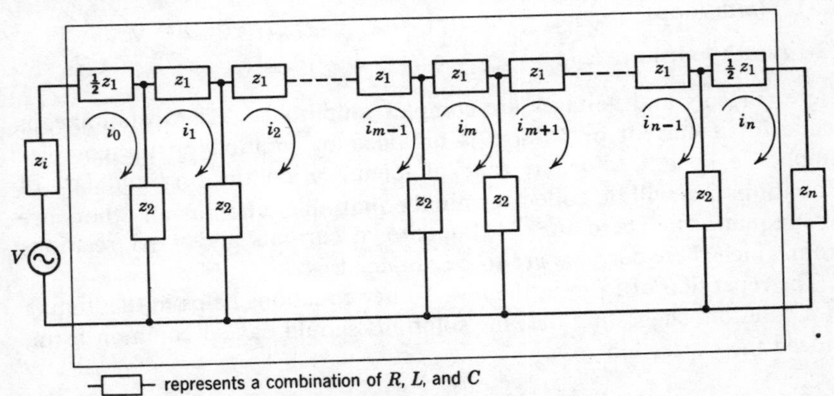

Figure 1. An elementary type of filter.

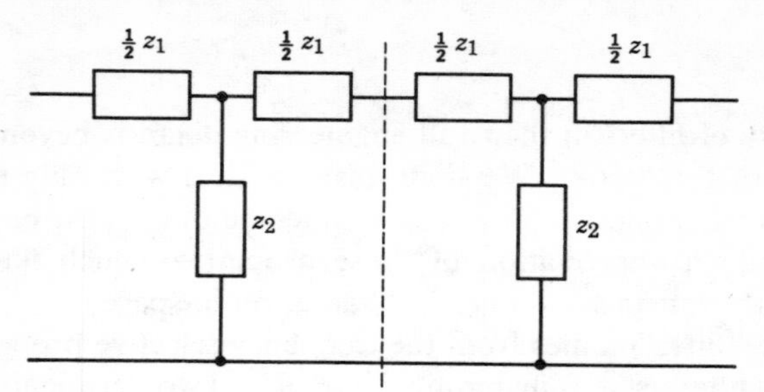

Figure 2. The circuit of Figure 1 is considered as a sequence of T-sections (two shown).

frequency ω, and internal impedance z_i. The output terminals on the right connect to a load impedance z_n. Within the filter itself, the impedances z_1 (all alike except for the first and last) are called series elements, and the impedances z_2 are shunt elements.

It is sometimes helpful to think of this filter circuit as being composed of T-sections as indicated in Figure 2. These sections are numbered from 1 to n, with m the general symbol to identify the number of a section. Maxwell loop currents are used in the analysis, these currents having subscripts running from 0 to n. The current i_m is the current in the loop between the mth and the $(m+1)$th section; i_0 is the current in the loop containing the generator, and i_n is the current in the loop containing the load. As we saw in Chapter 4, the use of Maxwell loop currents eliminates the need to write equations for Kirchhoff's point rule. Kirchhoff's loop rule yields the following:

(2-1*a*) (zeroth loop) $i_0z_i + \frac{1}{2}i_0z_1 - z_2(i_1 - i_0) = V$

(2-1*b*) (first loop) $z_2(i_1 - i_0) + z_1i_1 - z_2(i_2 - i_1) = 0$

(2-1*c*) (mth loop) $z_2(i_m - i_{m-1}) + z_1i_m - z_2(i_{m+1} - i_m) = 0$

(2-1*d*) (nth loop) $z_2(i_n - i_{n-1}) + (\frac{1}{2}z_1 + z_n)i_n = 0$

Here currents and voltages are complex amplitudes. The instantaneous value for a current or voltage is obtained by multiplying the complex amplitude by $e^{j\omega t}$. For the sake of identification the first and last of Equations 2-1 will be called terminal equations, whereas all others are filter equations. There are $n + 1$ unknown currents and $n + 1$ relations from which these currents are to be obtained.

The repetition of form in the $n - 1$ filter equations helps in the finding of a solution, suggesting that the solutions should have a common form. Indeed the expression

$$i_m = Ae^{\Gamma m} \tag{2-2}$$

is found to satisfy Equation 2-1*c* for any value of *A* provided

$$\frac{1}{2}(e^{\Gamma} + e^{-\Gamma}) = \cosh \Gamma = \frac{z_1}{2z_2} + 1 \tag{2-3}$$

Since, moreover, $\cosh \Gamma = \cosh(-\Gamma)$, we find that $i_m = Be^{-\Gamma m}$ also satisfies Equation 2-1*c* and, therefore; a complete solution has the form:

$$i_m = Ae^{\Gamma m} + Be^{-\Gamma m} \tag{2-4}$$

for arbitrary values of *A* and *B*. But only for special values of *A* and *B* will this solution also satisfy Equations 2-1*a* and 2-1*d*. These terminal equations in a sense constitute boundary conditions to fix the constants *A* and *B*.

Since, in general, z_1 and z_2 are complex numbers, Equation 2-3 shows that Γ will also in general be a complex number, to be written as

$$\Gamma = \alpha + j\beta \tag{2-5}$$

where α and β are real numbers to be determined from 2-3 by expanding the hyperbolic cosine of a sum:

$$\begin{aligned} \cosh(\alpha + j\beta) &= \cosh \alpha \cosh(j\beta) + \sinh \alpha \sinh(j\beta) \\ &= \cosh \alpha \cos \beta + j \sinh \alpha \sin \beta \end{aligned} \tag{2-6}$$

3. Transmission Characteristics

Equation 2-4 shows the total current i_m as the sum of two parts, $Ae^{\Gamma m}$ and $Be^{-\Gamma m}$. To obtain a feeling for the physical nature of these parts, let us first look at the *B*-term, writing it in its instantaneous form by including the factor $e^{j\omega t}$. (As in Chapter 8, we indicate instantaneous values with primes.)

$$i_{mB}' = Be^{-\alpha m}e^{j(\omega t - \beta m)} \tag{3-1}$$

Thus i_{mB}' lags behind the corresponding part of the input current

$$i_{0B}' = Be^{j\omega t}$$

by the phase angle βm, and behind the immediately preceding *B* current by an angle β. The complex amplitude of this part of the current is $Be^{-\alpha m}$ and thus decreases by a factor $e^{-\alpha}$ from one section to the next. In form, Equation 3-1 is a traveling wave function (Section 6) describing a wave traveling in the direction of increasing *m*, but the picture of a traveling wave must be qualified by the fact that the independent variable *m* takes on only integer values instead of being continuous.

The A term on the other hand, in its instantaneous form

$$i_{mA}' = Ae^{\alpha m}e^{j(\omega t+\beta m)} \tag{3-2}$$

is formally a wave traveling from the load back to the generator, similarly attenuated in the direction of travel. Presently it will be shown that the A-term is a current "wave" resulting from reflection at the load, just as light may be reflected from a surface on which it falls. And just as in some cases light may be totally absorbed, giving rise to no reflection, so in the case of the filter we shall see that some loads totally absorb the energy from the filter and produce no reflected wave. In this case the constant A is zero.

Should the constant α be zero, there will be no attenuation of the currents along the filter, and we speak of the filter as providing complete transmission. For $\alpha \neq 0$, $e^{-\alpha m}$ is a rather powerful attenuating function.

A physical feeling for the significance of Equation 2-3 is most readily obtained by examining the behavior of an ideal filter—one in which z_1 and z_2 consist of pure reactances with no resistance. z_1 and z_2 then each contain j as a factor and the right side of 2-3 is a pure real number, with the result that:

$$\sinh \alpha \sin \beta = 0 \tag{3-3a}$$

$$\cosh \alpha \cos \beta = \frac{z_1}{2z_2} + 1 \tag{3-3b}$$

Equation 3-3*a* requires either that $\alpha = 0$ or that β is an integer number of π. Which of the two choices holds depends on the magnitude of the right side of 3-3*b*. For should $\alpha = 0$, then $\cosh \alpha = 1$ and 3-3*b* becomes

$$\cos \beta = \frac{z_1}{2z_2} + 1 \qquad \text{for } (\alpha = 0) \tag{3-4}$$

A relation that can only hold if

$$-1 < \left(\frac{z_1}{2z_2} + 1\right) < +1 \tag{3-5}$$

On the other hand, if β is an integer number of π, then $\cos \beta = \pm 1$ and

$$\cosh \alpha = \pm\left(\frac{z_1}{2z_2} + 1\right) \tag{3-6}$$

Whether the positive or negative sign holds depends on whether $\left(\frac{z_1}{2z_2} + 1\right)$ is positive or negative since $\cosh \alpha$ can only be positive.

We may summarize the above discussion in a diagram such as Figure 3. Since z_1 and z_2 are, in general, functions of the frequency, it is the

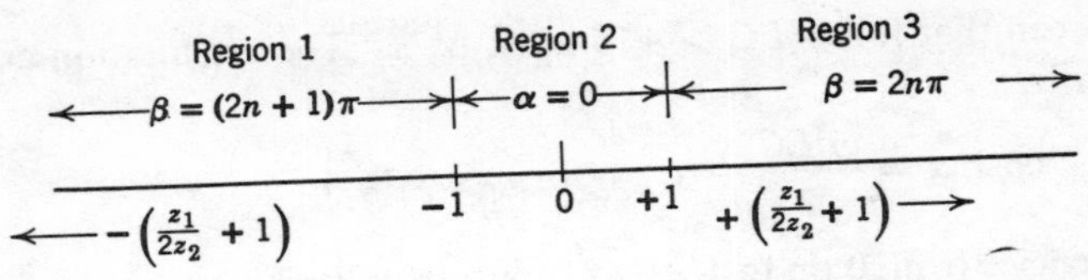

Figure 3. *Scales of* $\left(\frac{z_1}{2z_2} + 1\right)$ *showing required values of* α *and* β *in different ranges (ideal components).*

frequency which determines whether a given ideal filter is operating in region 1, 2, or 3. Because $\alpha = 0$ in region 2, those currents whose frequencies place $\left(\frac{z_1}{2z_2} + 1\right)$ in region 2 are transmitted to the load without attenuation, and hence these frequencies are in the "pass" region. For currents with frequencies such that $\left(\frac{z_1}{2z_2} + 1\right)$ is in region 1 or 3 α is greater than zero and the currents are attenuated. Regions 1 and 3, therefore, are called "stop" regions, although the word "stop" is not to be interpreted literally. It is a convenient term, implying "attenuated at least to some extent."

The inequality 3-5 expresses the condition for a frequency to be in a pass band or region. It is commonly rearranged into the following equivalent form:

$$0 < \left(-\frac{z_1}{z_2}\right) < 4 \tag{3-7}$$

As an example, consider a filter system consisting of series elements z_1, which are inductors of inductance L, and shunt elements z_2 which are capacitors of capacitance C. Then:

$$-\frac{z_1}{z_2} = -\frac{j\omega L}{-(j/\omega C)} = \omega^2 LC \tag{3-8}$$

Those frequencies for which $\omega^2 LC$ has values between 0 and 4 are in the pass region. The filter shows no attenuation for

$$\frac{0}{LC} < \omega^2 < \frac{4}{LC} \tag{3-9}$$

All other frequencies are attenuated to some degree, and hence fall within a stop region. For this type of filter, Equation 3-3*b* takes the form:

$$\cosh \alpha \cos \beta = \frac{j\omega L}{2[-(j/\omega C)]} + 1 = 1 - \tfrac{1}{2}(\omega^2 LC), \tag{3-10}$$

and it is seen that for $\omega^2 LC > 4$, β must be π. The attenuation constant α is given by:

$$\cosh \alpha = \frac{\omega^2 LC}{2} - 1 \qquad \text{for } \omega^2 LC > 4 \tag{3-11}$$

All frequencies from 0 up to $1/\pi\sqrt{LC}$ are transmitted without attenuation, while those above this limit are attenuated, the attenuation increasing with increasing frequency. Such a system is called a low-pass filter. One very

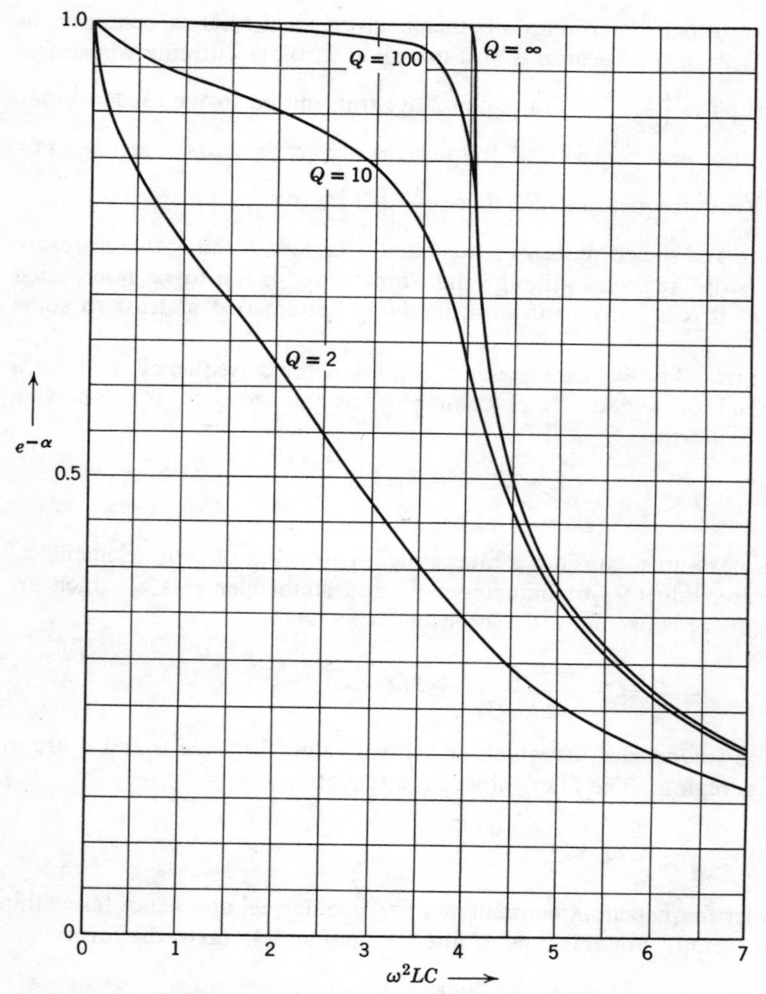

Figure 4. Attenuation of a low pass filter versus $\omega^2 LC$ for different Q's of the coils, assuming ideal capacitors.

common application is in a rectified power supply. The output of a rectifier consists of half-wave pulses which in a Fourier analysis are equivalent to a current of zero frequency plus currents of frequencies which are multiples of the power line frequency. If the components of a low-pass filter are so chosen that the upper limit of the pass region falls well below the power line frequency, all the alternating components are attenuated and there results an approximation to pure d-c in the load.

In so far as the filter components depart from ideal elements, this analysis is approximate. The effect of resistance in the filter components is primarily to increase α above zero in the pass region. The graph of Figure 4 shows how the exponential attenuation factor varies with frequency for a low-pass filter for various values of Q for the inductance (assuming an ideal capacitor).

Within the limitation that all coils have the same inductance L and all capacitors the same capacitance C, three other simple types of ideal filters can be constructed:

(*a*) High pass, in which z_1 is a capacitor C and z_2 is an inductor L.

(*b*) Band pass, in which z_1 is a series combination of L and C, and z_2 is a parallel combination of L and C.

(*c*) Band stop, in which z_1 is a parallel combination of L and C, and z_2 is a series combination of L and C.

The reader should analyze each of these to see the justification for the name given to it.

4. The Effect of the Terminations

We now return to a real filter system (nonideal elements) and determine the current amplitudes. The constants A and B are evaluated by requiring that Equation 2-4 must also satisfy the terminal equations of the zeroth and the nth loop of the system. By inserting Equation 2-4 into 2-1*a* and 2-1*d* we get, after grouping of terms (using $m = 0$ in 2-1*a* and $m = n$ in 2-1*d*):

$$A[z_i + \tfrac{1}{2}z_1 - z_2(e^{\Gamma} - 1)] + B[z_i + \tfrac{1}{2}z_1 - z_2(e^{-\Gamma} - 1)] = V \tag{4-1a}$$

$$Ae^{\Gamma n}[z_n + \tfrac{1}{2}z_1 - z_2(e^{-\Gamma} - 1)] + Be^{-\Gamma n}[z_n + \tfrac{1}{2}z_1 - z_2(e^{\Gamma} - 1)] = 0 \tag{4-1b}$$

These are linear equations in the two unknown quantities, A and B. They are somewhat simplified if we make the following substitutions:

$$z_k = -\tfrac{1}{2}z_1 + z_2(e^{\Gamma} - 1) = \tfrac{1}{2}z_1 - z_2(e^{-\Gamma} - 1) = \sqrt{(z_1{}^2/4) + z_1z_2} \tag{4-2}$$

(The reader should show that the three forms are equivalent.) With 4-2, Equations 4-1 become

$$A(z_i - z_k) + B(z_i + z_k) = V \tag{4-3}$$

$$Ae^{\Gamma n}(z_n + z_k) + Be^{-\Gamma n}(z_n - z_k) = 0$$

from which are obtained the rather symmetrical expressions:

$$A = \frac{-Ve^{-\Gamma n}(z_n - z_k)}{e^{\Gamma n}(z_i + z_k)(z_n + z_k) - e^{-\Gamma n}(z_i - z_k)(z_n - z_k)} \tag{4-4}$$

$$B = \frac{Ve^{\Gamma n}(z_n + z_k)}{e^{\Gamma n}(z_i + z_k)(z_n + z_k) - e^{-\Gamma n}(z_i - z_k)(z_n - z_k)}$$

Physically, the constants A and B are the two parts of the current at the input to the filter (the zeroth loop).

A comment is in order about the quantity z_k defined in Equation 4-2. z_k has the dimensions of impedance, is only a function of the components of the filter itself and not of the terminations, is in general a complex quantity, and is a function of the frequency. It is called variously "characteristic impedance," "iterative impedance," or "surge impedance" in the literature. We shall use the first.

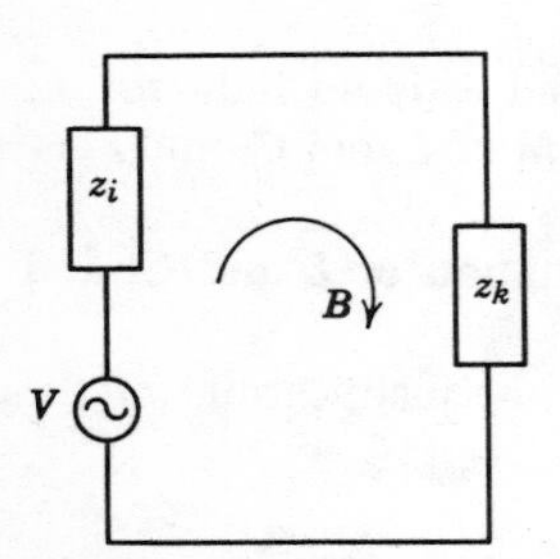

Figure 5. Equivalent input circuit of a filter with matched termination.

At a particular frequency we may compute z_k from z_1 and z_2. We may then construct an impedance equal to z_k out of inductors, capacitors, and resistors, connecting this impedance as a load to the output terminals. We are thus constructing a special case where $z_n = z_k$, in consequence of which, from 4-4:

$$A = 0 \qquad B = \frac{V}{z_i + z_k} \qquad (z_n = z_k) \tag{4-5}$$

The value for B in Equation 4-5 corresponds to the current in the circuit of Figure 5. We conclude that, if an impedance equal to z_k is connected across the output of the filter, the current in the generator is the same as if the same impedance were connected directly across its terminals with no filter present. Or, put in another way, if we measure the impedance at the input terminals (with an impedance bridge) when the output impedance is z_k, the result of our measurement will be z_k. Note that this result is independent of the number of sections in the filter. The reader should show that z_k is the only terminating impedance for which this is true.

There is for this particular load no reflected current wave ($A = 0$). All the energy which reaches the load from the filter is absorbed in the load and none is reflected back toward the generator.

Insert the constants A and B from Equation 4-4 into 2-4 for the total current at the mth section in terms of the applied voltage, the constants of the filter, and the terminations:

$$i_m = \frac{V[(z_n + z_k)e^{\Gamma(n-m)} - (z_n - z_k)e^{-\Gamma(n-m)}]}{e^{\Gamma n}(z_i + z_k)(z_n + z_k) - e^{-\Gamma n}(z_i - z_k)(z_n - z_k)} \tag{4-6}$$

By letting m be zero in Equation 4-6 and remembering that the voltage applied to the input terminals of the filter is $V - i_0 z_i$, we obtain a general expression for the impedance at the input to the filter:

$$z_{in} = z_k \frac{(z_n + z_k)e^{\Gamma n} + (z_n - z_k)e^{-\Gamma n}}{(z_n + z_k)e^{\Gamma n} - (z_n - z_k)e^{-\Gamma n}} \tag{4-7}$$

which is seen to reduce to z_k when $z_n = z_k$.

Two special cases are of particular interest:

CASE 1. $z_n = 0$. The load is a perfect short circuit. Call the input impedance for this case z_{sc}:

$$z_{sc} = z_k \frac{e^{\Gamma n} - e^{-\Gamma n}}{e^{\Gamma n} + e^{-\Gamma n}} = z_k \tanh \Gamma n \tag{4-8}$$

CASE 2. $z_n = \infty$. The load is a perfect open circuit. Call the input impedance for this case z_{oc}:

$$z_{oc} = z_k \frac{e^{\Gamma n} + e^{-\Gamma n}}{e^{\Gamma n} - e^{-\Gamma n}} = z_k \coth \Gamma n \tag{4-9}$$

By combining the results of Cases 1 and 2, we find that

$$z_k^2 = z_{sc} z_{oc} \tag{4-10}$$

Here is a relation that has considerable experimental importance. The characteristic impedance of a filter is an important property and Equation 4-10 shows that this property can be measured experimentally without knowledge of the components of the filter. Indeed, we may have a black box with two pairs of terminals leading from it and no knowledge of its contents. By making one impedance measurement at a terminal pair with the other pair open and another measurement with the other pair shorted, we have the information needed to compute the characteristic impedance. Curiously, this property of a filter, which seems to be closely associated with a special load, is readily obtained from two measurements, in each of which there is no load at all.

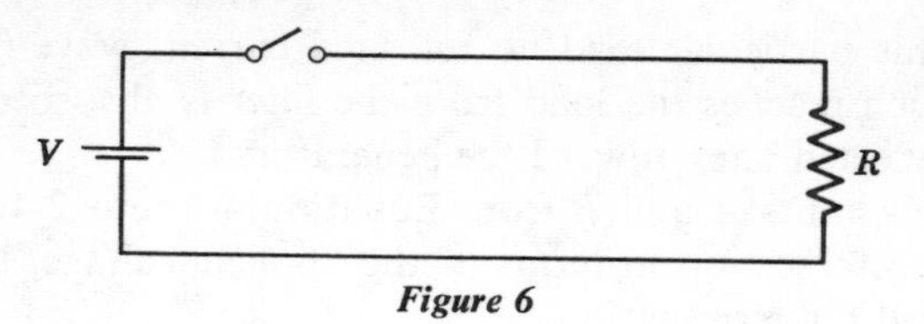

Figure 6

5. Transmission Lines

Implicit in the discussion of circuits has been the assumptions that connecting wires introduce no resistance, inductance, or capacitance. We now consider circumstances in which connecting wires cannot be ignored.

Consider, as an example, a battery and switch connected to a resistance by means of say, two long parallel wires. When the switch is closed, current does not start instantaneously in the resistance nor does the battery voltage appear instantaneously across the resistance. Instead, there is a step-pulse of voltage that travels from the battery to the resistance with a finite speed. If the switch is alternately opened and closed, several square-wave pulses may appear on the line before the first reaches the end. The speed with which such disturbances travel is, of course, very high, being usually equal to or very close to the speed of light. Consequently, if the time of travel is very short compared to the rate at which changes occur at the source, it can be neglected, whereas if the time of travel is appreciable (long wires) or if the frequency of change at the source is high, such a simplification is not justified.

As with the filter, so with continuous transmission lines it is not our purpose to enter into detailed discussion of engineering detail or applications, although we shall find that a discussion of some applications are warranted as illustrations of the theory. The subject of transmission lines is, however, a very practical one for the student who will be concerned with the instrumentation of modern physics, and the treatment given here is designed to develop a fundamental understanding on which he may readily build in the future.

6. Basic Transmission Line Equations

A pair of conductors connecting two points, A and B, with two other points, C and D, as in Figure 7, constitute a transmission line. Although such conductors may be of arbitrary shape and varying separation, we shall confine ourselves to parallel conductors of circular cross section whose properties are manageable with the analytical tools at hand. The points A and B are the input terminals of the line and are connected to a source of electrical energy which we shall indicate as an alternating current generator of angular frequency ω and internal impedance z_i. The points C and D are the output terminals of the line and are connected to an impedance, z_s, the load impedance. This impedance may, of course, be of any form, including conceivably the input terminals of another transmission line. For our discussion we shall assume that z_s is passive (contains no sources of emf) and linear (current in it is proportional to the voltage applied to it—no iron-core inductors, no rectifiers, etc.).

The significant properties of the line for our purposes are the resistance, the inductance, and the capacitance of the parallel pair of conductors, and the conductivity of the dielectric medium between the conductors. These properties are best expressed per unit length of line, since they are continuously distributed along the line. Thus R (ohms per meter) is the resistance of the conductors per unit length; L (henrys per meter) the inductance per unit length (cf.: Chapter 6, Problem 8); C (farads per meter) the capacitance per unit length (cf: Chapter 2, Problem 13), and G (mhos per meter) the conductance of the dielectric between the parallel pair. Note that these symbols now have an altered meaning from that given them in a-c circuit theory.

Over an infinitesimal length of the line, dx (Figure 7), the resistance will be $R\,dx$ and the inductance, $L\,dx$, so that there will be a series impedance within dx of $z\,dx = R\,dx + j\omega L\,dx$, where we identify

$$z = R + j\omega L \quad \text{(ohms per meter)} \tag{6-1}$$

as the series impedance per unit length. Within dx the dielectric will have a conductivity of $G\,dx$ and there will be a capacitance between the conductors of $C\,dx$. Consequently, there is a shunt admittance between the conductors within dx of $y\,dx = G\,dx + j\omega C\,dx$ and we identify

$$y = G + j\omega C \quad \text{(mhos per meter)} \tag{6-2}$$

as the shunt admittance per unit length.

The generator emf is $Ve^{j\omega t}$, so that at the point x there is a voltage across the transmission line of $v' = ve^{j\omega t}$, where v is a complex amplitude. Similarly, there is a current in the line at the point x given by $i' = ie^{j\omega t}$,

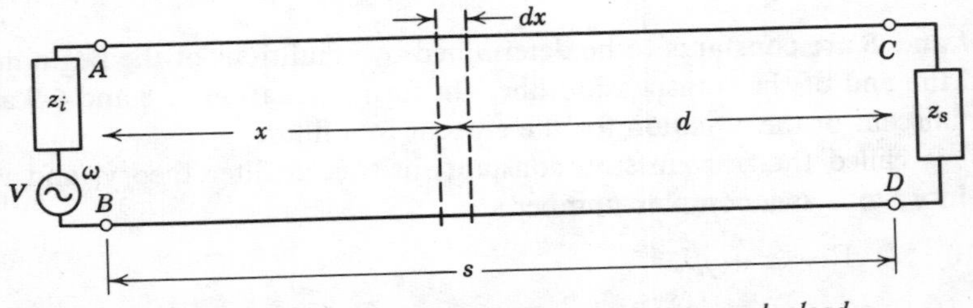

Figure 7. A transmission line between a generator and a load.

where i is the complex amplitude of the current. Both v' and i' are functions of x. The current i' in the series impedance of the line at x produces a change over dx of the voltage across the line, whose complex amplitude is

$$dv = -iz\,dx \tag{6-3}$$

where the negative sign indicates that for positive i the voltage decreases as x increases. Analogously, the voltage across the shunt admittance produces a shunting current, which is equal to the amount di by which the current in the conductors changes over dx:

$$di = -vy\,dx \tag{6-4}$$

The solution of these two differential equations describes the distribution of voltage and current amplitudes as a function of distance along the transmission line.

Separation of variables between Equations 6-3 and 6-4 is accomplished by differentiation and cross-substitution, yielding the two equations:

$$\frac{d^2v}{dx^2} = zyv \quad \text{and} \quad \frac{d^2i}{dx^2} = zyi \tag{6-5}$$

for the first of which, as a trial solution, we consider:

$$v = Ae^{\Gamma x} \tag{6-6}$$

which substitution shows is satisfactory provided:

$$\Gamma^2 = zy \tag{6-7}$$

Since only the square of Γ is determined, either $+\Gamma$ or $-\Gamma$ may be used in Equation 6-6, and therefore the complete solution has the form:

$$v = Ae^{\Gamma x} + Be^{-\Gamma x} \tag{6-8}$$

From Equation 6-4 the current distribution is

$$i = -A\frac{\Gamma}{z}e^{\Gamma x} + B\frac{\Gamma}{z}e^{-\Gamma x} \tag{6-9}$$

A and B are constants to be determined by conditions at the beginning and the end of the transmission line. In form, Equations 6-8 and 6-9 are reminiscent of the equation for the current in a filter.

Γ is called the transmission constant, just as in filter theory, and we again write it as a complex number

$$\Gamma = \alpha + j\beta \tag{6-10}$$

since z and y are in general complex.

Equations 6-8 and 6-9 are the amplitudes of the voltage and current as functions of x. The instantaneous values of these quantities are:

$$v' = Ae^{\alpha x}e^{j(\omega t+\beta x)} + Be^{-\alpha x}e^{j(\omega t-\beta x)} \tag{6-11}$$

$$i' = -A\sqrt{y/z}\, e^{\alpha x}e^{j(\omega t+\beta x)} + B\sqrt{y/z}\, e^{-\alpha x}e^{j(\omega t-\beta x)} \tag{6-12}$$

The traveling wave nature of the terms in this expression can be seen by examining in detail the real part of, say, $e^{j(\omega t-\beta x)}$, which is $\cos(\omega t - \beta x)$. By plotting this function against x for several successive values of t, it is seen that the cosinusoidal pattern moves in the direction of increasing space variable as time increases. (When the argument is $(\omega t + \beta x)$, the pattern moves in the direction of decreasing space variable.) Consider two of the curves in Figure 8 for which the separation in time is, say, Δt, and consider corresponding points on the two curves, where the curves cross the abscissa. These two intercepts are Δx apart and represent the same value of the cosine argument. Thus,

$$\omega t - \beta x = \omega(t + \Delta t) - \beta(x + \Delta x) \tag{6-13}$$

from which we find that

$$\frac{\Delta x}{\Delta t} = \frac{\omega}{\beta} \tag{6-14}$$

But $\Delta x/\Delta t$ is the velocity with which the cosine pattern moves along the x-axis and is, therefore, the wave velocity.

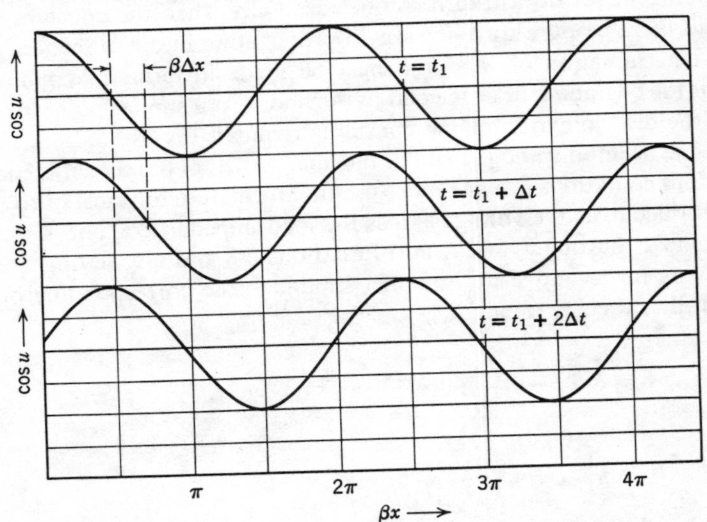

Figure 8. A traveling wave; cos u versus βx, where $u = \omega t - \beta x$.

The wavelength is defined as the distance between two corresponding points of the wave at a given instant of time, such as the distance between two successive peaks. If this distance is λ, then the cosine argument for the second peak can be written in two ways, if the argument for the first peak is $\omega t - \beta x$:

$$\omega t - \beta(x + \lambda) = \omega t - \beta x - 2\pi \tag{6-15}$$

from which

$$\lambda = \frac{2\pi}{\beta} \tag{6-16}$$

For this reason, β is sometimes called the wavelength constant.

The second terms of Equations 6-11 and 6-12 are recognized as representations of traveling wave patterns progressing toward increasing values of x and the first terms are patterns progressing toward decreasing values of x. Each pattern is exponentially attenuated in the direction of propagation. (In filter theory, the fact that the "space" variable, m, changed only by discrete intervals meant that we could not strictly speak of a traveling wave, although the concept was nevertheless useful.) In general, the velocity of propagation of the wave (Equation 6-14) depends on the frequency, unless β happens to be directly proportional to the frequency. Equations 6-11 and 6-12 show that in general there will be simultaneously existing waves propagating in both directions and, hence, that there will be standing waves on a transmission line. Only when the constant A is zero will there be pure traveling waves propagating in one direction.

The determination of A and B may be made in terms of conditions either at the beginning or at the end of the line. (At a similar point in filter theory the determination of the constants required the use of conditions at both the beginning and the end of the line. Wherein lies the difference?) The output conditions provide more useful information for most purposes. We introduce v_s as the voltage across the load impedance z_s and i_s as the current in z_s. Putting v_s and i_s into Equations 6-8 and 6-9, letting $x = s$, and solving for A and B in a manner similar to the corresponding operation in filter theory, we get:

$$A = \tfrac{1}{2}\left[v_s - i_s\left(\frac{z}{y}\right)^{1/2}\right]e^{-\Gamma s} \tag{6-17}$$

$$B = \tfrac{1}{2}\left[v_s + i_s\left(\frac{z}{y}\right)^{1/2}\right]e^{\Gamma s}$$

from which we get the complete expressions for the voltage and current amplitudes at a point x on the transmission line:

$$(6\text{-}18a) \qquad v = \tfrac{1}{2}v_s[e^{\Gamma(s-x)} + e^{-\Gamma(s-x)}] + \tfrac{1}{2}i_s\left(\frac{z}{y}\right)^{1/2}[e^{\Gamma(s-x)} - e^{-\Gamma(s-x)}]$$

$$= v_s \cosh \Gamma d + i_s\left(\frac{z}{y}\right)^{1/2} \sinh \Gamma d$$

$$(6\text{-}18b) \qquad i = \tfrac{1}{2}i_s[e^{\Gamma(s-x)} + e^{-\Gamma(s-x)}] + \tfrac{1}{2}v_s\left(\frac{y}{z}\right)^{1/2}[e^{\Gamma(s-x)} - e^{-\Gamma(s-x)}]$$

$$= i_s \cosh \Gamma d + v_s\left(\frac{y}{z}\right)^{1/2} \sinh \Gamma d$$

Here d, replacing $s - x$, is the distance from the load end to the point at which v or i is evaluated. Because much of our interest lies in the effect which the load has on the voltage and current distributions, Equations 6-18 are frequently of more interest than the comparable expressions found in terms of the voltage and current at the generator.

The impedance at the point d is defined as the ratio of the voltage to the current at d:

$$(6\text{-}19) \qquad z_d = \frac{v}{i} = \left(\frac{z}{y}\right)^{1/2} \frac{z_s \cosh \Gamma d + (z/y)^{1/2} \sinh \Gamma d}{(z/y)^{1/2} \cosh \Gamma d + z_s \sinh \Gamma d}$$

Evidently a load impedance equal to $(z/y)^{1/2}$ produces an impedance at any point of the line (including, of course, the input end) equal to $(z/y)^{1/2}$. Just as in filter theory, this is the only load for which input and output impedances are equal. This impedance, fixed by the properties of the transmission line, is called as in filter theory the characteristic impedance of the line.

The characteristic impedance z_k can also be looked upon as the input impedance which is approached as the length of the line becomes very large, approaching infinity, as is readily demonstrated from Equation 6-19. This interpretation is interesting but somewhat less fruitful than the one given immediately above.

The characteristic impedance of a transmission line is in general a function of the frequency, as is seen by expressing it in terms of R, G, L, and C:

$$(6\text{-}20) \qquad z_k = \left(\frac{R + j\omega L}{G + j\omega C}\right)^{1/2}$$

It is a quite different dependence, however, than was found for the characteristic impedance of a filter. This difference is most markedly brought out in a comparison of an ideal filter and an ideal transmission line, in each of which the resistances of conductors and the conductivities of dielectrics are zero. Equation 6-20 shows that the ideal transmission line characteristic impedance is a pure real number, and hence a pure resistance, independent of frequency; whereas the ideal filter characteristic impedance not only depends on frequency, but is a pure resistance over only a limited region of the spectrum, and is a pure reactance in other regions.

But, although the dependence of characteristic impedance in filters and transmission lines on frequency is considerably different, the basic meaning of the concept is the same. In filters, in transmission lines, and as we shall see later, in the propagation of waves in space, the concept of characteristic impedance always is associated with certain features—the input impedance of a system terminated in z_k is equal to z_k; in a system terminated in z_k there is no reflection of energy so that all energy reaching the load is absorbed in the load, and as a consequence there are no standing waves in the system resulting from interference of forward and backward traveling waves.

The meanings attributed to the impedance concept are meanings which arise out of our thinking in terms of circuit concepts, and impedance has value wherever such circuit concepts can legitimately be applied. It is surprising to what extent circuit concepts can be profitably used—even in dealing with waves in space—but these ideas have their limitations and they need to be carefully examined when they are applied to new situations. These limitations will become particularly apparent when we come to studying guided electromagnetic waves (Chapter 12).

7. Further Considerations about the Transmission Constant

By combining Equation 6-10 with 6-1, 6-2, and 6-7, the transmission constant, Γ, becomes

$$\Gamma = \alpha + j\beta = \sqrt{zy} = \sqrt{(R + j\omega L)(G + j\omega C)} \tag{7-1}$$

Squaring and equating the real and imaginary terms yields two algebraic equations from which α and β are obtained:

$$\alpha^2 = \tfrac{1}{2}\sqrt{(R^2 + \omega^2 L^2)(G^2 + \omega^2 C^2)} + \tfrac{1}{2}(RG - \omega^2 LC) \tag{7-2a}$$

$$\beta^2 = \tfrac{1}{2}\sqrt{(R^2 + \omega^2 L^2)(G^2 + \omega^2 C^2)} - \tfrac{1}{2}(RG - \omega^2 LC) \tag{7-2b}$$

Equation 7-1 is quite different from the equation defining Γ in filter theory and results in a different dependence of α and β on the frequency. The hyperbolic cosine of Γ in filter theory led to pass regions and stop regions in the frequency spectrum, regions which were, in ideal filters, sharply defined. No such distinct regions in the frequency spectrum exist in transmission line theory. Indeed, if the transmission line is ideal—conductors without resistance and perfect insulation—we find that

$$\alpha = 0 \qquad \beta = \omega\sqrt{LC} \tag{7-3}$$

An ideal transmission line transmits all frequencies without attenuation, the phase shift per meter is proportional to the angular frequency, and consequently, from Equation 6-14, all frequencies are propagated with the same velocity. The time required for each frequency component in a wave form of arbitrary shape—a pulse, say—to travel from source to load is the same and the transmission of the wave form is accomplished without distortion.

As an example of a nonideal transmission line, consider a cable connecting a microphone with a receiver. The conductors in such a cable have finite resistance and the insulation, though good, is not perfect. Voice sounds are composites of many frequencies and the electrical waves corresponding to these sounds are distorted by the time they reach the receiver because the different frequencies travel at different rates. If such a dispersion in velocities is great enough, the received sound becomes unintelligible. Ways of reducing the velocity dispersion in such a line are of practical interest in communication work. Three cases are noted here; the first two are exact and the third is a practical approximation.

CASE 1. $R = G = 0$. This is the case already mentioned of the ideal transmission line in which $\alpha = 0$ and $\beta = \omega\sqrt{LC}$, precisely the conditions required for uniform transmission velocity and no phase distortion. The received signal amplitude is the same as the sending amplitude for a matched termination, and there is no absorption of energy in the line itself.

CASE 2. $R/L = G/C$. Here neither the conductors nor the insulation is perfect. β has the same value as in Case 1, but α is not zero because electrical energy is converted to heat in the resistance of the line and in the insulation. Nevertheless, the value of α is independent of frequency, so that the amplitudes of currents of different frequencies bear the same relation to each other at the receiver as at the microphone. This case is only of theoretical interest, however, because the ratios required are difficult to achieve.

CASE 3. $G \cong 0$ and $\omega L \gg R$. Here we get a reduction in distortion but not necessarily perfect transmission. Since good quality insulation is readily available, it is possible to make G negligible, but the second condition is approximated by increasing the inductance, rather than by reducing the resistance. Coils of lumped inductance are inserted at regular intervals. With properly designed coils more inductive reactance is added than resistance, and so long as the coils are spaced at intervals which are small compared to the shortest wavelength to be used on the transmission line, their lumped nature does not significantly disturb the uniformity of the line. In such an inductively loaded line, Equation 7-2*a* becomes

$$\alpha^2 \cong \frac{R^2 C}{4L} \tag{7-4}$$

and 7-2*b* becomes

$$\beta^2 \cong \omega^2 LC\left(1 + \frac{R^2}{4\omega^2 L^2}\right) \cong \omega^2 LC \tag{7-5}$$

provided $(R/\omega L)^2$ can be neglected compared to unity. β is thus very closely proportional to ω; and α, although not zero, is very nearly independent of ω, as is necessary for low distortion.

8. Relation of Transmission Lines to Ordinary Circuits

We can now examine the relevance of the study of transmission lines to ordinary circuit theory. Let us consider a pair of wires which complete a closed circuit between circuit element 1 and circuit element 2. This closed loop may be only part of a more involved arrangement of components; nevertheless, in the application of Kirchhoff's laws, using the method of loop currents, the assumption is made that the loop current is the same at all parts of the loop at a given instant. But looking upon the connecting wires as a transmission line between element 1 and element 2, we see that this assumption is a usable approximation only when the length of the transmission line is a negligible fraction of the line wavelength at the frequency used. When this is not the case, the current may be in one sense at element 1 and even in the opposite sense at element 2, and its magnitude will vary continuously from point to point in between.

We may also visualize the situation in terms of the velocity of propagation of a signal, which is essentially the velocity of light for parallel conductors separated by air. The alternations of voltage at element 1 require a finite time to appear at element 2; whether this time is significant depends on whether it represents a significant fraction of the period of the voltage alternations.

Alternating current circuit theory as developed in Chapter 8 must therefore be used with care at higher frequencies. With the velocity of propagation close to 3×10^8 meters per second, the wavelength at 1 megacycle is 300 meters, compared to which the dimensions of a circuit on, say, a laboratory table is negligible. At 1000 megacycles, on the other hand, the wavelength is 30 centimeters, a quite different situation. Should ordinary circuit theory predict that the currents in two parts of the system 15 centimeters apart are in phase and equal, they may, in fact, differ greatly in both magnitude and phase because their separation amounts to a half-wavelength.

9. Short Transmission Lines—The Ideal Case

In addition to being used for communication purposes, transmission lines in high frequency devices of various sorts are used both as connectors and as simulators of inductance and capacitance. Occasionally they may be parallel wires, but more often they are coaxial, a geometry in which the electric and the magnetic fields are restricted to a well-defined region.

On short lengths of transmission line made of good conducting material the effect of attenuation is usually negligible, and it is possible to simplify the analysis by making the assumptions that there is no resistance and that the insulation is perfect—in short, we take α to be zero. Then the arguments of the hyperbolic functions involved in Equations 6-18 are pure imaginaries. Since

$$\sinh(j\beta d) = j \sin \beta d \quad \text{and} \quad \cosh(j\beta d) = \cos \beta d \tag{9-1}$$

the voltage and current equations become:

$$v = v_s \cos \beta d + j i_s z_k \sin \beta d \tag{9-2a}$$

$$i = i_s \cos \beta d + j \frac{v_s}{z_k} \sin \beta d \tag{9-2b}$$

and the expression for the impedance at a distance d from the end of the line becomes:

$$z_d = z_k \frac{z_s \cos \beta d + j z_k \sin \beta d}{z_k \cos \beta d + j z_s \sin \beta d} \tag{9-3}$$

When the impedance at the end of a transmission line is a dead short, Equation 9-3 becomes:

$$z_{sc} = j z_k \tan \beta d \tag{9-4}$$

where z_{sc} means "impedance at d when the load is short-circuited." On the other hand, when the impedance at the end of the line is infinite, Equation 9-3 becomes:

$$z_{oc} = -j z_k \cot \beta d \tag{9-5}$$

where z_{oc} means "impedance at d when load is an open circuit."

The product of Equations 9-4 and 9-5 is:

(9-6) $$z_{sc}z_{oc} = z_k^2$$

an expression in which the length of the line no longer appears. Just as with the comparable situation in filter theory, Equation 9-6 provides a method by which the characteristic impedance of a transmission line may be experimentally measured. The techniques by which such impedances are measured will be discussed in Section 12.

From the definition of characteristic impedance, $z_k = \sqrt{z/y}$, and from the assumption that in an ideal line $R = G = 0$, it follows that the characteristic impedance of an ideal transmission line is $\sqrt{L/C}$, a pure real number and hence a pure resistance.

10. The Ideal Line as a Reactance Element

The impedance of an ideal line with a shorted termination (Equation 9-4) is a pure reactance which is inductive or capacitative, depending on the sign of tan βd. Figure 9 is a plot of z_{sc} against βd. When $\beta d = \pi$, the length of the line is a half wave-length; when $\beta d = 2\pi$, the length is a whole wavelength, etc. Thus, a shorted line which is less than a quarter-wavelength long has an inductive input impedance. If the length is between a quarter- and a half-wavelength, the input impedance is capacitative, etc.

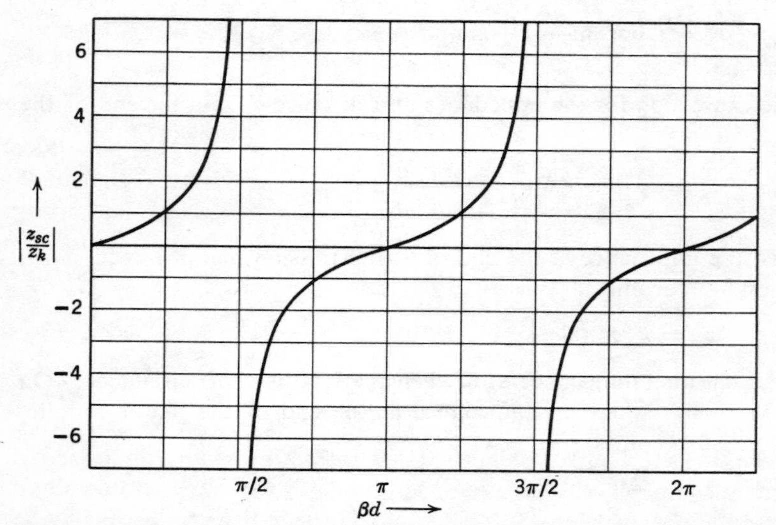

Figure 9. Input reactance of a shorted ideal transmission line versus length of line.

For a line length of exactly one-quarter wavelength, the impedance is infinite, a situation closely analogous to the impedance of a parallel combination of an ideal inductor and an ideal capacitor at resonance. For a line length of exactly one-half wavelength, the impedance is zero, analogous to a series combination of a perfect inductor and a perfect capacitor at resonance. (βd, which is $2\pi d/\lambda$, can be varied either by changing the physical length of the line or by changing the frequency.)

The impedance of a parallel resonant circuit is inductive (positive) for frequencies below resonance and capacitative above resonance. An examination of Figure 9 shows the same to be true for a shorted line. The comparison, however, is only qualitative, since the exact functional dependence of impedance on frequency is different for the two cases. For a series resonant circuit and for a half-wavelength shorted line the impedance is capacitative below and inductive above the resonant frequency, and the comparison is similarly qualitative. It is, however, often convenient in analyzing a circuit containing a transmission line as an element to treat it as a resonant *L-C* element.

A quarter wavelength shorted line has an interesting application. Suppose it is desired to support above ground but well insulated from ground a transmission line carrying a high-frequency current of wavelength λ. This line could be supported on ceramic insulators, but out of doors such insulators get wet and their surfaces become conducting. The supports can, instead, be made out of copper posts of such a length that the supported conductors are $\lambda/4$ meters above ground. Although this seems to provide just the opposite of good insulation, by joining the bottoms of the copper posts with a good shorting bar, the two posts become a quarter-wavelength transmission line whose impedance at the top is infinite. Anything, moreover, that improves the conductivity of the posts improves the approximation to an ideal line and improves their behavior as insulators! The usefulness of this arrangement is limited to one frequency (or odd multiples of it). At other frequencies the supporting posts, in effect, place either a capacitance or an inductance across the supported conductors. At twice the design frequency, the posts constitute a perfect short circuit!

11. Measurements on a Transmission Line

The impedance at a point on a transmission line distant d meters from the end of the line was defined in Section 6 as the ratio of the voltage across the line at the point d to the current in the line at d. Such voltages and currents could conceivably be measured by inserting the proper meters in and across the line. As with any measurements in physics, the question that must be faced is to what extent do the measuring instruments disturb the quantity being measured? On a long transmission line used at very low frequencies and hence long wavelengths, such an insertion of meters is possible, but is very quickly ruled out with decreasing wavelength. The experimental difficulties become prohibitive where a coaxial transmission line, instead of parallel wires, is used. On such coaxial lines at high frequencies, it is difficult to measure the absolute voltage across the line, but with a particular construction that we shall now discuss, it is possible rather simply to measure a quantity proportional to the voltage and hence to obtain voltage ratios. Current ratios may also be obtained by a similar method but the experimental difficulties are somewhat greater and they are usually inferred from their theoretical relation to the voltage ratios.

To obtain voltage ratios it is necessary to have a specially constructed coaxial line in which the outer conductor carries a narrow slot running parallel to the axis. Into this slot, but insulated from the outer conductor, is inserted a small probe, essentially a small antenna, on which is generated a voltage proportional to the electric field which exists within the line. This electric field is proportional to the voltage across the line, and the probe acts as an elementary voltmeter whose calibration is usually not known. By mounting the probe on a carriage which moves parallel to the axis and by conecting one of several types of detectors to it, measurements proportional to the voltage as a function of position on the line are obtained. Such a probe is illustrated in Figure 10.

The presence of the slot and the probe necessarily disturb conditions on the transmission line. However, since the slot runs parallel to the sense of the current in the outer conductor, the disturbance of the distribution of the current over this conductor is minimized by making the slot as narrow as possible. The presence of the probe in the electric field between the conductors alters the shape of the field, and this disturbance is minimized by minimizing the distance the probe is inserted. Here a compromise is reached between the sensitivity of the detecting instruments and the disturbance that can be tolerated.

In practice, when measurements are to be made in a coaxial system, a slotted line section of the same characteristics as the coaxial conductors is inserted at some junction. One of the simplest observations that can then be made is to determine whether the system is terminated in its characteristic impedance or not. For if it is, the absence of a reflected wave is shown by the fact that the voltage amplitude does not have maxima or minima as the probe is moved along the slot. This condition is unique, existing only for a termination equal to z_k.

In the next section we shall see, if voltage maxima and minima exist, how their location and their relative magnitudes make possible the

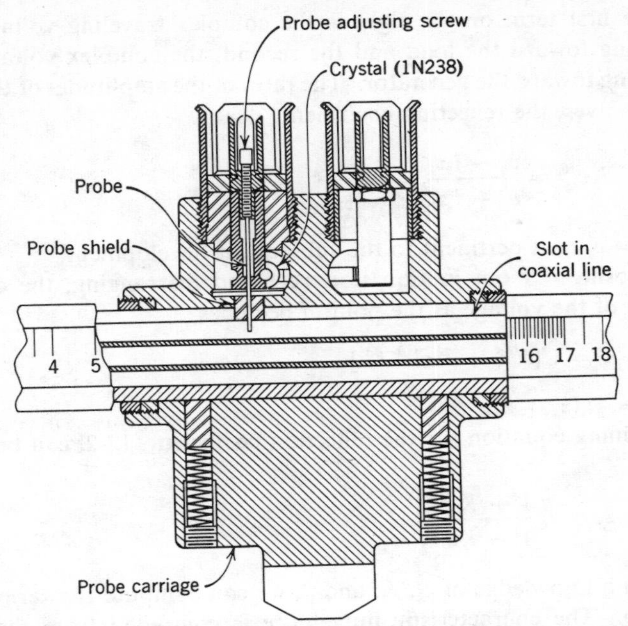

***Figure 10.** Diagram of a movable probe assembly on a slotted line. One outlet takes an adjustable shorted stub for resonating the probe system to the signal frequency. (Courtesy of General Radio Company.)*

computation of the load impedance. The method to be described, using the coaxial slotted line, is not the only one available—there is, in fact, a method that measures admittance rather than impedance, developed by the General Radio Company, that has advantages in experimental convenience—but the analysis of the coaxial slotted line as an impedance measuring device illustrates properties of the transmission line and is a good exercise in the application of transmission line equations. Such a slotted line is also a prototype of a similar device used in measurement on waveguides (Chapter 12).

12. Impedance Measurements with a Slotted Line

An ideal coaxial transmission line with a characteristic impedance z_k is terminated with an impedance z in which there is a voltage v and a current i . It is being used at a frequency and the traveling waves on the line have a wavelength $\lambda = 2\pi/\beta$. From Section 6 the instantaneous voltage at a point d meters from the load is given by

$$(12\text{-}1) \qquad v' = \tfrac{1}{2}(v_s + i_s z_k)e^{j(\omega t+\beta d)} + \tfrac{1}{2}(v_s - i_s z_k)e^{j(\omega t-\beta d)}$$

where the first term on the right is the complex traveling voltage wave propagating toward the load and the second, the complex voltage wave propagating toward the generator. The ratio of the amplitudes of these two traveling waves, the reflection coefficient,

$$Ke^{j\theta} = \frac{v_s - i_s z_k}{v_s + i_s z_k} = \frac{z_s - z_k}{z_s + z_k} \tag{12-2}$$

where $z_s = v_s/i_s$, is pertinent to the subsequent development.

By factoring $e^{j\omega t}$ out in Equation 12-1 and rearranging, the complex amplitude of the voltage at the point d becomes

$$v = C[e^{j\beta d} + Ke^{j(\theta - \beta d)}] \tag{12-3}$$

where $C = \frac{1}{2}i_s(z_s + z_k)$.

The defining equation for the reflection coefficient, 12-2, can be solved for z_s:

$$z_s = z_k \frac{1 + Ke^{j\theta}}{1 - Ke^{j\theta}} \tag{12-4}$$

Thus, with a knowledge of z_k, K, and θ, we can compute the terminating impedance. The characteristic impedance is computed from the cross-sectional dimensions of the coaxial line and the dielectric constant of the insulating medium (see Problem 11). In what follows we show how K and θ are obtained from measurements made with a slotted line.

The signal obtained from a probe on a slotted line is proportional to the real amplitude of the voltage. From Equation 12-3 the square of this amplitude is

$$vv^* = CC^*[1 + K^2 + 2K \cos (\theta - 2\beta\, d)] = |v|^2 \tag{12-5}$$

where the starred quantities are complex conjugates, obtained by changing j to $-j$, wherever it appears. The maximum voltage amplitude

$$|v|^2_{\max} = CC^*(1 + K)^2 \tag{12-6}$$

occurs at

$$\theta - 2\beta d = 2n\pi \tag{12-7}$$

and the minimum voltage

$$|v|^2_{\min} = CC^*(1 - K)^2 \tag{12-8}$$

occurs at

$$\theta - 2\beta d = (2n + 1)\pi \tag{12-9}$$

where n is an integer. Since the experimental measurements provide only relative values of the voltage, we must express the theory in terms of ratios. The *voltage standing wave ratio*, ρ, is the ratio of $|v|_{max}$ to $|v|_{min}$, and, from Equations 12-6 and 12-8, is

$$\rho = \frac{|v|_{max}}{|v|_{min}} = \frac{1+K}{1-K} \tag{12-10}$$

K is thus obtained from the experimentally measured value of ρ:

$$K = \frac{\rho - 1}{\rho + 1} \tag{12-11}$$

The next step is obtaining θ. Call the location of the point at which a voltage minimum occurs, d_m. From Equation 12-9, we get

$$\frac{\theta}{2} = \beta\, d_m + (n + \tfrac{1}{2})\pi \tag{12-12}$$

from which

$$\tan\frac{\theta}{2} = -\cot \beta d_m = -\cot\left(\frac{2\pi d_m}{\lambda}\right) \tag{12-13}$$

The wavelength is obtained from the generator frequency, if known, or it may be measured directly, using the fact that two successive voltage minima are separated by a half-wavelength.

θ is thus obtained from a determination of d_m and, together with ρ, it is used in Equation 12-4 to obtain z_s:

$$z_s = z_k \frac{\rho \csc^2 (2\pi d_m/\lambda) - j(\rho^2 - 1)\cot(2\pi d_m/\lambda)}{1 + \rho^2 \cot^2 (2\pi d_m/\lambda)} \tag{12-14}$$

This is the relation for obtaining the terminating impedance from measurements taken with a slotted line probe. The reader will learn in the laboratory that certain additional operations are introduced to refine the precision of measurement, but they contribute little to the illustration of the theory and will not be considered here. It may well occur to the reader that the numerical computations involved in Equation 12-14 can be burdensome if many impedances are to be determined. To reduce these computations, several ingenious charts have been devised upon which, with a ruler and compass, the impedance can be quickly located in terms of the data. At least one of these charts is also available in the form of a circular slide rule.*

* P. A. Smith, "Transmission Line Calculator," *Electronics*, 1944. The slide rule is manufactured by the Emeloid Company, Hillside, N.J.

13. Some Concluding Remarks

Transmission line theory has been developed in terms of the voltages and the currents on the conductors because such a development follows naturally after a study of a-c circuit theory. But transmission line behavior can also be described in terms of the electric and the magnetic fields in the space surrounding the conductors, since these fields are intimately associated with the currents and voltages on the conductors. A description of the properties of a transmission line in terms of these fields is a more elegant and a more general description, and shows some features that are not evident in the circuit picture of currents and voltages. Our analysis in this chapter nowhere indicated that there are circumstances in which waves of a more complicated nature may exist on a line, whereas such waves are predicted in the more fundamental field analysis developed more fully in Chapter 12. In practical applications, however, such higher order waves usually interfere with the desired functioning of transmission lines, and systems employing transmission lines are ordinarily designed to avoid the appearance of these waves. The simplest wave, as described in this chapter is readily handled in terms of circuit language, with which most physicists and engineers have an easy facility.

Discussion Question

1. The statement following Equation 4-10, suggesting that this equation may be applied to any kind of four-terminal network, may seem too broad inasmuch as Equation 4-10 was derived for the special case of a simple T-section filter. Discuss possible justifications for this generalization.

Problems

1. Locate the cut-off frequencies in terms of the circuit elements in the three types of simple filters described at the end of Section 3.

2. Obtain Equation 4-7 from Equation 4-6.

3. Show that the three expressions for z_k given in Equation 4-2 are indeed equivalent.

4. Show that in an ideal filter the characteristic impedance is a pure resistance in a pass band and a pure reactance in a stop band.

5. Draw a graph similar to Figure 9 for the reactance of an open-circuited ideal transmission line as a function of βd.

6. Show that the differential equation

$$\frac{\partial^2 v}{\partial x^2} = M^2 \frac{\partial^2 v}{\partial t^2}$$

has a solution in the form:

$$v = Af(p) + Bg(q)$$

where $p = x - ut$ and $q = x + ut$, u being a constant that has the dimensions of velocity. Show further that $f(p)$ represents a pattern which moves along the x-axis in the positive direction as time increases, and that $g(q)$ is a pattern which moves along the x-axis in the negative direction as time increases, and that M has the dimensions of reciprocal velocity. Functions of the type f or g are called travelling wave functions.

Show further that

$$\frac{\partial^2 v}{\partial x^2} = -M^2\omega^2 v$$

is a special case of the original differential equation when the time dependence is in the form $e^{j\omega t}$.

7. A three-section band-pass filter is constructed from ideal coils of 0.1 henry inductance and ideal capacitors of 0.76 μF. A generator develops 100 volts at the input terminals at a frequency of 1000 Hz and a 200-ohm pure resistor is connected to the output terminals. What is the magnitude of the current at the load, the magnitude of the voltage at the load, the phase of the output voltage relative to the input voltage, and the phase of the output current relative to the input current? What power is dissipated in the resistor and how does this compare with the power at the input terminals?

DISCUSSION. How is it that, although there is no attenuation in the filter of Problem 7, the output and the input currents are not the same in magnitude? What would be the situation if the load resistance equalled the characteristic impedance of the filter?

8. In the text, the constants of integration of the transmission line equations are evaluated in terms of the current and the voltage at the end of the line. Evaluate them in terms of the generator voltage, generator impedance, and current at the input to the line.

9. Show, in Equations 7-2, that when the relation $R/L = G/C$ holds, that $\beta = \omega\sqrt{LC}$. Also find α for this case. Similarly, when $R^2 \ll \omega^2 L^2$ and $G \cong 0$, show that $\beta \cong \omega\sqrt{LC}$ and that $\alpha \cong \sqrt{R^2C/4L}$.

10. Compute the inductance per unit length, the capacitance per unit length, and hence the characteristic impedance of an ideal parallel wire transmission line for which the separation of the wires is d and the radius of each wire is a. Make the assumption that $a \ll d$. What will be the velocity of propagation on this line?

11. Compute the inductance per unit length, the capacitance per unit length, and hence the characteristic impedance of an ideal coaxial transmission line for which the inner conductor has a radius a and the outer conductor has an inside radius b. Assume the currents are essentially surface currents on the conductors. What will be the velocity of propagation on this line?

12. In each of the following cases decide whether the characteristic impedance of the transmission line is pure resistive, pure inductive, pure capacitative, complex inductive, or complex capacitative.

(*a*) An ideal transmission line, with perfect insulation and no resistance in the conductors.

(*b*) A transmission line with finite resistance in the conductors, but with perfect insulation.

(*c*) A transmission line with zero resistance in the conductors, but with imperfect insulation.

13. An ideal coaxial line has an inner conductor of diameter 1.0 cm and an outer conductor of inner diameter 2.3 cm. The dielectric is air. At the input is a generator which produces a voltage at the input terminals of 50 volts at a frequency of 300 megaHz. As a load there is a pure resistance of 100 ohms, located 0.15 meter from the input. What is the magnitude of the current and the voltage in the load, the phase difference of the voltage from the input voltage, and the phase difference of the current from the input current? What power is received by the line from the generator and what power is received from the line by the load?

14. An ideal transmission line of characteristic impedance 70 ohms has a load whose impedance is $80 + 20j$ ohms. What is the amplitude of the reflected traveling wave of voltage compared to the forward traveling wave? What is the voltage standing wave ratio?

15. An ideal transmission line has a characteristic impedance z_k. It is desired that power shall be supplied to a load resistance R_s without reflection. Another ideal transmission line is obtained, whose length is made equal to one-quarter of a wavelength at the frequency to be employed. This quarter-wave line is inserted between R_s and the terminus of the original line. Show that, if the characteristic impedance of the quarter-wave line is $z_c^2 = z_k R_s$, there is no reflected wave in the original line. (There are, however, reflections within the quarter-wave element.)

16. A slotted line is supplied with power at 500 megaHz. The location of the first voltage minima from four different load impedances was found to be (measured from the load): (*a*) 22 cm; (*b*) 30 cm; (*c*) 16 cm; (*d*) 5.0 cm. Decide in each case whether the load impedance is capacitative or inductive in nature. The answer can be obtained from a study of the graph of Figure 9.

17. The following data were taken on a slotted line ($z_k = 50$ ohms) terminated with an unknown impedance: relative amplitude of a voltage maximum, 80, located 25 cm from the load; relative amplitude of a voltage minimum, 25, located 15 cm from the load; locations of other minima (of the same relative amplitude) at 35 cm, 55 cm, and 75 cm from the load. Find the value of the terminating impedance.

18. Derive relations similar to Equations 12-5 to 12-9 inclusive for the current on a transmission line.

19. An ideal transmission line has a characteristic impedance z_k. It is terminated in a complex impedance. Express the ratio of the voltage to the current at a point on the line where there is a voltage minimum in terms of the voltage standing wave ratio. In terms of the ratio, what is the impedance at a voltage maximum?

20. An ideal transmission line of 50 ohms characteristic impedance is terminated with a resistance of 35 ohms and is being used at a frequency of 500 megaHz. It is desired that there be no reflected wave returning to the gener-

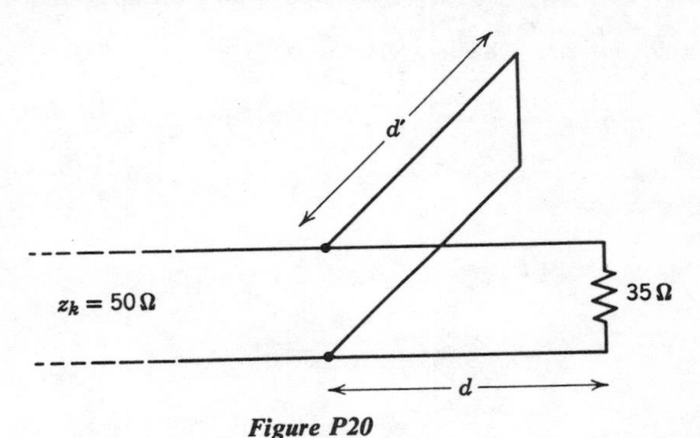

Figure P20

ator. To do this a length of the same kind of transmission line is attached in parallel to the main line at a point *d* meters from the load. The second line is terminated in a short, the length from the point of attachment to the short being *d'* meters. Determine *d* and *d'* in order that there be no reflected wave.

21. If a transmission line is regarded as a sequence of infinitesimal filter sections in which the series elements have an impedance $z\,dx$ and the shunt elements an impedance $1/(y\,dx)$, show that the expression for the characteristic impedance of a filter

$$z_k = \sqrt{z_1^2/4 + z_1 z_2}$$

reduces to the expression for a transmission line

$$z_k = \sqrt{z/y}$$

22. An ideal transmission line whose characteristic impedance is z_1 feeds into another transmission line whose characteristic impedance is z_2. The second line is either infinitely long or is terminated in a load equal to z_2. Show that in the first transmission line

$$\frac{v_r}{v_f} = \frac{z_2 - z_1}{z_2 + z_1} \qquad \frac{i_r}{i_f} = -\frac{z_2 - z_1}{z_2 + z_1}$$

$$\frac{v_t}{v_f} = \frac{2z_2}{z_2 + z_1} \qquad \frac{i_t}{i_f} = \frac{2z_1}{z_2 + z_1} = \frac{z_1}{z_2}\frac{2z_2}{z_2 + z_1}$$

where the subscript *f* applies to the forward wave in the first line; *r* to the reflected wave; and *t* to the wave in the second line.

REFERENCES

1. Michels, *Electrical Measurements and Their Applications* (Van Nostrand), 1957. Chapters 8 and 9.
2. Ryder, *Networks, Lines, and Fields* (Prentice-Hall), 1955. Chapters 4, 5, 6, 7, 8.
3. Ramo and Whinnery, *Fields and Waves In Modern Radio* (Wiley), 2nd ed., 1953. Chapters 1 and 9.
4. Page and Adams, *Principles of Electricity* (Van Nostrand), 2nd ed., 1949. Chapter XV.
5. Harnwell, *op. cit.*, Chapter XIV.

10

Maxwell's field equations

Relations among the electromagnetic field variables based on Coulomb's law, Ampère's law, Faraday's law, and Ohm's law were summarized at the end of Chapter 7. The developments of this chapter will be primarily based on the four integral relations:

(*A*) $$\int \mathbf{D} \cdot d\mathscr{A} = \int \rho \, d\mathscr{V}$$

(*B*) $$\int \mathbf{B} \cdot d\mathscr{A} = 0$$

(*C*) $$\oint \mathbf{E} \cdot d\mathbf{l} = -\frac{d}{dt}\int \mathbf{B} \cdot d\mathscr{A}$$

(*D*) $$\oint \mathbf{H} \cdot d\mathbf{l} = \int \mathbf{J} \cdot d\mathscr{A}$$

and their point or differential counterparts:

(*a*) $$\nabla \cdot \mathbf{D} = \rho$$

(*b*) $$\nabla \cdot \mathbf{B} = 0$$

(*c*) $$\nabla \times \mathbf{E} = -\dot{\mathbf{B}}$$

(*d*) $$\nabla \times \mathbf{H} = \mathbf{J}$$

where $\mathbf{J}$ is the true current density and ρ is the free charge density. The defining relations for **D** and **H** are

$$(e) \qquad \mathbf{D} = \epsilon_0 \mathbf{E} + \mathbf{P}$$

$$(f) \qquad \mathbf{H} = \frac{\mathbf{B}}{\mu_0} - \mathbf{M}$$

and inasmuch as we shall very frequently restrict our attention to linear, homogeneous, isotropic media, we shall reduce (*e*) and (*f*) to the simpler forms:

$$(e') \qquad \mathbf{D} = \epsilon_0 \kappa \mathbf{E} = \epsilon \mathbf{E}$$

$$(f') \qquad \mathbf{B} = \mu_0 \kappa_m \mathbf{H} = \mu \mathbf{H}$$

But before we go into the implications of these equations, it is first necessary to identify and correct an inconsistency which primarily involves (*D*) and (*d*).

1. Maxwell's Modification of Ampère's Circuital Law

We saw in Chapter 5 that the integral form of Ampère's circuital law must hold for all surfaces whose perimeter is the closed line integral on the left of (*D*). The finding of one such surface for which the equality does not hold is sufficient to show that the relation is either entirely wrong or at least incomplete. In Figure 1 is shown a simple alternating current circuit consisting of a parallel plate capacitor across the terminals of a generator. Let us choose an instant of time while the capacitor is either charging or discharging, so that there is a true current in the connecting wires. From (*D*) the line integral of **H** around the circular path is equal to the integral of **J** over the surface *A* or *B* or *C*. The integral of **J** over either *A* or *B* is the total current in the wire; but the same integral evaluated over the surface *C*, which passes between the capacitor plates, is zero.

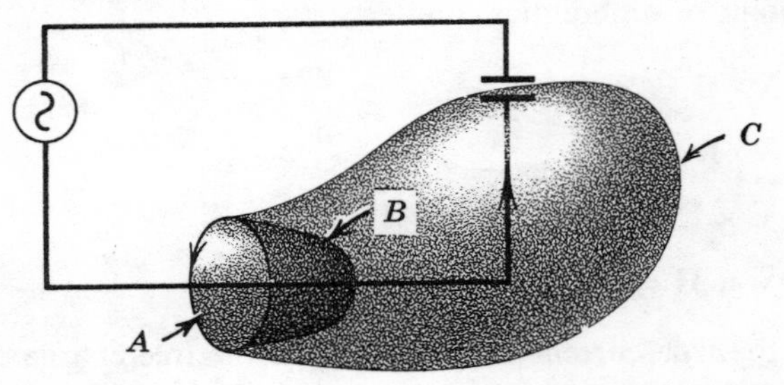

Figure 1. The integrals of **J** *over the surfaces B and C are not the same.*

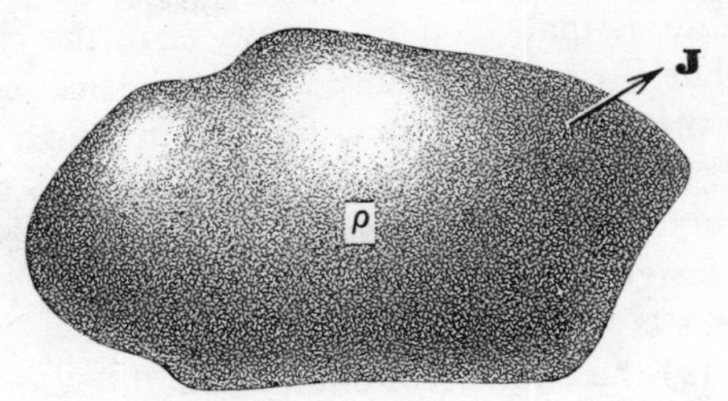

***Figure 2.** A rate of change of total charge within the volume is associated with a current density at the surface.*

This difficulty with the circuital law is brought out in a more general way, without reference to a special circuit, by taking the divergence of (*d*):

$$\nabla \cdot \nabla \times \mathbf{H} = \nabla \cdot \mathbf{J} = 0 \tag{1-1}$$

The conclusion from (*d*) and Equation 1-1 is that the divergence of **J** is always zero, a statement that we can readily show is not true. For let us imagine a closed surface of arbitrary shape in a region containing charges of density ρ (Figure 2). If these charges are moving, there is also a current density **J** in this region. The charge which flows through an element $d\mathscr{A}$ of the surface per second is $\mathbf{J} \cdot d\mathscr{A}$ and the total charge per second flowing out of this volume is $\int \mathbf{J} \cdot d\mathscr{A}$, the integral being taken over the closed surface. On the other hand the total charge within the volume at any instant is $\int \rho \, d\mathscr{V}$, whose rate of change within the volume is $d/dt \int \rho \, d\mathscr{V}$. If net electric charge can be neither created nor destroyed, then the rate of flow of charge out of the surface must be equal to the negative of the rate of change of charge within the surface, and it follows that;

$$\int \mathbf{J} \cdot d\mathscr{A} = -\frac{\partial}{\partial t} \int \rho \, d\mathscr{V} \tag{1-2}$$

By Gauss's theorem, the left side is converted to a volume integral and, on the right, the time derivative is put under the integral sign, provided the surface does not change shape or size with time. Thus

$$\int \nabla \cdot \mathbf{J} \, d\mathscr{V} = -\int \frac{\partial \rho}{\partial t} \, d\mathscr{V} \tag{1-3}$$

Since 1-3 is true for an arbitrary volume, it follows that the integrands are equal:

$$\nabla \cdot \mathbf{J} = -\frac{\partial \rho}{\partial t} \tag{1-4}$$

This is the charge continuity equation.

Whereas (*d*) requires that $\nabla \cdot \mathbf{J}$ be always zero, the charge continuity equation shows that $\nabla \cdot \mathbf{J}$ is zero only if charge densities do not change with time. Since there certainly are situations in which charge densities change with time, then either Ampère's law is incorrect as it stands or we must not assume that net charge cannot be created or destroyed. Maxwell had the insight to recognize this inconsistency and he proposed modifying Ampère's law.

With Equation (*a*) the charge continuity equation can be rewritten as

$$\nabla \cdot \mathbf{J} = -\frac{\partial}{\partial t}(\nabla \cdot \mathbf{D}) = -\nabla \cdot \dot{\mathbf{D}} \tag{1-5}$$

which, upon regrouping, becomes

$$\nabla \cdot (\mathbf{J} + \dot{\mathbf{D}}) = 0 \tag{1-6}$$

Thus it is the vector $(\mathbf{J} + \dot{\mathbf{D}})$, rather than $\mathbf{J}$ itself whose divergence is zero, indicating that the right side of Ampère's circuital law be modified by adding the term $\dot{\mathbf{D}}$:

$$\nabla \times \mathbf{H} = \mathbf{J} + \dot{\mathbf{D}} \tag{1-7}$$

Equation 1-7 is now consistent with the continuity of charge, and it is also valid, as the original circuital law was, for steady-state conditions. The term $\dot{\mathbf{D}}$ has the dimensions of a current density and is called the Displacement Current Density, often simply (and carelessly) the displacement current.

This modification of Ampère's circuital law has been motivated essentially by a desire for mutual consistency among the electromagnetic relations. The choice of the form of the correction was guided by an attachment to the law of conservation of charge which is comparable in its urgency to our attachment to the law of conservation of energy. Both laws are inductive laws, and their validity and that of arguments based on them depends on whether the predictions arising from their use can be verified. Most of the remainder of this book will be devoted to the development of these predictions.

2. The Wave Equations

The four fundamental differential equations now stand as:

(*a*) $\nabla \cdot \mathbf{D} = \rho$

(*b*) $\nabla \cdot \mathbf{B} = 0$

(*c*) $\nabla \times \mathbf{E} = -\dot{\mathbf{B}}$

(*d′*) $\nabla \times \mathbf{H} = \mathbf{J} + \dot{\mathbf{D}}$

All involve first derivatives. Several kinds of operations on these equations lead to interesting and informative relations. One in particular includes differentiating and intersubstituting to obtain second-order differential equations in the separated field variables, solutions to which turn out to have the form of traveling wave functions.

As a first step we take the curl of (c):

$$\text{(2-1)}\qquad \nabla \times (\nabla \times \mathbf{E}) = -\nabla \times \dot{\mathbf{B}} = -\frac{\partial}{\partial t}(\nabla \times \mathbf{B})$$

For a medium in which the relations $\mathbf{B} = \mu\mathbf{H}$ and $\mathbf{D} = \epsilon\mathbf{E}$ are valid, Equation (d') can be used to eliminate $\mathbf{H}$ from 2-1:

$$\text{(2-2)}\qquad \nabla \times (\nabla \times \mathbf{E}) = -\frac{\partial}{\partial t}(\mu\nabla \times \mathbf{H}) = -\mu\dot{\mathbf{J}} - \mu\epsilon\ddot{\mathbf{E}}$$

The further restriction is made that the medium in which these fields exist is uncharged and nonconducting, thus eliminating the term in $\mathbf{J}$. With the identity:

$$\text{(2-3)}\qquad \nabla \times (\nabla \times \mathbf{E}) = \nabla(\nabla \cdot \mathbf{E}) - \nabla \cdot (\nabla\mathbf{E})$$

Equation 2-2 becomes

$$\text{(2-4)}\qquad \nabla(\nabla \cdot \mathbf{E}) - \nabla \cdot (\nabla\mathbf{E}) = -\mu\epsilon\ddot{\mathbf{E}}$$

Since in such a medium the divergence of $\mathbf{E}$ is zero, the first term on the left goes out, and we are left with

$$\text{(2-5)}\qquad \nabla^2\mathbf{E} = \mu\epsilon\ddot{\mathbf{E}}$$

In Cartesian coordinates, and only in Cartesian coordinates, this vector equation separates into the three scalar differential equations for the three components of $\mathbf{E}$:

$$\text{(2-6}a\text{)}\qquad \nabla^2 E_x = \mu\epsilon\ddot{E}_x$$

$$\text{(2-6}b\text{)}\qquad \nabla^2 E_y = \mu\epsilon\ddot{E}_y$$

$$\text{(2-6}c\text{)}\qquad \nabla^2 E_z = \mu\epsilon\ddot{E}_z$$

These are wave equations (see Chapter 9, Problem 6), with solutions which are functions of the space and time variables in combinations such as $(r - vt)$ or $(r + vt)$. The velocity of propagation of the traveling waves is

$$\text{(2-7)}\qquad v = \frac{1}{\sqrt{\mu\epsilon}}$$

which for free space, where $\mu = \mu_0$ and $\epsilon = \epsilon_0$, has a value $\approx 3 \times 10^8$ meters per second, the velocity of light. This is the first of several indications that light is electromagnetic in nature.

By performing comparable operations on Ampère's circuital law (d'), we arrive at a similar wave equation for **H**:

$$\nabla^2 \mathbf{H} = \mu\epsilon\ddot{\mathbf{H}} \tag{2-8}$$

which, in Cartesian coordinates, reduces to

$$\nabla^2 H_x = \mu\epsilon\ddot{H}_x \tag{2-9a}$$

$$\nabla^2 H_y = \mu\epsilon\ddot{H}_y \tag{2-9b}$$

$$\nabla^2 H_z = \mu\epsilon\ddot{H}_z \tag{2-9c}$$

Equations 2-5 and 2-8 are valid with any type of coordinate system, Cartesian or curvilinear, but the terms of the Laplacian are not separately functions of single components of the field for systems other than Cartesian.

Maxwell's equations require that the spacial and temporal variations of an electric field or a magnetic field must always be so related as to be expressible in the form of a traveling wave. Moreover, the interrelation of the electric field and the magnetic field through Ampère's and Faraday's laws insure that where there is a traveling electric field wave, it is always associated with a traveling magnetic field wave.

3. Plane Waves

One particularly simple form of traveling wave is the plane wave, in which space variations occur with respect to only one variable. Like many other concepts in physics, this is an idealization which cannot be completely realized physically, but which simplifies the mathematical manipulation and provides valuable physical insight as well as a close approximation to many practical applications.

By starting only with the assumption that an electric field exists which is a function only of x and t, we can show that a traveling electromagnetic plane wave is a necessary consequence. The assumed electric field, which we write $\mathbf{E}(x, t)$ has, in general, three components: $E_x(x, t)$, $E_y(x, t)$ and $E_z(x, t)$. At a given instant over any plane parallel to the y-z plane, each of these components is constant.

We assume that the medium is homogeneous, isotropic, that there are no free charges, that the conductivity of the medium is zero, and that the relations $\mathbf{D} = \epsilon\mathbf{E}$ and $\mathbf{B} = \mu\mathbf{H}$ hold. The field equations then take the form:

$$\nabla \cdot \mathbf{E} = 0 \tag{3-1a}$$

$$\nabla \cdot \mathbf{H} = 0 \tag{3-1b}$$

$$\nabla \times \mathbf{E} = -\mu\dot{\mathbf{H}} \tag{3-1c}$$

$$\nabla \times \mathbf{H} = \epsilon\dot{\mathbf{E}} \tag{3-1d}$$

As was shown in Section 2, **E** and **H** must each satisfy the wave equation

$$\nabla^2\mathbf{E} = \mu\epsilon\ddot{\mathbf{E}} \tag{3-2a}$$

$$\nabla^2\mathbf{H} = \mu\epsilon\ddot{\mathbf{H}} \tag{3-2b}$$

Since $\partial E_y/\partial y$ and $\partial E_z/\partial z$ are zero, it follows from Equation 3-1*a* that $\partial E_x/\partial x$ is also zero. It therefore follows from 3-2*a* that

$$\frac{\partial^2 E_x}{\partial t^2} = 0 \quad \text{and} \quad E_x = a + bt \tag{3-3}$$

where a and b are constants. The x component of the electric field can at most be a linear function of the time. Also from 3-2*a*

$$\frac{\partial^2 E_y}{\partial x^2} = \mu\epsilon \frac{\partial^2 E_y}{\partial t^2} \tag{3-4a}$$

$$\frac{\partial^2 E_z}{\partial x^2} = \mu\epsilon \frac{\partial^2 E_z}{\partial t^2} \tag{3-4b}$$

showing that E_y and E_z must each be a traveling wave function such as:

$$E_y = E_{y0} f(x \pm vt) \tag{3-5a}$$

$$E_z = E_{z0} g(x \pm vt) \tag{3-5b}$$

where E_{y0} and E_{z0} are constants, independent of x and t.

E and **H** are related through the curl equations. From Equation 3-1*c* we obtain the time derivatives of **H**:

$$\frac{\partial H_x}{\partial t} = 0 \tag{3-6a}$$

$$\frac{\partial H_y}{\partial t} = \frac{1}{\mu}\frac{\partial E_z}{\partial x} \tag{3-6b}$$

$$\frac{\partial H_z}{\partial t} = -\frac{1}{\mu}\frac{\partial E_y}{\partial x} \tag{3-6c}$$

Then with 3-5, after integration with respect to time, the magnetic field components are found to be:

(3-7*a*) $$H_x = C_x(x, y, z)$$

(3-7*b*) $$H_y = \pm \frac{E_{z0}}{\mu v} g(x \pm vt) + C_y(x, y, z)$$

(3-7*c*) $$H_z = \mp \frac{E_{y0}}{\mu v} f(x \pm vt) + C_z(x, y, z)$$

where C_x, C_y and C_z are constants of integration which may be functions of position but not of time. Thus the magnetic field consists of two parts—a time dependent part with only y and z components and a time independent part given by $\mathbf{i}C_x + \mathbf{j}C_y + \mathbf{k}C_z$. This time independent part may be any constant field in the laboratory, such as the earth's field, or it may be zero. At any rate, from 3-1*b* and 3-2*b* it is readily shown that its Laplacian and its divergence are both zero, and its curl has only an x component which is ϵb from Equation 3-3.

Our interest is in the time dependent part:

(3-8) $$\mathbf{H} = \pm \mathbf{j} \frac{E_{z0}}{\mu v} g(x \pm vt) \mp \mathbf{k} \frac{E_{y0}}{\mu v} f(x \pm vt)$$

together with the electric field associated with it:

(3-9) $$\mathbf{E} = \mathbf{j}E_{y0} f(x \pm vt) + \mathbf{k}E_{z0} g(x \pm vt)$$

Evidently the y component of $\mathbf{H}$ depends on the z component of $\mathbf{E}$, and the z component of $\mathbf{H}$ depends on the y component of $\mathbf{E}$, according to

(3-10*a*) $$E_y = \mp\sqrt{\mu/\epsilon}\, H_z$$

(3-10*b*) $$E_z = \pm\sqrt{\mu/\epsilon}\, H_y$$

where the upper sign goes with $(x + vt)$ and the lower sign with $(x - vt)$, the reverse and forward directions of propagation, respectively, of the traveling wave. There is no relation between E_y and H_y, nor between E_z and H_z. There are thus two independent electromagnetic waves (two planes of polarization, in optical language), in each of which the electric field, the magnetic field, and the direction of propagation are mutually perpendicular. The two waves are quite independent of one another, except for such relationship as might have been established at their source.

The ratio of the electric to the magnetic field in a traveling plane wave is, from Equation 3-10, a constant determined by the medium and has units of (volts per meter)/(amperes per meter) = ohms. It is the characteristic impedance of the medium, and we shall see in the next chapter

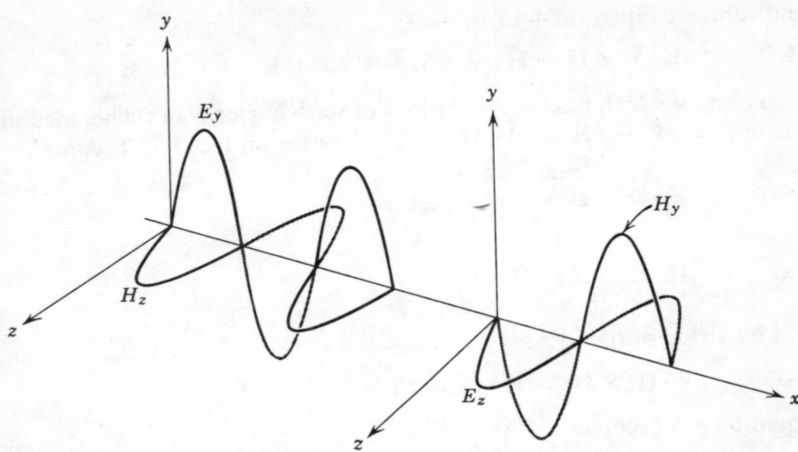

Figure 3. The two polarizations of a plane wave traveling in the positive x-direction.

that there is a close analogy between it and the impedance, z_k, of a transmission line.

The medium that has been assumed in this derivation has several formal similarities to an ideal transmission line. There is no dissipation of energy in it (it is uncharged and nonconducting); the velocity of propagation of a wave is independent of frequency. Thus, whatever the form of the functions f and g, they are propagated without distortion. Later we shall encounter media in which the characteristic impedance is a function of the frequency. In such a medium it is necessary to deal separately with the individual components of a complex wave form.

4. The Poynting Vector

We saw in Chapters 2 and 6 that energy density in the field is a function of the field intensity, and we should expect that out of the differential field equations we can obtain an energy relation consistent with the law of conservation of energy.

Starting with Ampère's circuital law and Faraday's law:

$$\nabla \times \mathbf{H} = \mathbf{J} + \dot{\mathbf{D}} \tag{4-1}$$

$$\nabla \times \mathbf{E} = -\dot{\mathbf{B}} \tag{4-2}$$

we perform a scalar multiplication of the first by $\mathbf{E}$ and of the second by $\mathbf{H}$:

$$\mathbf{E} \cdot \nabla \times \mathbf{H} = \mathbf{E} \cdot \mathbf{J} + \mathbf{E} \cdot \dot{\mathbf{D}} \tag{4-3}$$

$$\mathbf{H} \cdot \nabla \times \mathbf{E} = -\mathbf{H} \cdot \dot{\mathbf{B}} \tag{4-4}$$

and subtract Equation 4-4 from 4-3:

$$\mathbf{E} \cdot \nabla \times \mathbf{H} - \mathbf{H} \cdot \nabla \times \mathbf{E} = \mathbf{E} \cdot \mathbf{J} + \mathbf{E} \cdot \dot{\mathbf{D}} + \mathbf{H} \cdot \dot{\mathbf{B}} \tag{4-5}$$

each term of which has the dimensions of watts/meter^3. In such a medium where $\mathbf{D} = \epsilon\mathbf{E}$ and $\mathbf{B} = \mu\mathbf{H}$, the last two terms on the right become

$$\mathbf{E} \cdot \dot{\mathbf{D}} = \epsilon\mathbf{E} \cdot \dot{\mathbf{E}} = \frac{d}{dt}(\tfrac{1}{2}\epsilon E^2) \tag{4-6}$$

and

$$\mathbf{H} \cdot \dot{\mathbf{B}} = \mu\mathbf{H} \cdot \dot{\mathbf{H}} = \frac{d}{dt}(\tfrac{1}{2}\mu H^2)$$

and by virtue of the identity:

$$\nabla \cdot (\mathbf{E} \times \mathbf{H}) = \mathbf{H} \cdot (\nabla \times \mathbf{E}) - \mathbf{E} \cdot \nabla \times \mathbf{H} \tag{4-7}$$

Equation 4-5 becomes

$$-\nabla \cdot (\mathbf{E} \times \mathbf{H}) = \mathbf{E} \cdot \mathbf{J} + \frac{d}{dt}(\tfrac{1}{2}\epsilon E^2 + \tfrac{1}{2}\mu H^2) \tag{4-8}$$

The fact that this is a statement of the conservation of energy is more clearly brought out by integrating this point relation over a finite volume. The integral of the divergence on the left is changed, by Gauss's theorem, to an integral of the vector $\mathbf{E} \times \mathbf{H}$ over the surface, bounding the volume.

$$-\int (\mathbf{E} \times \mathbf{H}) \cdot d\mathscr{A} = \int \mathbf{E} \cdot \mathbf{J}\, d\mathscr{V} + \frac{d}{dt}\int \left(\tfrac{1}{2}\epsilon E^2 + \tfrac{1}{2}\mu H^2\right) d\mathscr{V} \tag{4-9}$$

Each term in Equation 4-9 is a rate of change of energy associated with the volume, and the whole equation expresses an energy balance. The first term on the right is the energy associated with the motion of free charges within the volume; the second term is the rate of change of the total energy in the fields within the volume. If power is being supplied to put charges in motion and if the energy of the fields is increasing with time, such power must be supplied to the volume through the bounding surface, a power given by the negative of the surface integral on the left. The surface integral itself is the rate at which energy is passing outward through the bounding surface, and it follows as a logical interpretation that $(\mathbf{E} \times \mathbf{H})$ represents the rate at which energy passes outward per unit area of this surface.

The vector $\mathbf{E} \times \mathbf{H}$ is called the Poynting vector, after John Henry Poynting, an English physicist of the Nineteenth century. It is identified by the symbol $\mathbf{S}$. The sense in which the energy flows is obtained from the right-hand rule for vector multiplication.

There will be many opportunities in which the interpretation of $\mathbf{E} \times \mathbf{H}$ as a rate of flow of energy per unit area will be profitable. In most cases of practical interest, such an interpretation is valid, although it must always be kept in mind that only the integral of **S** over a closed surface can be physically measured (cf: the physical interpretation of energy density, Chapter 2). One interesting example is a wire of length L and radius a carrying a direct current I (Figure 4). The electric field in the wire (and at its surface) is V/L, where V is the voltage applied over the length of the wire. The magnetic field at the surface of the wire is $I/2\pi a$. These two fields are mutually perpendicular. The Poynting vector is directed into the volume of the wire and has a magnitude

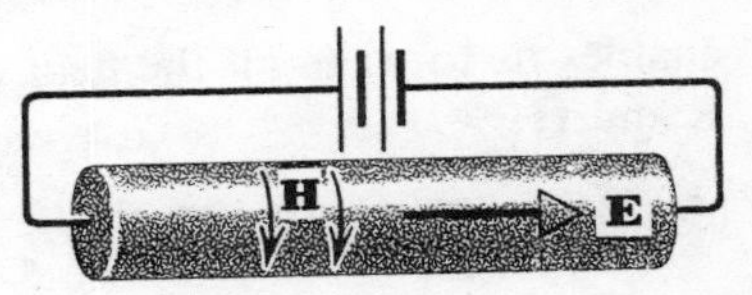

Figure 4. The electric field and the magnetic field at the surface of a wire of radius a carrying a direct current I.

$$S = \frac{V}{L}\frac{I}{2\pi a} \tag{4-10}$$

The fields are constant in time, so that Equation 4-9 becomes

$$\int \frac{V}{L}\frac{I}{2\pi a}\, d\mathscr{A} = \int \mathbf{E}\cdot\mathbf{J}\, d\mathscr{V} = VI \tag{4-11}$$

Thus the energy flowing into the wire through the sides equals the rate at which heat is developed within the wire. Just how it is that the connections to the energy source, say a battery, are at the ends of the wire, yet energy flows in through the sides, should be pondered by the reader.

That the interpretation of $\mathbf{E} \times \mathbf{H}$ as a flow of energy per unit area must be used with care is illustrated by the following example. Suppose the wire in Figure 4 is not connected to the battery, but instead carries a static charge. Suppose further that there is a static magnetic field in the vicinity, such as, say, the earth's field. At the surface of the wire the Poynting vector is in general nonzero, but we sense that the interpretation of **S** as a flow of energy is in this case strained. As noted before, the physically measurable quantity is the integral of **S** over the surface of the wire, and where **E** and **H** are unrelated this integral is, of course, zero.

5. The Scalar and Vector Potentials

In Chapter 5, Section 17, the concept of the vector potential was introduced, and in Chapter 6, Section 10, it was shown that Faraday's law, together with the vector potential enables us to write the electric field in terms of the vector and the scalar potential:

$$\mathbf{E} = -\dot{\mathbf{A}} - \nabla\phi \tag{5-1}$$

This relation together with the defining relation for **A**:

(5-2) $$\mathbf{B} = \nabla \times \mathbf{A}$$

enables us to write all the field equations in terms of **A** and ϕ instead of **E** and **H**:

(5-3*a*) $$\nabla \cdot (-\dot{\mathbf{A}} - \nabla\phi) = \frac{\rho}{\epsilon}$$

(5-3*b*) $$\nabla \cdot (\nabla \times \mathbf{A}) = 0$$

(5-3*c*) $$-\nabla \times \dot{\mathbf{A}} - \nabla \times (\nabla\phi) = -\nabla \times \dot{\mathbf{A}}$$

(5-3*d*) $$\frac{1}{\mu}\nabla \times \nabla \times \mathbf{A} = \mathbf{J} - \epsilon\ddot{\mathbf{A}} - \epsilon\nabla\dot{\phi}$$

Equations 5-3*b* and 5-3*c* are identities since they were used in obtaining the relations between **E**, **B**, **A**, and ϕ in the first place. The first and fourth can be expanded:

(5-4*a*) $$\nabla \cdot \dot{\mathbf{A}} + \nabla^2\phi = -\frac{\rho}{\epsilon}$$

(5-4*b*) $$\nabla(\nabla \cdot \mathbf{A}) - \nabla^2\mathbf{A} = \mu\mathbf{J} - \mu\epsilon\ddot{\mathbf{A}} - \mu\epsilon\nabla\dot{\phi}$$

In principle these two differential equations can be solved as they stand for **A** and ϕ in terms of ρ and **J**. But the solution is greatly simplified by noting that the expression $\mathbf{B} = \nabla \times \mathbf{A}$ does not define the function **A** unambiguously (for we can add to **A** the gradient of any scalar function, since the curl of a gradient is identically zero). However, a vector function is completely specified (except for possible additive constants) by specifying *both* its curl and its divergence over all space, provided there are no sources (charges or currents, in the case of the field functions we are dealing with) at infinity. Since only the already specified curl of **A** is needed in the field equations, we are at liberty to specify the divergence of **A** in whatever way will be most convenient for us. We shall choose the following, often called the Lorentz gauge condition:

(5-5) $$\nabla \cdot \mathbf{A} = -\mu\epsilon\dot{\phi}$$

This choice puts Equations 5-4 into the form:

(5-6*a*) $$\nabla^2\phi = -\frac{\rho}{\epsilon} + \mu\epsilon\ddot{\phi}$$

(5-6*b*) $$\nabla^2\mathbf{A} = -\mu\mathbf{J} + \mu\epsilon\ddot{\mathbf{A}}$$

resulting in separated equations in ϕ and **A**. Note that in a region in which ρ and **J** are each zero, these equations become simply wave equations for ϕ and **A**, for which the form of the solution can immediately be written down. In Cartesian coordinates the differential equation in **A** can be written as three separate scalar equations in the components of **A**. It would be surprising if we did not get wave equations for ϕ and A since they are associated with **E** and **H**, which are themselves wave functions. Under static conditions Equation 5-6*a* takes the form of Poisson's equation (Chapter 2, Section 13).

6. Solutions for the Potentials

In a rather ingenious fashion the solution for the scalar potential in Equation 5-6*a* can be found when the charge density giving rise to the potential function is defined over all space and goes to zero at infinity.

To do this, we first derive an extension of Gauss's theorem known as Green's theorem. Let p and q be two scalar functions of position and define two vector functions:

$$\mathbf{R} = p\nabla q \quad \text{and} \quad \mathbf{S} = q\nabla p \tag{6-1}$$

Gauss's theorem applied to **R** and to **S** is:

$$\int \mathbf{R} \cdot d\mathscr{A} = \int \nabla \cdot \mathbf{R}\, d\mathscr{V} \qquad \int \mathbf{S} \cdot d\mathscr{A} = \int \nabla \cdot \mathbf{S}\, d\mathscr{V} \tag{6-2}$$

We form the difference of the two equations in 6-2.

$$\int (p\nabla q - q\nabla p) \cdot d\mathscr{A} = \int [\nabla \cdot (p\nabla q) - \nabla \cdot (q\nabla p)]\, d\mathscr{V} \tag{6-3}$$

Using the vector identity

$$\nabla \cdot (s\mathbf{T}) = (\nabla s) \cdot \mathbf{T} + s(\nabla \cdot \mathbf{T}) \tag{6-4}$$

where s is a scalar and **T** a vector function, Equation 6-3 becomes

$$\begin{aligned} \int (p\nabla q - q\nabla p) \cdot d\mathscr{A} &= \int [\nabla q \cdot \nabla p + p\nabla^2 q - \nabla p \cdot \nabla q - q\nabla^2 p]\, d\mathscr{V} \\ &= \int (p\nabla^2 q - q\nabla^2 p)\, d\mathscr{V} \end{aligned} \tag{6-5}$$

which is Green's theorem. The functions p and q need only be continuous, single-valued, with continuous derivatives. By making a judicious choice for these functions, we shall be able to find a solution to Equation 5-6*a*, at least when the time dependence of the charge density goes as $e^{j\omega t}$.

Let us choose for p the space part of the potential function in 5-6*a*, where we take the charge density as going to zero at infinity and having the form

$$\rho = \rho' e^{j\omega t} \tag{6-6}$$

where ρ' is that part of the charge density function that depends only on space variables. This is, of course, a special case where the time variation in all parts of space is the same. From 5-6*a* it is seen that the potential ϕ can then be written as

$$\phi = \phi' e^{j\omega t} \tag{6-7}$$

where ϕ' is the space dependent part of the potential function.

For the function q, let us choose

$$q = \frac{e^{-jmr}}{r} \tag{6-8}$$

where m is a constant. Then

$$\nabla q = -\frac{\mathbf{r}}{r^3} e^{-jmr}(jmr + 1) \tag{6-9}$$

$$\nabla^2 q = -m^2 \frac{e^{-jmr}}{r} \tag{6-10}$$

and Equation 6-5 becomes

$$\int e^{-jmr}\left[\frac{\phi'\mathbf{r}}{r^3} + \frac{jm\phi'\mathbf{r}}{r^2} + \frac{\nabla\phi'}{r}\right] \cdot d\mathscr{A} = \int \frac{e^{-jmr}}{r} [m^2\phi' + \nabla^2\phi'] \, d\mathscr{V} \tag{6-11}$$

We seek a value for ϕ' at the point P, which we shall take as the origin of coordinates. Because of the denominators, the integrands have singularities at the origin. We therefore take as the volume of integration all space except that within a sphere of radius a about the origin. The surface of integration on the left is the surface at infinity and the surface of the sphere of radius a (whose normal points inward toward P).

Over the surface at infinity the area integral on the left of 6-11 is zero, and over the surface of the sphere of radius a it becomes:

$$-e^{-jma}\left\{\frac{1}{a^2}\int \phi' \, d\mathscr{A} + \frac{jm}{a}\int \phi' d\mathscr{A} + \frac{1}{a}\int \frac{\partial \phi'}{\partial r} \, d\mathscr{A}\right\} \tag{6-12}$$

We now make use of the fact that the average value of ϕ' over the sphere of radius a is

$$\overline{\phi'} = \frac{1}{4\pi a^2}\int \phi' \, d\mathscr{A} \tag{6-13}$$

with similar reasoning for $\partial\phi'/\partial r$ (cf: Chapter 2, Problem 22). Thus 6-12 becomes

$$-e^{-jma}\left\{\frac{4\pi a^2\overline{\phi'}}{a^2}+\frac{jm4\pi a^2\overline{\phi'}}{a}+\frac{4\pi a^2}{a}\overline{\frac{\partial\phi'}{\partial r}}\right\} \tag{6-14}$$

As we let the radius a approach zero, $\overline{\phi'}\rightarrow\phi_P'$, the value of ϕ' at P, and 6-14 approaches $-4\pi\phi_P'$, so that 6-11 is:

$$-4\pi\phi_P'=\int\frac{e^{-jmr}}{r}(m^2\phi'+\nabla^2\phi')\,d\mathscr{V} \tag{6-15}$$

Now we return to the right side of 6-11. Putting 6-6 and 6-7 into 5-6a, we see that

$$\nabla^2\phi'+\omega^2\mu\epsilon\phi'=-\frac{\rho'}{\epsilon} \tag{6-16}$$

Let $m^2=\omega^2\mu\epsilon$, so that Equation 6-15 becomes

$$-4\pi\phi_P'=\int\frac{e^{-jmr}}{r}\left(-\frac{\rho'}{\epsilon}\right)d\mathscr{V} \tag{6-17}$$

and reintroducing the time dependence, we get finally

$$\phi_P=\frac{1}{4\pi\epsilon}\int\frac{\rho' e^{j\omega(t-r\sqrt{\mu\epsilon})}}{r}\,d\mathscr{V} \tag{6-18}$$

which is a particular solution for Equation 5-6a when the charge density is defined over all space and goes to zero at infinity. It should be noted that 6-18 reduces to Equation 4-4 of Chapter 2 under static conditions ($\omega=0$). The derivation may be extended to conditions in which the charge density is defined only within a finite volume with potentials and potential gradients defined over the surface of this volume. In this case the surface integral is not over a surface at infinity, but is taken over the surface of the finite volume. We shall not develop this derivation further, however, since the implications of 6-18 are adequate for our present purposes.

Equation 5-6a is a differential equation for the scalar function ϕ. In Cartesian coordinates Equation 5-6b separates into three scalar equations for the three components of the vector potential **A**, each of which has the same form as 5-6a, and hence their solutions can be written down immediately by analogy:

$$A_x=\frac{\mu}{4\pi}\int\frac{J_x' e^{j\omega(t-r\sqrt{\mu\epsilon})}}{r}\,d\mathscr{V} \tag{6-19}$$

with similar expressions for the y and z components. Here J_x' is the space dependent part of the current density.

The physical picture which Equations 6-18 and 6-19 are describing can best be seen by first returning to the static case, Equation 4-4 and Figure 5 of Chapter 2:

$$\phi_P = V_P = \frac{1}{4\pi\epsilon_0}\int \frac{\rho\, d\mathscr{V}}{r} \tag{6-20}$$

The potential at the point P is the sum of the contributions from charges in the volume elements $d\mathscr{V}$ in the space surrounding the point P. But let us suppose that the charge density in the volume element changes in time. The electric field associated with the charge in $d\mathscr{V}$ changes and consequently the potential at P changes, but this change is not instantaneously observed at P because, as we have seen, an electric field change travels with a finite velocity given by $v = 1/\sqrt{\mu\epsilon}$. The potential observed at P at the time t is that due to the charge density that existed in $d\mathscr{V}$ at an earlier time, $t - r/v$, where r/v is the time required for the change to travel from $d\mathscr{V}$ to P. Therefore, the integrals in Equations 6-18 and 6-19 are over the charge and current densities existing at the appropriate earlier times. Such potentials are called "retarded potentials." If the charge density, for instance, is a function of time and space, $\rho(x, y, z, t)$, the integral in 6-20 is to be computed by using this function with the modified time, $t - r/v$, inserted wherever t appears.

To compute the electric and the magnetic field when these fields are time dependent, it is often easier to find A and ϕ and then to use Equations 5-1 and 5-2. But because of the Lorentz gauge condition of Equation 5-5, it is not necessary to determine both $\mathbf{A}$ and ϕ from the conditions of a problem. Usually the simplest procedure is to determine $\mathbf{A}$; then obtain ϕ from $\mathbf{A}$ through 5-5; obtain $\mathbf{E}$ from $\mathbf{A}$ and ϕ from 5-1; and then obtain $\mathbf{H}$ either from 5-2 or from its relation to $\mathbf{E}$ in the Maxwell equations. An example of this kind of computation follows.

7. Radiation from Antennas

The relationships of Section 6 will be applied to the determination of the electric and magnetic fields in the region surrounding a conductor in which there are changing currents. When such a conductor is intentionally designed for the purpose of radiating electromagnetic energy, it is called an antenna. (But radiation can also be an undesired effect whose influence on the functioning of a device may need to be known.)

We first write the vector potential $\mathbf{A}$ at a point in space a distance r away from the conductor in the form of Equation 6-19. The integration, although formally over all space, need, of course, only be carried out over the conductor itself, because the current density is zero everywhere else. Since, in general, different parts of the conductor are at different distances from the point where **A** is to be determined, and since r appears both in the wave function and in the denominator, performing the integration can well be difficult. Especially is this true for an extended antenna if a complete description of the electromagnetic field both close to and far from the antenna is desired. In most practical cases, however, it is the field at large distances which is of interest and the simplifications that can thus be introduced make the integration much easier.

In the case of those antennas which consist of a configuration of current carrying wires, the procedure includes determining the radiation field produced by a very small element of the antenna and then summing at the point in space the fields of all such elements of the radiating system. As an illustration of the uses of the vector potential in the determination of a radiation field we begin by obtaining the field produced by such a small element, the so-called differential antenna or Hertzian dipole.

8. The Differential Antenna

We imagine a very short conducting element of length Δz along the z-direction, with a cross-sectional area $\mathscr{A}$ (Figure 5). The point P at which we wish to determine the electromagnetic field is a distance r from the center of Δz, where r is large compared to Δz and makes an angle θ with the z-direction. Because of symmetry about the z-axis the only variables are θ and r.

The current in the element Δz is assumed to be of constant magnitude

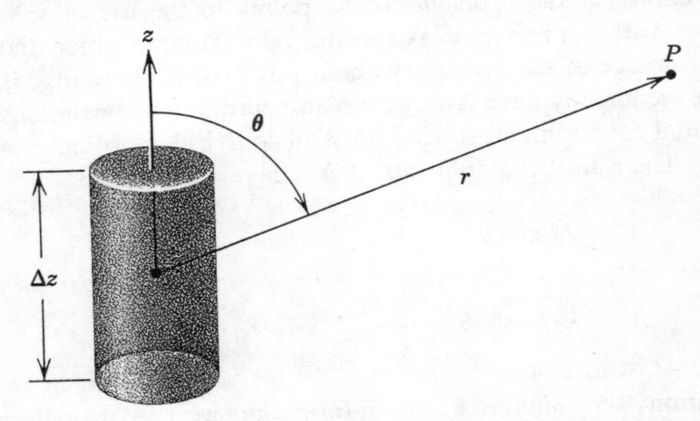

Figure 5. The differential antenna or Hertzian dipole. $\Delta z \ll \lambda$

along Δz, an approximation possible only if we imagine Δz as part of a much longer conductor. (Obviously, if Δz were an isolated element, the current at the ends would have to be zero.) We also assume the current to vary sinusoidally with time:

$$i' = Ie^{j\omega t} \tag{8-1}$$

where I is the real amplitude of the current i'. The current density is then

$$J' = \frac{I}{\mathscr{A}} e^{j\omega t} \tag{8-2}$$

In computing the vector potential from Equation 6-19 we need the current density which existed at a time r/v earlier than the time for which we wish the fields at P. Hence we must use in the integrand the expression:

$$J\left(t - \frac{r}{v}\right) = \frac{I}{\mathscr{A}} e^{j(\omega t - \beta r)} \tag{8-3}$$

where $\beta = \omega/v$.

The assumption that r is large compared to Δz allows the approximation that r is essentially constant over the volume of the antenna and that terms involving r may be taken outside the integral. What remains under the integral is the differential volume, which, integrated over the antenna element, is just $\mathscr{A}\Delta z$. Since, moreover, the current density has only a z component, it follows that the vector potential also has only a z component, and is given by the expression:

$$A_z = \frac{\mu I}{4\pi\mathscr{A}} \frac{e^{j(\omega t - \beta r)}}{r} \mathscr{A}\Delta z = \frac{\mu I \, \Delta z}{4\pi} \frac{e^{j(\omega t - \beta r)}}{r} \tag{8-4}$$

The nature of the problem clearly points to the use of spherical coordinates with the reference axis in the z-direction, in order that we may take advantage of the symmetry about this axis. This change is possible because we already have **A** in Cartesian coordinates. Beginning with the differential equation and solving for **A** in spherical coordinates is another matter. The r and θ components of **A** are

$$\begin{aligned} A_r &= M \cos\theta \, \frac{e^{j(\omega t - \beta r)}}{r} \\ A_\theta &= -M \sin\theta \, \frac{e^{j(\omega t - \beta r)}}{r} \end{aligned} \tag{8-5}$$

where $M = \mu I \Delta z / 4\pi$.

Equation 8-5 contains all the information we need to determine the electric and magnetic fields which exist at the point P.

In spherical coordinates the divergence of **A** is

$$\nabla \cdot \mathbf{A} = \frac{1}{r^2}\frac{\partial}{\partial r}(r^2 A_r) + \frac{1}{r \sin\theta}\frac{\partial}{\partial\theta}(\sin\theta A_\theta) + \frac{1}{r\sin\psi}\frac{\partial A_\psi}{\partial\psi} \tag{8-6}$$

and symmetry conditions enable us to eliminate derivatives with respect to ψ. We therefore obtain, in turn:

$$\nabla \cdot \mathbf{A} = -M\cos\theta\left(\frac{j\beta}{r} + \frac{1}{r^2}\right)e^{j(\omega t - \beta r)} \tag{8-7}$$

and

$$\dot{\phi} = -\frac{1}{\mu\epsilon}\nabla \cdot \mathbf{A} = \frac{\omega^2}{\beta^2}M\cos\theta\left(\frac{j\beta}{r} + \frac{1}{r^2}\right)e^{j(\omega t - \beta r)} \tag{8-8}$$

and

$$\phi = -\frac{j\omega}{\beta^2}M\cos\theta\left(\frac{j\beta}{r} + \frac{1}{r^2}\right)e^{j(\omega t - \beta r)} \tag{8-9}$$

In spherical coordinates the gradient is

$$\begin{aligned}\nabla\phi &= \mathbf{r}_1\frac{\partial\phi}{\partial r} + \boldsymbol{\theta}_1\frac{1}{r}\frac{\partial\phi}{\partial\theta} + \boldsymbol{\psi}_1\frac{1}{r\sin\theta}\frac{\partial\phi}{\partial\psi} \\ &= \frac{j\omega M}{\beta^2}\left\{\mathbf{r}_1\left[\frac{2}{r^3} + \frac{2j\beta}{r^2} - \frac{\beta^2}{r}\right]\cos\theta\right. \\ &\quad \left. + \boldsymbol{\theta}_1\left(\frac{1}{r^3} + \frac{j\beta}{r^2}\right)\sin\theta\right\}e^{j(\omega t - \beta r)}\end{aligned} \tag{8-10}$$

where $\mathbf{r}_1$, $\boldsymbol{\theta}_1$, and $\boldsymbol{\psi}_1$ are unit vectors in the directions of increasing r, θ, and ψ, respectively. Using the time derivative of the vector potential, together with Equation 8-10, we find the electric field at P to be

$$\begin{aligned}\mathbf{E} = M&\left\{\mathbf{r}_1\left(-\frac{2j\omega}{\beta^2 r^3} + \frac{2\omega}{\beta r^2}\right)\cos\theta\right. \\ &\left. + \boldsymbol{\theta}_1\left(-\frac{j\omega}{\beta^2 r^3} + \frac{\omega}{\beta r^2} + \frac{j\omega}{r}\right)\sin\theta\right\}e^{j(\omega t - \beta r)}\end{aligned} \tag{8-11}$$

From the curl of **A** in spherical coordinates

$$\begin{aligned}\nabla \times \mathbf{A} &= \frac{\mathbf{r}_1}{r\sin\theta}\left[\frac{\partial}{\partial\theta}(A_\psi\sin\theta) - \frac{\partial A_\theta}{\partial\psi}\right] \\ &\quad + \frac{\boldsymbol{\theta}_1}{r}\left[\frac{1}{\sin\theta}\frac{\partial A_r}{\partial\psi} - \frac{\partial}{\partial r}(rA_\psi)\right] + \frac{\boldsymbol{\psi}_1}{r}\left[\frac{\partial}{\partial r}(rA_\theta) - \frac{\partial A_r}{\partial\theta}\right]\end{aligned} \tag{8-12}$$

it is evident that H has only a ψ component:

$$\mathbf{H} = \boldsymbol{\psi}_1 \frac{M \sin \theta}{\mu}\left(\frac{1}{r^2} + \frac{j\beta}{r}\right)e^{j(\omega t - \beta r)} \tag{8-13}$$

The electric and magnetic fields of Equations 8-11 and 8-13 show a somewhat involved dependence on r, but these relations are simplified when we restrict our attention to fields at values of r very large compared to the wavelength, where the terms in $1/r^2$ and $1/r^3$ are negligibly small compared to the terms in $1/r$. At such distances the r component of $\mathbf{E}$ becomes negligible and we are left with:

$$E_\theta = \frac{j\omega M \sin \theta}{r} e^{j(\omega t - \beta r)} \tag{8-14a}$$

$$H_\psi = \frac{j\beta M \sin \theta}{\mu r} e^{j(\omega t - \beta r)} \tag{8-14b}$$

The electric and magnetic fields at great distances from the antenna are thus in phase, at right angles to each other, and normal to the direction of propagation of the radiation. Their ratio

$$\frac{E_\theta}{H_\psi} = \frac{\omega\mu}{\beta} = \sqrt{\mu/\epsilon} \tag{8-15}$$

is the same as was found in Section 3 to be characteristic of plane waves.

At much smaller values of r, on the other hand, where r is nevertheless large compared to Δz, the terms in $1/r^3$ are dominant in the electric field components and the term in $1/r^2$ in the magnetic field:

$$E_r = -\frac{2jI\,\Delta z \cos \theta}{4\pi\omega\epsilon r^3} e^{j(\omega t - \beta r)} \tag{8-16a}$$

$$E_\theta = -\frac{jI\,\Delta z \sin \theta}{4\pi\omega\epsilon r^3} e^{j(\omega t - \beta r)} \tag{8-16b}$$

$$H_\psi = \frac{I\,\Delta z \sin \theta}{4\pi r^2} e^{j(\omega t - \beta r)} \tag{8-16c}$$

Several points of interest are to be noted in this so-called "induction field." The electric field components are one-quarter cycle out of phase with the magnetic field components; the electric field components are of the same form as the field components from an electric dipole (Chapter 2, Equation 8-5) with a dipole moment of $-I\Delta z/\omega$; and the magnetic field component is that of a current element $I\Delta z$ (cf: Chapter 5, Equation 6-3).

9. Energy Radiated by the Differential Dipole

Before we analyze the radiation from a larger antenna, it is instructive to compute the power which a differential antenna radiates. This power is obtained by integrating the Poynting vector, **S**, over a sphere surrounding the antenna.

The Poynting vector for the differential antenna is

$$\mathbf{S} = \mathbf{r}_1 E_\theta H_\varphi - \boldsymbol{\theta}_1 E_r H_\varphi \tag{9-1}$$

since E_φ, H_r, and H_θ are zero.

It was shown in Chapter 8 that average power may be obtained directly by the product of the complex voltage with the complex conjugate of the current, where the amplitudes are in terms of effective values. In this present problem the electric field corresponds to voltage (volts/meter) and the magnetic field to current (amperes/meter). Since the current in the antenna was expressed in terms of the peak value I, we need to divide each field by $\sqrt{2}$ to convert to effective values. Then the time average of the Poynting vector is readily found to be

$$\bar{\mathbf{S}} = \frac{M^2}{2\mu}\left[\mathbf{r}_1 \sin^2\theta\left(\frac{\beta\omega}{r^2} - \frac{j\omega}{\beta^2 r^5}\right) + \boldsymbol{\theta}_1 \sin\theta\cos\theta\,\frac{2j\omega}{r^3}\left(\frac{1}{\beta^2 r^2} + 1\right)\right] \tag{9-2}$$

in which it is seen that in only the radial component is there a real part. Thus net power is directed only radially outward from the antenna. The instantaneous energy does have a θ component of flow, but it drops out in averaging over a whole number of cycles.

The total power radiated is found by integrating the real term in Equation 9-2 over a sphere of radius r:

$$\bar{P} = \int \bar{S}\, 2\pi r^2 \sin\theta\, d\theta \tag{9-3}$$

from which we get

$$\bar{P} = \frac{4\pi\beta\omega M^2}{3\mu} = \frac{\mu\sqrt{\mu\epsilon}\,\omega^2 I^2(\Delta z)^2}{12\pi} \tag{9-4}$$

showing, as we intuitively expect, that the radius of the sphere of integration drops out. The same power passes through any sphere centered on the antenna. By substituting for ω:

$$\omega = 2\pi f = 2\pi\frac{v}{\lambda} = \frac{2\pi}{\lambda\sqrt{\mu\epsilon}} \tag{9-5}$$

the average power becomes

$$\bar{P} = \frac{\pi}{3}\sqrt{\mu/\epsilon}\, I^2 \left(\frac{\Delta z}{\lambda}\right)^2 \qquad (9\text{-}6)$$

The radiated energy is directly proportional to the characteristic impedance of the surrounding medium and to the square of the ratio of the length of the antenna to the wavelength (so long as this ratio is small). It can readily be shown that only the far field terms are involved in the radiated power.

Engineers often find it useful to speak of the effective resistance which an antenna presents to the generator to which it is connected as a load. In terms of this radiation resistance, the average power is written as

$$\bar{P} = \frac{I^2 R_{\text{antenna}}}{2} \qquad (9\text{-}7)$$

Comparison of Equation 9-7 with 9-6 shows that

$$R_{\text{antenna}} = \tfrac{2}{3}\pi\sqrt{\mu/\epsilon}\left(\frac{\Delta z}{\lambda}\right)^2 \qquad (9\text{-}8)$$

For free space $\mu = \mu_0 = 4\pi \times 10^{-7}$ henry/meter and $\epsilon = \epsilon_0 = 10^{-9}/36\pi$ farads/meter, from which the value of the characteristic impedance is found to be 120π, resulting in a radiation resistance for the differential dipole antenna of

$$R_{\text{dipole}} = 80\pi^2\left(\frac{\Delta z}{\lambda}\right)^2 \cong 800\left(\frac{\Delta z}{\lambda}\right)^2 \qquad (9\text{-}9)$$

10. The Half-wave Dipole

As an example of the use of the differential dipole in antenna calculation we compute the electromagnetic field radiated by an antenna whose length is one-half wavelength, the driving generator being connected at the middle of the antenna. The current now is no longer uniform over the length of the antenna; it will in fact approximate very closely a sinusoidal standing wave with zeroes at the ends and a maximum in the middle.

We place the reference axis of spherical coordinates parallel to the antenna with the origin at the center of the antenna (Figure 6). Letting the z coordinate measure distance along the reference axis, the current on the antenna is given by

$$i' = I \cos\frac{2\pi z}{\lambda}\, e^{j\omega t} \qquad (10\text{-}1)$$

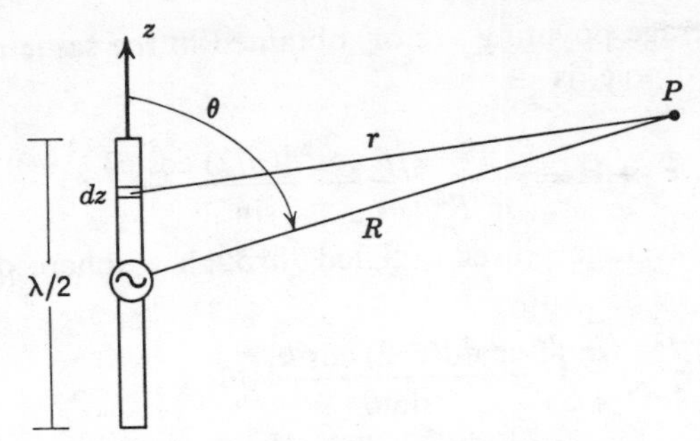

Figure 6. The half-wave dipole antenna.

We are interested in the distant field terms and we integrate Equations 8-14 over the length of the antenna (letting Δz now be dz). In this integration r is not constant but the variation in r is significant essentially only in the exponential term where it affects the phase differences in the fields coming from different parts of the antenna. In the denominator where small variations in r can be neglected we shall use the distance R from P to the center of the antenna. But the closer approximation $R - z\cos\theta$ is needed in the exponent. Thus the electric field at P from the element dz is

$$dE_\theta = \frac{j\omega\mu I e^{-j\beta R} e^{j\omega t} \cos \frac{2\pi z}{\lambda}\, e^{j\beta z \cos\theta} \sin\theta \, dz}{4\pi R} \tag{10-2}$$

and the total electric field, from integrating Equation 10-2 over the length of the antenna, is

$$E_\theta = \frac{j\omega\mu I \sin\theta \, e^{j(\omega t - \beta R)}}{4\pi R} \int_{-\lambda/4}^{\lambda/4} \cos \frac{2\pi z}{\lambda}\, e^{j\beta z \cos\theta} \, dz \tag{10-3}$$

$$= \frac{jI}{2\pi R} \sqrt{\mu/\epsilon}\, e^{j(\omega t - \beta R)} \frac{\cos[(\pi/2)\cos\theta]}{\sin\theta}$$

This integral can be found in a standard table of integrals. Similarly the magnetic field is

$$H_\psi = \frac{jI}{2\pi R} e^{j(\omega t - \beta R)} \frac{\cos[(\pi/2)\cos\theta]}{\sin\theta} \tag{10-4}$$

The ratio of electric to magnetic fields at a great distance from the half-wave dipole is the characteristic impedance of the medium, just as for the differential dipole.

The time average poynting vector, obtained in the same manner as for the differential dipole, is

$$\bar{\mathbf{S}} = \bar{\mathbf{E}} \times \bar{\mathbf{H}} = \frac{I^2}{8\pi^2 R^2}\sqrt{\frac{\mu}{\epsilon}}\,\frac{\cos^2[(\pi/2)\cos\theta]}{\sin^2\theta} \tag{10-5}$$

Using 10-5 the average power radiated through a sphere of radius R is found to be

$$\bar{P} = \frac{I^2}{4\pi}\sqrt{\frac{\mu}{\epsilon}}\int_0^{\pi}\frac{\cos^2[(\pi/2)\cos\theta]}{\sin\theta}\,d\theta \tag{10-6}$$

and the radiation resistance is

$$R = \frac{1}{2\pi}\sqrt{\frac{\mu}{\epsilon}}\int_0^{\pi}\frac{\cos^2[(\pi/2)\cos\theta]}{\sin\theta}\,d\theta \tag{10-7}$$

Computation of this integral by numerical methods gives a value of 73.09 ohms. A polar plot of field strength and power against angle θ is shown in Figure 7. (In the differential dipole the radiation resistance and the power were functions of the wavelength, whereas the wavelength does not appear in Equations 10-6 and 10-7. Why?)

It is beyond the scope of this book to go further into the subject of antennas. There are many forms of radiating systems designed for various purposes and all have directive properties of some sort. Those we have considered radiate uniformly about the z-axis but discriminate

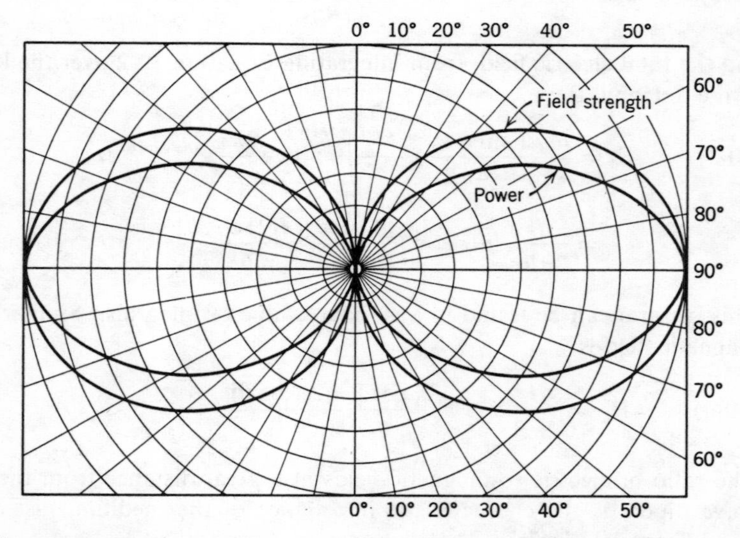

Figure 7. Field strength and power versus polar angle for the half-wave dipole.

against the z-direction. Others concentrate the radiation strongly along one or a few preferred directions. References 1 and 2 give good accounts of antennas and their properties.

Discussion Questions

1. Write the Maxwell field equations in terms of **E** and **B**, in terms of **E** and **H**, in terms of **D** and **H**, and in terms of **D** and **B**. Are there circumstances in which one or more of these sets may properly be used and others may not?

2. An electron rotates in a circular orbit. Discuss qualitatively what radiation might be expected, on the basis of electromagnetic theory, in the plane of such an orbit. How should it be polarized? Then consider the electric and the magnetic fields on the axis perpendicular to the plane of the orbit. How may these fields be expected to change with time? What is the nature of the radiation on this axis? What kind of polarization may it be expected to have?

3. The Lorentz condition, Equation 5-5, is a special case of gauge invariance. Are there other choices that might be made? Investigate and discuss the concept further. [See References: 2, p. 440ff; 3, p. 181; 4, p. 180ff.]

4. Does the functional choice in Equation 6-6 limit the generality of the results obtained? If so, to what extent?

Problems

1. Show that the displacement current between the plates of a parallel plate capacitor is equal to the conduction current in the wires leading to the plates. Make the approximation of ignoring fringing electric flux at the edges of the plates. (But see Reference 10.)

2. Relative to a set of Cartesian coordinates a certain direction has the cosine directors l, m, and n. Show that

$$\mathbf{E} = \mathbf{E}_0 f(lx + my + nz - vt)$$

represents a plane wave propagating in this direction with a speed v.

3. Show that, in general, the differential equation for a traveling wave of a field **F** is

$$\nabla^2 \mathbf{F} = \frac{1}{v^2}\frac{\partial^2 \mathbf{F}}{\partial t^2}$$

where v is the velocity of propagation of the wave.

4. The magnetic components of two electromagnetic waves have the same amplitude. One wave exists in empty space and the other in a medium of relative dielectric constant 4 and relative permeability 1. What is the ratio of the amplitudes of the electric components and what is the ratio of the energy densities associated with the two waves?

5. An electric field whose components are

$$E_x = A \sin \beta y \sin \omega t \qquad E_y = 0 \qquad E_z = 0$$

is present in an uncharged, nonconducting medium that is infinite in extent and has a dielectric constant of ϵ and a permeability of μ. Determine the associated magnetic field, evaluate β, show that each is a combination of traveling waves, and determine the directions of propagation.

6. Find the time-averaged value of the Poynting vector for the plane wave discussed in Sec. 3.

7. A wire of circular cross section has a resistance of 0.5 ohm per meter of length and carries a d-c current of 2 amperes. Determine the electric and magnetic fields at the surface of the wire, the magnitude of the Poynting vector at the surface, the integral of the Poynting vector over the surface of a meter length of the wire, and show that this integral is equal to the power dissipated into heat in the wire as found from Joule's law.

8. For plane waves in free space, show that the average energy per unit time crossing an area normal to the direction of propagation is equal to the average field energy within a cylinder whose base is this area and whose height is numerically equal to the velocity.

9. If, on the average, the rate at which energy reaches the earth as sunlight is 1.5 kilowatts per square meter, what is the amplitude of the electric vector and the amplitude of the magnetic vector? Treat as a plane wave. What is the energy per cubic meter in this sunlight?

10. What fraction of the power from a differential dipole is radiated within 10° of the equatorial plane?

11. Integrate Equation 10-7 to verify the value of the radiation resistance for a halfwave antenna. Use graphical integration.

12. Show that at a given point in space far from a differential dipole and at a given instant the energy density in the electric field and the energy density in the magnetic field are equal. Show that the same is true for the half-wave dipole antenna.

13. A circular current loop forms a magnetic dipole whose moment is $m = I\mathscr{A}$, where $\mathscr{A}$ is the area of the loop and I the current. Find the magnetic vector potential, **A**, at a point (r, θ, ϕ), where r is very large compared to the radius of the loop. The normal to the plane of the loop through the center is the polar axis of coordinates. Consider the current as being the same magnitude at all parts of the loop and varying with time as $e^{j\omega t}$. Make approximations similar to those made for the differential dipole. In addition, assume that the scalar potential is uniform at all parts of the loop, and that it may be taken as the zero reference. From this vector potential, obtain the electric and magnetic fields in the radiation from this loop. Such radiation is called magnetic dipole radiation. [Reference 2, p. 474ff is helpful.]

REFERENCES

1. Ramo and Whinnery, *Fields and Waves in Modern Radio* (Wiley), 2nd ed., 1953. Chapters 4 and 12.
2. Corson and Lorain, *Introduction to Electromagnetic Fields and Waves* (Freeman), 1962. Chapters 8, 9, and 13.
3. Jackson, *op. cit.* Chapter 9.
4. Whitmer, *op. cit.* Chapter 11.
5. Panofsky and Phillips, *op. cit.* Chapters 9 and 11.
6. Slater and Frank, *Electromagnetism* (McGraw-Hill), 1947. Chapter II, Section 4, Chapters VIII and XII.
7. Smythe, *op. cit.* Chapters XIII and XIV.
8. Scott, *op. cit.* Chapter 10.
9. Reitz and Milford, *op. cit.* Chapter 15.
10. French and Tassman, "Displacement Currents and Magnetic Fields," *Am. Jour. Phys.*, **31,** 201 (1963).
11. Harrison *et al.*, "Possibility of Observing the Magnetic Charge of an Electron," *Am. Jour. Phys.*, **31,** 249 (1963).

11

Plane waves—the influence of the medium

The electromagnetic wave sufficiently far from its source is plane and transverse, consisting of mutually perpendicular electric and magnetic field components. So long as there is no conductivity in the medium, the wave propagates without attenuation and the electric and magnetic components remain in phase. In a conducting medium both the relative phase and the amplitudes of the components change. In an anisotropic medium the dielectric constant is not a scalar number but becomes a tensor. When the properties of the medium change along the path of the wave, there are changes in direction as well as in phase and amplitude. How the electromagnetic wave is affected by such conductivity, anisotropy, or discontinuity in the medium—these are the subjects to be discussed in this chapter.

1. Conductivity of the Medium

Let us assume that the medium, instead of being a pure dielectric, as in Chapter 10, Section 2, obeys Ohm's law according to the relation

(1-1) $\mathbf{J} = \sigma \mathbf{E}$

where σ is the conductivity of the medium. We continue to assume the medium to be uncharged, homogeneous, and isotropic. Of the Maxwell equations, only Ampère's circuital law:

(1-2) $\nabla \times \mathbf{H} = \mathbf{J} + \dot{\mathbf{D}}$

is affected by the presence of conductivity in the medium. How it is affected is best shown by writing it first for a nonconductor and then for a medium of conductivity σ, where both media are linear media and where the time dependence of the fields is as $e^{j\omega t}$, so that wherever the operator d/dt appears it is replaced by $j\omega$. For these two kinds of media, Equation 1-2 becomes:

$$\text{(1-3}a\text{)} \qquad \nabla \times \mathbf{H} = j\omega\epsilon_0\kappa\mathbf{E} \qquad \text{(nonconductor)}$$

$$\text{(1-3}b\text{)} \qquad \nabla \times \mathbf{H} = \sigma\mathbf{E} + j\omega\epsilon_0\kappa\mathbf{E}$$

$$= j\omega\epsilon_0\left(\kappa - j\frac{\sigma}{\omega\epsilon_0}\right)\mathbf{E} \qquad \text{(conductor)}$$

A comparison of these two forms shows that so long as we are dealing with linear media (media which do not fit this restriction are uncommon), the effect of conductivity in the medium is that, instead of a simple real number for the relative dielectric constant, as in nonconductors, we must use a complex number. All the formal operations which have been and will be carried out for pure dielectrics can also be applied to conductors, provided only that we make this change in the dielectric constant. The effect of a complex dielectric constant is to introduce an attenuation on the wave amplitude (as did complex impedances in filters and transmission lines) and to change the characteristic impedance of the medium from a pure real number to a complex number, making it a function of the frequency.

2. Attenuation Resulting from Conductivity

Instead of writing out the complex dielectric constant in full, we shall often find it convenient to indicate it with a subscript c:

$$\text{(2-1)} \qquad \epsilon_c = \epsilon_o[\kappa - j(\sigma/\omega\epsilon_0)]$$

and similarly to use the same subscript on the wavelength constant and the characteristic impedance. Thus

$$\text{(2-2)} \qquad j\beta_c = j\omega\sqrt{\mu\epsilon_c} = j\omega\sqrt{\mu\epsilon_0[\kappa - (j\sigma/\omega\epsilon_0)]} = m + jn$$

where m and n are real numbers which may be obtained from 2-2:

$$\text{(2-3}a\text{)} \qquad m^2 = \frac{\omega^2\mu}{2}[|\epsilon_c| - \epsilon]$$

$$\text{(2-3}b\text{)} \qquad n^2 = \frac{\omega^2\mu}{2}[|\epsilon_c| + \epsilon]$$

Here $|\epsilon_c|$ is the absolute value of the complex number, ϵ_c, and $\epsilon = \kappa\epsilon_0$.

Including 2-2, the electric field component of an electromagnetic plane wave traveling in the z-direction takes the form:

$$E = E_0 e^{j(\omega t - \beta_c z)} = E_0 e^{-mz} e^{j(\omega t - nz)} \tag{2-4}$$

where E_0 is the amplitude of the field at $z = 0$. The magnetic field component is similar in form. The wave amplitude is attenuated as it propagates into a conducting medium, the attenuation governed by the exponential e^{-mz}. In a nonconducting medium $m = 0$ and Equation 2-4 reduces to the form for pure dielectrics.

In a distance $1/m$ the amplitude of the electromagnetic wave decreases by a factor $1/e$. This distance is of particular interest in connection with good conductors, where it is ordinarily small. It is called the skin depth and is assigned the symbol δ. A good conductor is a substance in which $\sigma/\omega\epsilon_0 \gg \kappa$, so that in effect κ can be neglected and ϵ_c becomes a pure imaginary. In this case

$$m^2 = n^2 = \frac{\omega\mu\sigma}{2} = \pi f\mu\sigma = \frac{1}{\delta^2} \tag{2-5}$$

where $f = \omega/2\pi$. Then Equation 2-4 becomes

$$\mathbf{E} = \mathbf{E}_0 e^{-z/\delta} e^{j[\omega t - (z/\delta)]} \tag{2-6}$$

Not only has the amplitude decreased by $1/e$ in the distance δ, but the phase of the wave at $z = \delta$ differs by just one radian from the phase at $z = 0$.

δ can be related to the dissipation of the energy of an electromagnetic wave in a good conductor in a rather simple way. The current density, being proportional to $\mathbf{E}$, has the same form as Equation 2-6:

$$\mathbf{J} = \mathbf{J}_0 e^{-z/\delta} e^{j[\omega t - (z/\delta)]} \tag{2-7}$$

Suppose there is in a semi-infinite conductor (Figure 1) an electromagnetic plane wave traveling from the surface into the medium. Let the direction of propagation be z with the electric field and the current density in the x-direction. We now imagine a rectangular parallelepiped in the medium, with base at $z = 0$, of sides w and L, where L is parallel to J, extending into the medium a height D. The charge which passes per second through an area wD is

$$i' = w\int_0^D J_x\,dz = wJ_0\int_0^D e^{-[(z/\delta)+(jz/\delta)]} e^{j\omega t}\,dz \tag{2-8}$$

which, upon letting D approach infinity, becomes

$$i' = wJ_0 \frac{\delta}{1+j} e^{j\omega t} = \frac{wJ_0\delta}{2}(1-j)e^{j\omega t} = \frac{wJ_0\delta}{\sqrt{2}} e^{j[\omega t - (\pi/4)]} \tag{2-9}$$

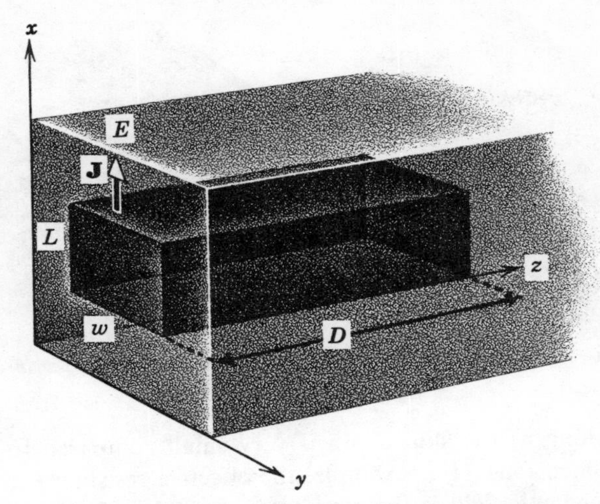

Figure 1. An imaginary volume of base wL extending into a conducting medium.

where J_0 is the peak value of J at $z = 0$. The effective value of this current is obtained by dividing the amplitude by $\sqrt{2}$:

$$I_{\text{eff}} = \frac{wJ_0\delta}{2} \tag{2-10}$$

The heat developed per second in this parallelepiped of infinite height is

$$\begin{aligned} \bar{P} &= \int_0^D (J_x w\, dz)^2 \frac{L}{\sigma w\, dz} \\ &= \frac{wL}{2\sigma} J_0^2 \int_0^\infty e^{-2z/\delta}\, dz \\ &= \frac{wLJ_0^2\delta}{4\sigma} \end{aligned} \tag{2-11}$$

where σ is the conductivity of the medium. Thus the resistance of the medium under the area wL is obtained from

$$R = \frac{\bar{P}}{I_{\text{eff}}^2} = \frac{1}{\sigma}\frac{L}{w\delta} \tag{2-12}$$

This resistance is equal to the direct current resistance of a parallelepiped of the same base and of height δ. The current which an electromagnetic wave of the form 2-6 produces in an infinitely thick conductor develops the same heat which a direct current of the same effective value develops in a thickness δ. Although the reasoning has been applied to the simple geometry of a semiinfinite body, the results apply very closely for other

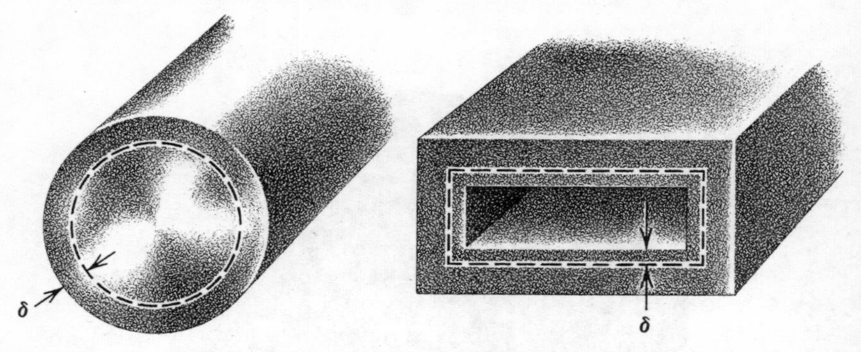

Figure 2. *Skin depth in a cylindrical rod.*

Figure 3. *Skin depth on the inside of a rectangular tube.*

shapes, so long as the skin depth is very small compared to the cross-sectional dimensions. For example, the effective resistance of a rod for an electromagnetic field is the resistance of a tube of thickness δ at the surface of the rod (Figure 2).

The resistance of a pipe of rectangular cross section within which there is an electromagnetic field is the resistance of a shell of depth δ on the inside surface of the pipe (Figure 3).

These considerations have practical significance in connection with the use of waveguides, the theory of which is to be treated in the next chapter.

3. Characteristic Impedance in a Conductor

As the symbol for the characteristic impedance of the medium for electromagnetic waves we use $\eta = (\mu/\epsilon)^{1/2}$, which for a conducting material, with a complex dielectric constant, becomes

$$\eta_c = \left(\frac{\mu}{\epsilon_c}\right)^{1/2} = \left[\frac{\mu}{\epsilon_0[\kappa - (j\sigma/\omega\epsilon_0)]}\right]^{1/2} = p + jq = |\eta_c|\, e^{j\varphi} \tag{3-1}$$

where p, q, $|\eta_c|$, and φ are

$$p = \frac{n}{\omega\,|\epsilon_c|} \tag{3-2a}$$

$$q = \frac{m}{\omega\,|\epsilon_c|} \tag{3-2b}$$

$$|\eta_c| = \sqrt{p^2 + q^2} = \sqrt{\mu/|\epsilon_c|} \tag{3-2c}$$

$$\tan\varphi = \frac{q}{p} = \frac{m}{n} = \left(\frac{|\epsilon_c| - \epsilon}{|\epsilon_c| + \epsilon}\right)^{1/2} \tag{3-2d}$$

Using the exponential form in Equation 3-1, the magnetic and the electric fields of the wave are related by

$$H = \frac{E}{|\eta_c|} e^{-j\varphi} \tag{3-3}$$

Thus in a conducting medium not only is the wave attenuated but the electric and the magnetic components are no longer in phase, as they were in pure dielectrics. In a very good conductor, where, from Equation 2-5, $m = n$ and hence $p = q$, then $\tan \varphi = 1$, so that the electric and magnetic fields are 45° out of phase with each other.

The impedances of such good conductors as copper, silver, gold, and even lead are very small fractions of an ohm. An electromagnetic wave in a dielectric approaching a good conducting surface is almost completely reflected—in circuit language the conductor presents an almost perfect short circuit at the boundary—analogous to a transmission line wave approaching a shorted termination. In the limiting approximation of an ideal conductor to be used in discussing waveguides, we can assume 100% reflection at the surface of the conducting sides of the guide. Clearly such an ideal conductor will absorb no energy from the wave.

4. Waves in Anisotropic Media

We have been discussing the behavior of waves in isotropic homogeneous materials in which **D** and **E** are parallel vectors related through the scalar constant ϵ. The fact that ϵ is a scalar quantity implies that the x component of **D** is related only to the x component of **E**, the y component of **D** only to the y component of **E**, etc., with the constant of proportionality the same for each component. In an anisotropic medium in which electrical properties are different in different directions, a component of **D** is in general a function of all three components of **E**. Thus

$$\begin{aligned} D_x &= \epsilon_{xx}E_x + \epsilon_{xy}E_y + \epsilon_{xz}E_z \\ D_y &= \epsilon_{yx}E_x + \epsilon_{yy}E_y + \epsilon_{yz}E_z \\ D_z &= \epsilon_{zx}E_x + \epsilon_{zy}E_y + \epsilon_{zz}E_z \end{aligned} \tag{4-1}$$

in which the ϵ_{ij}'s are elements of a tensor and their values depend on the orientation of the reference axes. For any set of axes it turns out that $\epsilon_{xy} = \epsilon_{yx}$, $\epsilon_{xz} = \epsilon_{zx}$, and $\epsilon_{yz} = \epsilon_{zy}$; that is, the tensor is symmetrical. Happily there is a special choice of axes for such a symmetrical tensor in which $\epsilon_{ij} = 0$ for $i \neq j$ and $\epsilon_{ij} = \epsilon_i$ for $i = j$. Such a set of axes can always be found for a symmetrical tensor without loss of generality. Then Equation 4-1 becomes

$$D_x = \epsilon_x E_x \qquad (4\text{-}2)$$
$$D_y = \epsilon_y E_y$$
$$D_z = \epsilon_z E_z$$

Having found a set of three mutually perpendicular directions for which 4-2 holds, we still have freedom to choose the labeling of the axes. It is convenient to label the axes in such a way that:

$$\epsilon_x < \epsilon_y < \epsilon_z \qquad (4\text{-}3)$$

We adopt the exponential form of a sinusoidal traveling wave function for the displacement, but we can no longer confine our treatment to propagation along the z-axis because we wish now to compare wave behaviors in different directions. Figure 4 shows a wavefront propagating in the direction fixed by the unit vector **m**, the distance of the wavefront from the origin being given by the variable M. The vector **r** locates an arbitrary position in the wavefront, and the plane of the wavefront is fixed by the relation $\mathbf{m} \cdot \mathbf{r} = M$. The displacement vector is

$$\mathbf{D} = \mathbf{D}_0 e^{j(\omega t - \mathbf{m} \cdot \mathbf{r}\, \beta)} = \mathbf{D}_0 e^{j(\omega t - \beta M)} \qquad (4\text{-}4)$$

The Maxwell field equations insure that if **D** is in the form of 4-4, then **B**, **H**, and (from 4-2) **E** have similar forms.

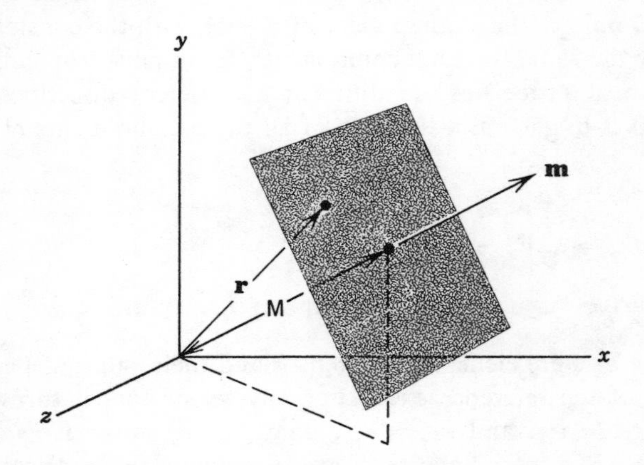

Figure 4. *A wave front distant M from the origin and propagating in the direction* **m**.

From Equation 4-4 the differential operators with respect to time and distance become

$$\frac{\partial}{\partial t} = j\omega \tag{4-5a}$$

$$\frac{\partial}{\partial x} = \frac{\partial M}{\partial x}\frac{\partial}{\partial M} = \mathbf{m}\cdot\frac{\partial \mathbf{r}}{\partial x}\frac{\partial}{\partial M} = \mathbf{m}\cdot\mathbf{i}\frac{\partial}{\partial M} = -j\beta m_x \tag{4-5b}$$

$$\frac{\partial}{\partial y} = -j\beta m_y \tag{4-5c}$$

$$\frac{\partial}{\partial z} = -j\beta m_z \tag{4-5d}$$

so that

$$\nabla = \mathbf{i}\frac{\partial}{\partial x} + \mathbf{j}\frac{\partial}{\partial y} + \mathbf{k}\frac{\partial}{\partial z} = -j\beta\mathbf{m} \tag{4-5e}$$

where m_x, m_y, and m_z are the components of **m**, the cosine directors of the direction of propagation. The four Maxwell equations become, using Equations 4-5:

$$\nabla\cdot\mathbf{D} = 0 \qquad -j\beta\mathbf{m}\cdot\mathbf{D} = 0 \tag{4-6a}$$

$$\nabla\cdot\mathbf{B} = 0 \qquad -j\beta\mathbf{m}\cdot\mathbf{B} = 0 \tag{4-6b}$$

$$\nabla\times\mathbf{H} = \dot{\mathbf{D}} \qquad -j\beta\mathbf{m}\times\mathbf{H} = j\omega\mathbf{D} \tag{4-6c}$$

$$\nabla\times\mathbf{E} = -\dot{\mathbf{B}} \qquad -j\beta\mathbf{m}\times\mathbf{E} = -j\omega\mathbf{B} \tag{4-6d}$$

Equations 4-6*a* and 4-6*b* show that **D** and **B** are perpendicular to the direction of propagation of the wavefront. Equation 4-6*c* shows that **D** is perpendicular to **H** and hence to **B** (we assume no anisotropy in the magnetic properties of the medium). And Equation 4-6*d* shows that **B** and **E** are perpendicular. But since the divergence of **E** is not zero, **E** and **D** are not parallel. Nevertheless, **D**, **E**, and **m** are coplanar since all three are perpendicular to **B**. The spacial relationship of these vectors is illustrated in Figure 5.

The interesting result is that the direction of propagation of the energy, which from the Poynting vector, is normal to the **E-H** plane, is not the same as the direction of propagation of the wavefront. The direction of the Poynting vector, fixed by the unit vector **s**, makes an angle α with **m**, the same as the angle between **E** and **D**:

$$\cos\alpha = \mathbf{m}\cdot\mathbf{s} = \frac{\mathbf{E}\cdot\mathbf{D}}{|E|\,|D|} \tag{4-7}$$

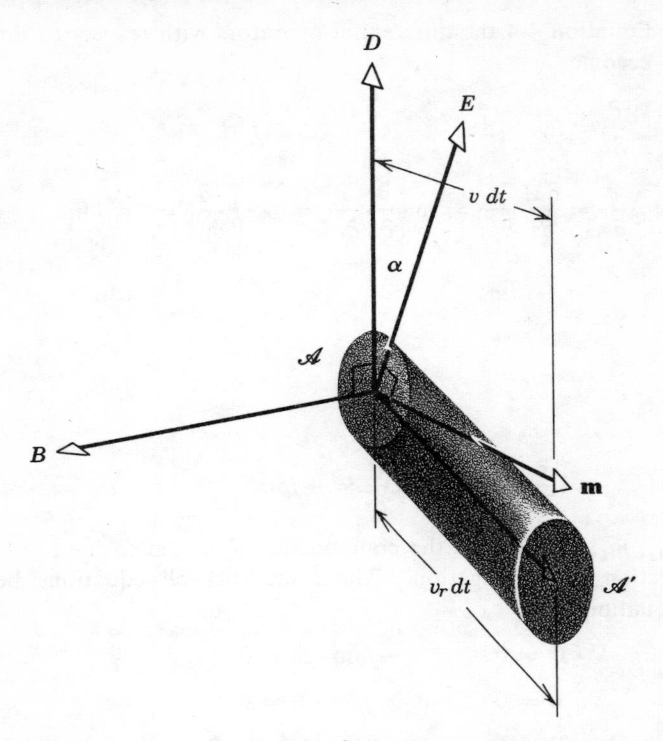

Figure 5. Field vectors in an anisotropic medium and the direction of energy propagation.

Let us picture two positions of a plane wavefront which are separated by a time dt (Figure 5). On the first plane imagine an area $\mathscr{A}$. The energy which is carried by the fields at $\mathscr{A}$ travels through the cylinder whose other end is $\mathscr{A}'$ and whose generators make an angle α with $\mathbf{m}$. The direction of energy flow is called the ray direction and its velocity is the ray velocity. The volume of this cylinder is

$$(\mathbf{v}_r\, dt) \cdot \mathscr{A} = v_r\, dt\, \mathscr{A} \cos \alpha \tag{4-8}$$

If w is the energy density within this region, the total energy in the cylindrical volume is

$$w v_r\, dt\, \mathscr{A} \cos \alpha \tag{4-9}$$

The energy which flows out through $\mathscr{A}'$ (or has come in through $\mathscr{A}$) in a time dt is

$$S\mathbf{s}\, dt \cdot \mathscr{A} = S\, dt\, \mathscr{A} \cos \alpha \tag{4-10}$$

Equating 4-9 and 4-10 gives the energy velocity or ray velocity:

(4-11) $$v_r = \frac{S}{w}$$

The ray velocity is related to the velocity of the wavefront by writing the energy density and Poynting's vector in terms of the fields. The energy density in the electric field is $\frac{1}{2}\mathbf{D}\cdot\mathbf{E}$ and the energy density in the magnetic field is $\frac{1}{2}\mathbf{B}\cdot\mathbf{H}$. From Equations 4-6*c* and 4-6*d* these become:

(4-12*a*) $$w_E = \tfrac{1}{2}\mathbf{D}\cdot\mathbf{E} = \tfrac{1}{2}\left(-\frac{\mathbf{m}\times\mathbf{H}}{v}\right)\cdot\mathbf{E} = \frac{1}{2}\frac{\mathbf{m}\cdot\mathbf{S}}{v}$$

(4-12*b*) $$w_H = \tfrac{1}{2}\mathbf{B}\cdot\mathbf{H} = \frac{1}{2}\left(\frac{\mathbf{m}\times\mathbf{E}}{v}\right)\cdot\mathbf{H} = \frac{1}{2}\frac{\mathbf{m}\cdot\mathbf{S}}{v}$$

where use has been made of the fact that $\beta = \omega/v$. These two energy densities are equal as they always are in a nonabsorptive medium. Their sum is

(4-13) $$w = \frac{1}{v}\,\mathbf{m}\cdot(\mathbf{E}\times\mathbf{H}) = \frac{1}{v}\,\mathbf{m}\cdot(S\mathbf{s}) = \frac{1}{v}\,S\cos\alpha$$

Putting Equation 4-13 into 4-11, we obtain

(4-14) $$v = v_r\cos\alpha$$

The ray velocity is thus greater than the wavefront velocity.

We now compute the velocity of propagation of the wavefront. The wave equation is obtained by the procedure used in Chapter 10, Section 2. We take the curl of 4-6*d* and substitute from 4-6*c* to get:

(4-15) $$\nabla^2\mathbf{E} - \nabla(\nabla\cdot\mathbf{E}) = \mu\ddot{\mathbf{D}}$$

Since $\nabla\cdot\mathbf{E} \neq 0$, this equation cannot be reduced to the simple form of Equation 2-5, Chapter 10. With 4-5*e*, Equation 4-15 becomes

(4-16) $$\mathbf{E} - \mathbf{m}(\mathbf{m}\cdot\mathbf{E}) = \mu v^2\mathbf{D}$$

which yields the three scalar equations in the components of **D**:

(4-17*a*) $$\frac{D_x}{\epsilon_x} - m_x(\mathbf{m}\cdot\mathbf{E}) = \mu v^2 D_x$$

(4-17*b*) $$\frac{D_y}{\epsilon_y} - m_y(\mathbf{m}\cdot\mathbf{E}) = \mu v^2 D_y$$

(4-17*c*) $$\frac{D_z}{\epsilon_z} - m_z(\mathbf{m}\cdot\mathbf{E}) = \mu v^2 D_z$$

where use has been made of Equation 4-2 to replace E_x, E_y, and E_z with D_x, D_y, and D_z, respectively. Solving for the components of **D**, we have

$$D_x = \frac{m_x(\mathbf{m}\cdot\mathbf{E})}{\mu(v_x^2 - v^2)} \tag{4-18a}$$

$$D_y = \frac{m_y(\mathbf{m}\cdot\mathbf{E})}{\mu(v_y^2 - v^2)} \tag{4-18b}$$

$$D_z = \frac{m_z(\mathbf{m}\cdot\mathbf{E})}{\mu(v_z^2 - v^2)} \tag{4-18c}$$

where

$$v_x^2 = \frac{1}{\mu\epsilon_x} \qquad v_y^2 = \frac{1}{\mu\epsilon_y} \qquad v_z^2 = \frac{1}{\mu\epsilon_z} \tag{4-19}$$

Note that v_x, v_y, and v_z in 4-19 are not components of v, although they do have the dimensions of velocity. They are quantities associated with the three directions in the medium.

We now put Equations 4-18 into the divergence equation, 4-6*a*:

$$\frac{\mathbf{m}\cdot\mathbf{E}}{\mu}\left(\frac{m_x^2}{v_x^2 - v^2} + \frac{m_y^2}{v_y^2 - v^2} + \frac{m_z^2}{v_z^2 - v^2}\right) = 0 \tag{4-20}$$

The expression in brackets is zero since $\mathbf{m}\cdot\mathbf{E}$ is not zero. Equation 4-20 is to be solved for the velocity v of a wave front propagating in a direction fixed by the unit vector **m**. Since 4-20 is a quadratic in v^2, the solution is double valued:

$$v^2 = \tfrac{1}{2}\Big\{m_x^2(v_y^2 + v_z^2) + m_y^2(v_z^2 + v_x^2) + m_z^2(v_x^2 + v_y^2) \pm \sqrt{\begin{matrix}[m_x^2(v_y^2 + v_z^2) + m_y^2(v_z^2 + v_x^2) + m_z^2(v_x^2 + v_y^2)]^2 \\ - 4[m_x^2v_y^2v_z^2 + m_y^2v_z^2v_x^2 + m_z^2v_x^2v_y^2]\end{matrix}}\Big\} \tag{4-21}$$

What is the physical significance of there being two velocities of propagation for each direction? From Equation 4-18 we see that there is a displacement vector associated with each velocity. Let the two values of v be v_1 and v_2. Then the corresponding $\mathbf{D}_1$ and $\mathbf{D}_2$ are

$$\mathbf{D}_1 = \frac{\mathbf{m}\cdot\mathbf{E}_1}{\mu}\left(\mathbf{i}\frac{m_x}{v_x^2 - v_1^2} + \mathbf{j}\frac{m_y}{v_y^2 - v_1^2} + \mathbf{k}\frac{m_z}{v_z^2 - v_1^2}\right) \tag{4-22a}$$

$$\mathbf{D}_2 = \frac{\mathbf{m}\cdot\mathbf{E}_2}{\mu}\left(\mathbf{i}\frac{m_x}{v_x^2 - v_2^2} + \mathbf{j}\frac{m_y}{v_y^2 - v_2^2} + \mathbf{k}\frac{m_z}{v_z^2 - v_2^2}\right) \tag{4-22b}$$

It is not difficult to show that the two displacement vectors are mutually perpendicular. The dot product of $\mathbf{D}_1$ and $\mathbf{D}_2$ is

$$\text{(4-23)} \qquad \mathbf{D}_1 \cdot \mathbf{D}_2 = \frac{(\mathbf{m} \cdot \mathbf{E}_1)(\mathbf{m} \cdot \mathbf{E}_2)}{\mu} \sum_{i=x,y,z} \frac{m_i^2}{(v_i^2 - v_1^2)(v_i^2 - v_2^2)}$$

By multiplying and dividing by $(v_1^2 - v_2^2)$ and noting that

$$\text{(4-24)} \qquad \frac{m_i^2(v_1^2 - v_2^2)}{(v_i^2 - v_1^2)(v_i^2 - v_2^2)} = \frac{m_i^2}{(v_i^2 - v_1^2)} - \frac{m_i^2}{(v_i^2 - v_2^2)}$$

we can transform Equation 4-23 into

$$\text{(4-25)} \qquad \mathbf{D}_1 \cdot \mathbf{D}_2 = \frac{(\mathbf{m} \cdot \mathbf{E}_1)(\mathbf{m} \cdot \mathbf{E}_2)}{\mu} \frac{1}{v_1^2 - v_2^2}\left[\sum_i \frac{m_i^2}{v_i^2 - v_1^2} - \sum_i \frac{m_i^2}{v_i^2 - v_2^2}\right]$$

Comparison with 4-20 shows that v_1 and v_2 are just the values which make each summation in Equation 4-25 zero, and hence $\mathbf{D}_1$ is normal to $\mathbf{D}_2$. Since these two components have different velocities, they have different indices of refraction and the corresponding waves are refracted by different amounts at a discontinuity in the medium. Thus the two components of an unpolarized light beam (in a calcite crystal, for example) propagate with different velocities and are, in general, refracted through different angles at an interface with another medium. Each of the refracted beams is polarized. The Nicol prism, often used for producing polarized light, makes use of this property.

A further interesting property of an anisotropic medium may be deduced from Equation 4-21. If the radical is zero for a particular direction of $\mathbf{m}$, then the two velocities are equal. We find this direction in terms of its direction cosines, m_x, m_y, and m_z by using the relation

$$\text{(4-26)} \qquad m_x^2 + m_y^2 + m_z^2 = 1$$

to eliminate one of the components of $\mathbf{m}$, say, m_z. Equating the terms under the radical in Equation 4-21 to zero and solving for m_x, we get

$$\text{(4-27)} \qquad m_x^2 = -\left[\frac{(v_x^2 - v_y^2) + m_y^2(v_z^2 - v_y^2)}{(v_z^2 - v_x^2)}\right] \pm \frac{2m_y}{v_z^2 - v_x^2}\sqrt{(v_z^2 - v_y^2)(v_x^2 - v_y^2)}$$

Equations 4-3 and 4-19 show that $v_x > v_y > v_z$, and hence the term under the radical in 4-27 is negative. The only possibility of a real value for m_x is if $m_y = 0$, showing that a direction along which the two velocities are equal must lie in the xz-plane, with cosine directors:

$$\text{(4-28)} \qquad m_x = \pm\left(\frac{v_x^2 - v_y^2}{v_x^2 - v_z^2}\right)^{1/2} \qquad m_y = 0 \qquad m_z = \pm\left(\frac{v_y^2 - v_z^2}{v_x^2 - v_z^2}\right)^{1/2}$$

There are thus two such directions in the xz-plane. These directions are called the optic axes. The velocity of propagation along an optic axis is

$$v = v_y \tag{4-29}$$

v_y is a special velocity of propagation normal to the y-axis. If $v_x = v_y > v_z$, then $m_x = 0$ and $m_z = 1$, and there is only one optic axis, namely, the z-axis. For $v_x > v_y = v_z$, $m_x = 1$ and $m_z = 0$ and the single optic axis is along the x-axis.

Anisotropy which leads to the above behavior requires a crystalline structure in the medium although not all crystals show this effect. In optics, this property in crystals is called birefringence, and the crystals are called birefringent crystals.

5. Boundary Conditions

The changes in electric and magnetic fields at a discontinuity in the medium are limited by conditions imposed by the Maxwell equations. Such boundary conditions have already been discussed in Chapter 3 for the electric field and the displacement vector when these fields are independent of time. Here we find that for time-varying fields the boundary conditions are not greatly different, and the argument is extended to include **B** and **H** as well. The discussion is confined to plane waves and to abrupt changes in the medium. What happens when the medium changes continuously can be found in more advanced texts on optics.

To establish the boundary conditions, we use the integral forms of the field equations:

$$\int \mathbf{D} \cdot d\mathscr{A} = \int \rho \, d\mathscr{V} \tag{5-1a}$$

$$\int \mathbf{B} \cdot d\mathscr{A} = 0 \tag{5-1b}$$

$$\oint \mathbf{E} \cdot d\mathbf{l} = -\int \dot{\mathbf{B}} \cdot d\mathscr{A} \tag{5-1c}$$

$$\oint \mathbf{H} \cdot d\mathbf{l} = \int \mathbf{J} \cdot d\mathscr{A} + \int \dot{\mathbf{D}} \cdot d\mathscr{A} \tag{5-1d}$$

Let M-N in Figure 6 represent a surface of discontinuity between medium 1 and medium 2. A Gaussian volume is constructed in the form of a right cylinder with generators normal to the boundary and with end areas on opposite sides of the boundary and infinitesimally close together. Let the end areas be $\mathscr{A}$, sufficiently small so that the fields are effectively constant over their extent.

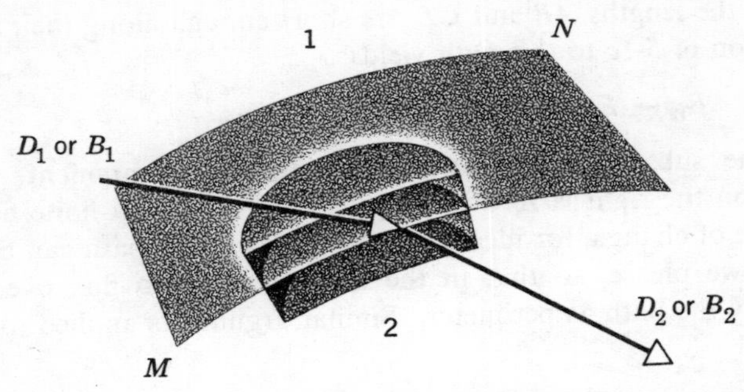

Figure 6. A Gaussian volume at a surface of discontinuity between medium 1 and medium 2.

The left side of Equation 5-1a integrates the normal component of **D** over the area of the cylinder, while the right side is just the surface charge enclosed by the cylinder. Thus,

$$D_{n2}\mathscr{A} - D_{n1}\mathscr{A} = \sigma\mathscr{A} \tag{5-2}$$

or

$$D_{n2} - D_{n1} = \sigma$$

where the subscript n indicates the normal component, and σ is the surface charge density. By the same argument, Equation 5-1b yields

$$B_{n2} - B_{n1} = 0 \tag{5-3}$$

At this same interface let us describe a closed path, $ABCD$ (Figure 7) in which AB and CD are parallel to each other and to the interface, infinitesimally separated and on either side of the interface. The integral on the left of Equation 5-1c integrates the component of **E** parallel to $d\mathbf{l}$; that is, the component tangential to the surface, around the closed

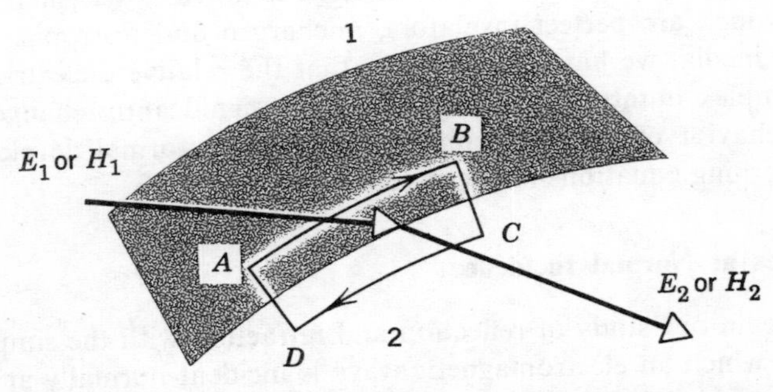

Figure 7. A closed path of a line integral at a surface of discontinuity.

path. If the lengths AB and CD are short enough along their extent, the application of 5-1c to this path yields

$$E_{t2} - E_{t1} = 0 \tag{5-4}$$

where the subscript t indicates the tangential component. The area integral on the right is zero so long as the field **B** is a finite field with a finite rate of change, for ideally the two sides of the path can be made as close as we please, so that in the limit there is no flux over the area defined by the path as perimeter. Similar arguments applied to Equation 5-1d yield

$$H_{t2} - H_{t1} = J_s \tag{5-5}$$

where J_s is a surface density of current, ideally a layer of current of no thickness. Such a concept is of value when we come to use ideal conductors on the surface of which there may be infinite volume current densities.

The boundary conditions to which electromagnetic field components are subject can be summarized as follows:

The normal component of the displacement changes across a surface of discontinuity by the surface density of free charge (Equation 5-2). At an uncharged surface the normal component of **D** is continuous.

The tangential component of the electric field across an interface between two media is continuous (Equation 5-4).

The normal component of the flux density is continuous across a surface of discontinuity (Equation 5-3).

The tangential component of the magnetic field changes across a surface of discontinuity by the surface current density (Equation 5-5). At an interface between nonconductors the tangential component of **H** is continuous.

These boundary conditions are now employed in the study of the reflection and refraction of electromagnetic waves. We shall deal with media which are perfect insulators, uncharged and isotropic. For conducting media, we have already seen that the relative dielectric constant is a complex number. Although this may significantly change the ultimate behavior of the electromagnetic wave, the formal development of the governing equations is not affected.

6. Waves at Normal Incidence

We begin our study of reflection and refraction with the simplest case, that for which an electromagnetic wave is incident normally at a surface between two dielectric media. In Figure 8 is shown an incident plane

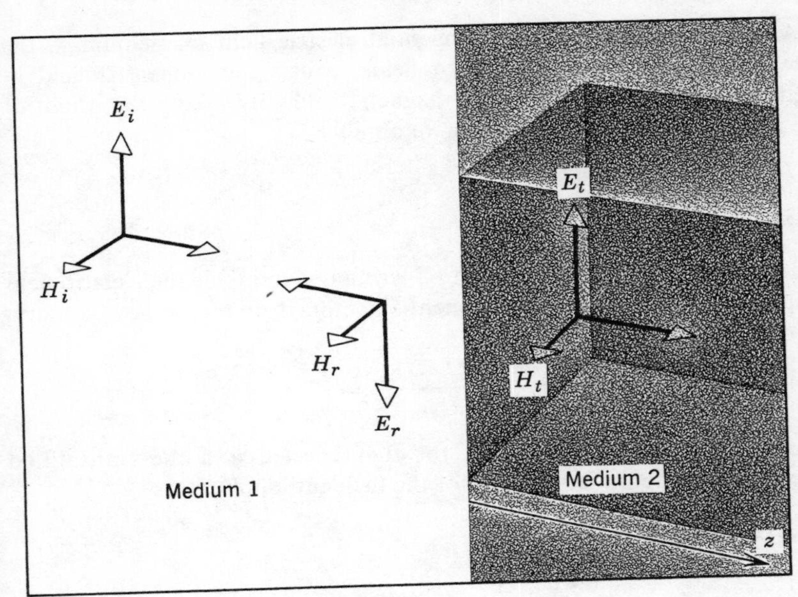

Figure 8. Plane wave normally incident at a plane boundary between two dielectrics.

wave in medium 1, with mutually perpendicular $\mathbf{E}_i$ and $\mathbf{H}_i$, traveling normally toward the boundary of medium 2. A reflected wave, $\mathbf{E}_r$-$\mathbf{H}_r$ is traveling away from the boundary in medium 1, and a transmitted wave, $\mathbf{E}_t$-$\mathbf{H}_t$, away from the boundary in medium 2.

We take the components of the incident wave to be

$$E_i = E_{0i}e^{j(\omega t - \beta_1 z)} \qquad H_i = H_{0i}e^{j(\omega t - \beta_1 z)} \tag{6-1a}$$

Similarly the reflected and transmitted fields are

$$E_r = E_{0r}e^{j(\omega t + \beta_1 z)} \qquad H_r = H_{0r}e^{j(\omega t + \beta_1 z)} \tag{6-1b}$$

$$E_t = E_{0t}e^{j(\omega t - \beta_2 z)} \qquad H_t = H_{0t}e^{j(\omega t - \beta_2 z)} \tag{6-1c}$$

where

$$H_i = \frac{E_i}{\eta_1} \qquad H_r = -\frac{E_r}{\eta_1} \qquad H_t = \frac{E_t}{\eta_2} \tag{6-2}$$

$$\eta_1 = \sqrt{\mu_1/\epsilon_1} \quad \eta_2 = \sqrt{\mu_2/\epsilon_2} \quad \beta_1 = \omega\sqrt{\mu_1\epsilon_1} \quad \beta_2 = \omega\sqrt{\mu_2\epsilon_2}$$

and where the subscripts 1 and 2 identify the medium. Note the change of sign required in the reflected wave because of the reversal of the direction of propagation.

At the boundary the total tangential electric field in medium 1, the sum of the incident and reflected fields, equals the tangential field in medium 2, and similarly for the magnetic field. By taking the origin of the coordinate z at the boundary, we obtain

$$E_i + E_r = E_t \qquad \text{(6-3}a\text{)}$$

$$H_i + H_r = H_t \qquad \text{(6-3}b\text{)}$$

But Equations 6-3*a* and 6-3*b* can be written in terms of the electric field amplitudes by using 6-2 (the exponential factors drop out at the boundary, $z = 0$):

$$E_{0i} + E_{0r} = E_{0t} \qquad \frac{E_{0i} - E_{0r}}{\eta_1} = \frac{E_{0t}}{\eta_2} \qquad \text{(6-4)}$$

These two relations are solved for the reflected and the transmitted electric field amplitudes in terms of the incident amplitude:

$$E_{0r} = E_{0i}\frac{\eta_2 - \eta_1}{\eta_2 + \eta_1} \qquad \text{(6-5}a\text{)}$$

$$E_{0t} = E_{0i}\frac{2\eta_2}{\eta_2 + \eta_1} \qquad \text{(6-5}b\text{)}$$

The corresponding magnetic field amplitudes are

$$H_{0r} = -\frac{E_{0r}}{\eta_1} = -H_{0i}\frac{\eta_2 - \eta_1}{\eta_2 + \eta_1} \qquad \text{(6-6}a\text{)}$$

$$H_{0t} = \frac{E_{0t}}{\eta_2} = H_{0i}\frac{2\eta_1}{\eta_2 + \eta_1} \qquad \text{(6-6}b\text{)}$$

If medium 2 has a greater impedance than medium 1 (this usually means that $\epsilon_2 < \epsilon_1$, since except for ferromagnetic materials the change in the permeability from one medium to another is negligible), then the reflected electric field is in the same direction at the boundary as the incident electric field, whereas the magnetic field reverses. If medium 2 has a smaller impedance, it is the electric field that reverses. This situation has a close analogy to the behavior of voltage and current at the junction of two transmission lines, the analogy becoming exact with ideal lines having pure real characteristic impedances.

Now consider what happens when medium 2 is a perfectly conducting medium. We found in Section 3 that the characteristic impedance of such a medium is zero. From Equations 6-5 it follows that

$$E_{0r} = -E_{0i} \quad \text{and} \quad E_{0t} = 0 \qquad \text{(6-7)}$$

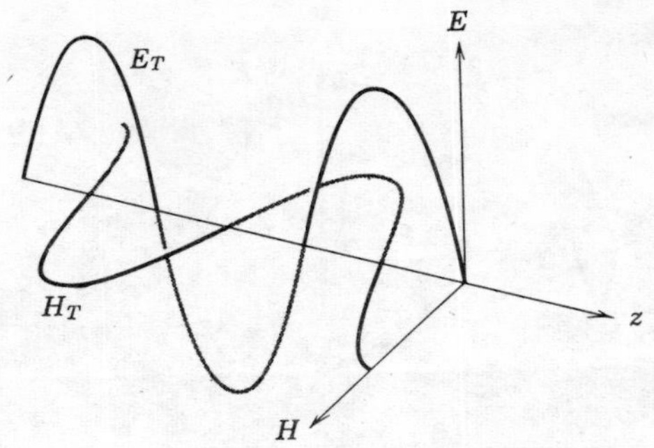

Figure 9. Plots of real parts of E_T and H_T versus z at $\omega t = \pi/4$.

while from 6-6 we get

$$H_{0r} = H_{0i} \qquad \text{and} \qquad H_{0t} = 2H_{0i} \tag{6-8}$$

There is thus no propagating wave in the second medium. The total electric field in medium 1 (the sum of the forward and the reflected waves at any instant) is

$$\begin{aligned} E_T &= E_{0i}e^{j(\omega t - \beta_1 z)} - E_{0i}e^{j(\omega t + \beta_1 z)} \\ &= -2jE_{0i}e^{j\omega t} \sin \beta_1 z \\ &= 2E_{0i}e^{j[\omega t - (\pi/2)]} \sin \beta_1 z \end{aligned} \tag{6-9}$$

whereas the total magnetic field is

$$\begin{aligned} H_T &= H_{0i}e^{j(\omega t - \beta_1 z)} + H_{0i}e^{j(\omega t + \beta_1 z)} \\ &= 2H_{0i}e^{j\omega t} \cos \beta_1 z \end{aligned} \tag{6-10}$$

showing that the total electric and magnetic fields are standing waves, with the two fields one-quarter cycle out of phase both in time and in space (Figure 9).

7. Reflection at Oblique Incidence—Snell's Laws

Reflection and transmission of an electromagnetic wave at other than normal incidence on a dielectric interface introduces some interesting new problems. A plane wave, $\mathbf{E}_i$-$\mathbf{H}_i$, is incident on the boundary between medium 1 and medium 2 with its direction of propagation making an angle θ_i with the normal to the boundary. Let $\mathbf{m}_i$ be a unit vector in the direction of propagation of this incident wave, and let $\mathbf{r}$ be a vector drawn from an origin in the boundary to a point on a phase plane of $\mathbf{E}_i$-$\mathbf{H}_i$ (Figure 10). The distance of this phase plane from the origin is

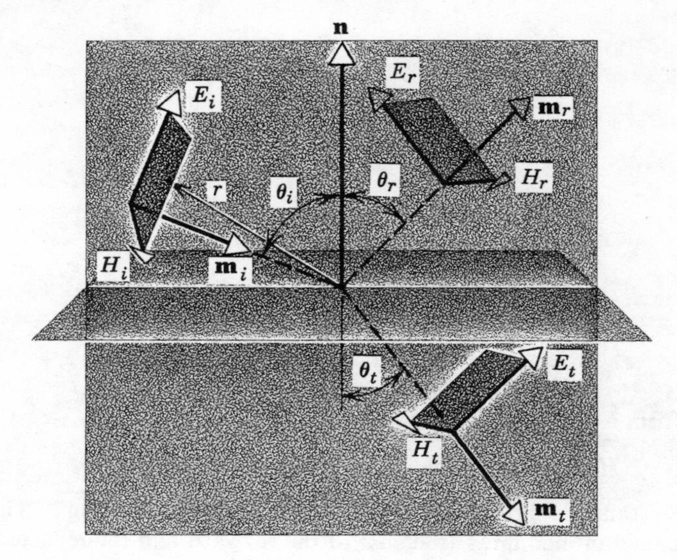

Figure 10. *Waves at oblique incidence on a boundary.*

$\mathbf{m}_i \cdot \mathbf{r}$, so that the electric and magnetic components of the incident wave are:

(7-1*a*) $$E_i = E_{0i} e^{j(\omega t - \beta_1 \mathbf{m}_i \cdot \mathbf{r})}$$

(7-1*b*) $$H_i = H_{0i} e^{j(\omega t - \beta_1 \mathbf{m}_i \cdot \mathbf{r})} = \frac{E_{0i}}{\eta_1} e^{j(\omega t - \beta_1 \mathbf{m}_i \cdot \mathbf{r})}$$

where η_1 and β_1 were given in 6-2.

In like manner we may treat the reflected and transmitted waves, $\mathbf{E}_r$-$\mathbf{H}_r$, and $\mathbf{E}_t$-$\mathbf{H}_t$, respectively. Let $\mathbf{m}_r$ be a unit vector in the direction of propagation of the reflected wave (without assuming that this wave and the incident wave make equal angles with the normal to the interface), and let $\mathbf{m}_t$ be a unit vector in the direction of propagation of the transmitted wave in medium 2. Then the field components of the reflected and transmitted electromagnetic waves can be written:

(7-2*a*) $$E_r = E_{0r} e^{j(\omega t - \beta_1 \mathbf{m}_r \cdot \mathbf{r})}$$

(7-2*b*) $$H_r = H_{0r} e^{j(\omega t - \beta_1 \mathbf{m}_r \cdot \mathbf{r})} = \frac{E_{0r}}{\eta_1} e^{j(\omega t - \beta_1 \mathbf{m}_r \cdot \mathbf{r})}$$

(7-3*a*) $$E_t = E_{0t} e^{j(\omega t - \beta_2 \mathbf{m}_t \cdot \mathbf{r})}$$

(7-3*b*) $$H_t = H_{0t} e^{j(\omega t - \beta_2 \mathbf{m}_t \cdot \mathbf{r})} = \frac{E_{0t}}{\eta_2} e^{j(\omega t - \beta_2 \mathbf{m}_t \cdot \mathbf{r})}$$

For pictorial purposes the three wave planes have been drawn widely separated, but we are interested in those phase planes of the waves which intersect at a common point in the boundary, where the boundary conditions on the fields can be used to establish relations among the fields. Let $\mathbf{r}_0$ be a vector lying in the boundary plane and terminating at such a point. Equation 5-4 fixes a relation among the tangential components of the electric field at this point:

$$E_{0i(\tan)}e^{j(\omega t-\beta_1\mathbf{m}_i\cdot\mathbf{r}_0)} + E_{0r(\tan)}e^{j(\omega t-\beta_1\mathbf{m}_r\cdot\mathbf{r}_0)} = E_{0t(\tan)}e^{j(\omega t-\beta_2\mathbf{m}_t\cdot\mathbf{r}_0)} \tag{7-4}$$

a relation that must hold for any instant t and at any point in the boundary. Equation 7-4 is independent of t and $\mathbf{r}_0$, provided the exponents of the exponential factors are all equal and thus can be canceled. This boundary condition therefore requires, as has been assumed in the writing of the relation, that the frequencies of the reflected and the refracted waves be the same as that of the incident wave. It also requires that

$$\beta_1\mathbf{m}_i\cdot\mathbf{r}_0 = \beta_1\mathbf{m}_r\cdot\mathbf{r}_0 = \beta_2\mathbf{m}_t\cdot\mathbf{r}_0 \tag{7-5}$$

The first equality in 7-5 may be rewritten as

$$\beta_1(\mathbf{m}_i - \mathbf{m}_r)\cdot\mathbf{r}_0 = 0 \tag{7-6}$$

which says that the vector difference between $\mathbf{m}_i$ and $\mathbf{m}_r$ is a vector normal to $\mathbf{r}_0$ and hence normal to the boundary plane, from which we conclude that $\mathbf{m}_i$, $\mathbf{m}_r$, and $\mathbf{n}$, the normal to the boundary can be drawn in a single plane. Furthermore, since $\mathbf{m}_i\cdot\mathbf{r}_0 = -r_0\sin\theta_i$ and $\mathbf{m}_r\cdot\mathbf{r}_0 = -r_0\sin\theta_r$, we conclude that

$$\sin\theta_i = \sin\theta_r \tag{7-7}$$

corresponding to the well-known Snell's law of reflection in optics.

The second equality in Equation 7-5 may now be rewritten as

$$(\beta_1\mathbf{m}_i - \beta_2\mathbf{m}_t)\cdot\mathbf{r}_0 = 0 \tag{7-8}$$

from which we conclude that $\mathbf{m}_i$ and $\mathbf{m}_t$ can also be drawn in a plane with $\mathbf{n}$, and furthermore that

$$\beta_1\sin\theta_i = \beta_2\sin\theta_t \tag{7-9}$$

which becomes, on substituting for the β's:

$$\sqrt{\mu_1\epsilon_1}\sin\theta_i = \sqrt{\mu_2\epsilon_2}\sin\theta_t \tag{7-10}$$

Snell's law of refraction in optics is

$$n_1\sin\theta_i = n_2\sin\theta_t \tag{7-11}$$

where n_1 and n_2 are relative indices of refraction. The index of refraction of a medium is inversely proportional to the velocity of light in that medium, so that

$$\frac{v_2}{v_1} = \frac{n_1}{n_2} \tag{7-12}$$

But the radicals in Equation 7-10 are the reciprocal velocities in media 1 and 2 and hence are proportional to the refractive indices. In all media in which optical properties can be tested, the permeability differs negligibly from that of free space, so that, if light is to be regarded as being an electromagnetic wave, there should be a correspondence between the optically measured index of refraction and the electrically measured square root of the relative dielectric constant. Both quantities are functions of the frequency, and when both are measured at the same frequency, the equivalence is experimentally confirmed. Obtaining Snell's optical laws from electromagnetic theory in this manner further confirms the electromagnetic nature of light.

8. Parallel and Perpendicular Polarization

The derivation of Snell's laws in Section 7 required no specification or restriction of the orientation of the electric and magnetic fields in the incident, reflected, or refracted waves. Only the fact that the boundary conditions required the exponential factor to drop out was used. But when we seek relations among the amplitudes of the beams, it will be necessary to consider separately the components of the fields which are in the plane of incidence—the plane defined by the boundary normal and the vector $\mathbf{m}_i$—and the components which are perpendicular to the plane of incidence.

When the incident wave has its electric field in (and the magnetic field normal to) the plane of incidence, it is said to be polarized parallel to the plane of incidence; whereas when the electric field is normal to the plane of incidence, the wave is said to be polarized normal to the plane of incidence. This identification is conventional to electromagnetic theory, but it is unhappily opposite to that which by tradition has come to be used in optics.

Since the fields in the incident wave can always be resolved into components parallel and normal to the plane of incidence, we may regard any such wave as the sum of two waves. In one the electric field is in the plane of incidence. In the other the magnetic field is in the plane of incidence. We shall find that the reflected and transmitted wave amplitudes are not the same for the two types of polarization. After the parallel and normal parts of the reflected and transmitted wave have been found, the resultant waves are obtained by combining the two polarizations vectorially.

9. Oblique Incidence—Parallel Polarization

Figure 11 shows an incident wave approaching a surface of discontinuity between two media, with the electric field in the plane of incidence and the magnetic field normal to it. The direction of propagation makes an angle θ_i with the normal to the boundary. In accordance with Snell's law of reflection, the angle of reflection is also θ_i and the angle of refraction is θ_t.

The continuity of the tangential components of the electric field at the boundary has already been expressed in Equation 7-4, which we now use again, but we shall henceforth no longer write the exponential factors since they have been shown to drop out at the boundary. If the incident amplitude of the electric field is E_{0i}, then its component parallel to the boundary is $E_{0i} \cos \theta_i$, with similar forms for the reflected and transmitted fields. Thus Equation 7-4 becomes

$$E_{0i} \cos \theta_i - E_{0r} \cos \theta_i = E_{0t} \cos \theta_t \tag{9-1}$$

The magnetic fields in the three waves are entirely parallel to the boundary, and hence the continuity of the magnetic field at the boundary is

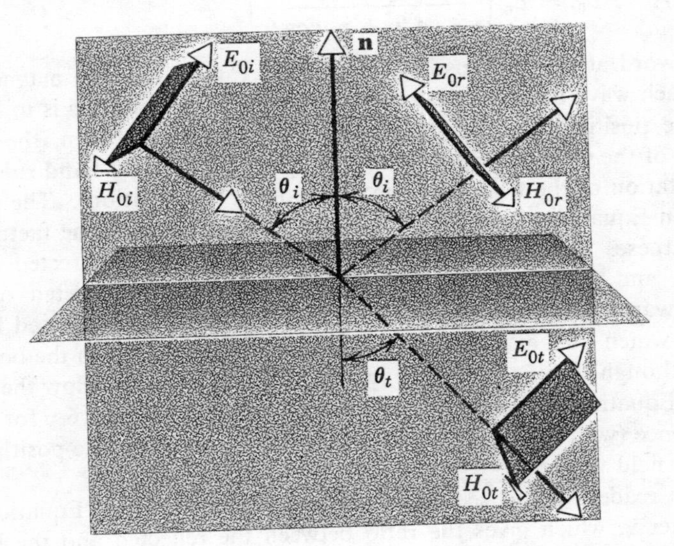

Figure 11. Reflection and refraction for parallel polarization.

given by

$$H_{0i} + H_{0r} = H_{0t} \tag{9-2}$$

Using

$$\eta_1 = \frac{E_{0i}}{H_{0i}} = \frac{E_{0r}}{H_{0r}} \qquad \eta_2 = \frac{E_{0t}}{H_{0t}} \tag{9-3}$$

Equation 9-1 becomes

$$\eta_1 \cos \theta_i (H_{0i} - H_{0r}) = \eta_2 H_{0t} \cos \theta_t \tag{9-4}$$

From Equations 9-2 and 9-4 we solve for H_{0r} and H_{0t} in terms of H_{0i}, the characteristic impedances of the media, and the angles of incidence and refraction. Thus

$$H_{0r} = -H_{0i}\left(\frac{\eta_2 \cos \theta_t - \eta_1 \cos \theta_i}{\eta_2 \cos \theta_t + \eta_1 \cos \theta_i}\right) \tag{9-5a}$$

$$H_{0t} = H_{0i}\left(\frac{2\eta_1 \cos \theta_i}{\eta_2 \cos \theta_t + \eta_1 \cos \theta_i}\right) \tag{9-5b}$$

from which, with the aid of 9-3, we get

$$E_{0r} = -E_{0i}\left(\frac{\eta_2 \cos \theta_t - \eta_1 \cos \theta_i}{\eta_2 \cos \theta_t + \eta_1 \cos \theta_i}\right) \tag{9-6a}$$

$$E_{0t} = E_{0i}\left(\frac{2\eta_2 \cos \theta_i}{\eta_2 \cos \theta_t + \eta_1 \cos \theta_i}\right) \tag{9-6b}$$

A word about the sense of the fields in Figure 11. The magnetic field for each wave has been drawn out from the paper and this is to be taken as the positive sense for this field. For each beam, then, the positive sense of the electric field is obliquely upward (cf: right-hand rule for the orientation of the fields and the direction of propagation). The negative sign in Equations 9-5*a* and 9-6*a* shows that (so long as the factor in the parentheses is positive) the reflected magnetic field is directed into the paper and necessarily the reflected electric field is directed obliquely downward. It should be kept in mind that the fields discussed here are those which exist at a given instant and at a given point on the boundary, even though they have been drawn at points above and below the boundary. Equations 9-5 and 9-6 reduce to Equations 6-5 and 6-6 for normal incidence (with due regard for the conventions used for the positive sense of the field vectors).

The reader should see the analogy between 9-6*a* and Equation 12-2, Chapter 9, which gives the ratio between the reflected and the forward wave amplitudes on a transmission line.

10. Oblique Incidence—Normal Polarization

For this component of polarization we take the electric field normal to the plane of incidence, drawn outward from the plane of the page, and the magnetic field obliquely downward, for propagation in the directions indicated in Figure 12. The continuity of the tangential components of H require that

$$-H_{0i} \cos \theta_i + H_{0r} \cos \theta_i = -H_{0t} \cos \theta_t \tag{10-1}$$

while the continuity of the tangential components of E yields

$$E_{0i} + E_{0r} = E_{0t} \tag{10-2}$$

After introducing the electric field amplitudes into Equation 10-1 from Equations 9-3, we can solve for the electric field amplitudes to get

$$E_{0r} = E_{0i} \left(\frac{\dfrac{\eta_2}{\cos \theta_t} - \dfrac{\eta_1}{\cos \theta_i}}{\dfrac{\eta_2}{\cos \theta_t} + \dfrac{\eta_1}{\cos \theta_i}} \right) \tag{10-3a}$$

$$E_{0t} = E_{0i} \left(\frac{\dfrac{2\eta_2}{\cos \theta_t}}{\dfrac{\eta_2}{\cos \theta_t} + \dfrac{\eta_1}{\cos \theta_i}} \right) \tag{10-3b}$$

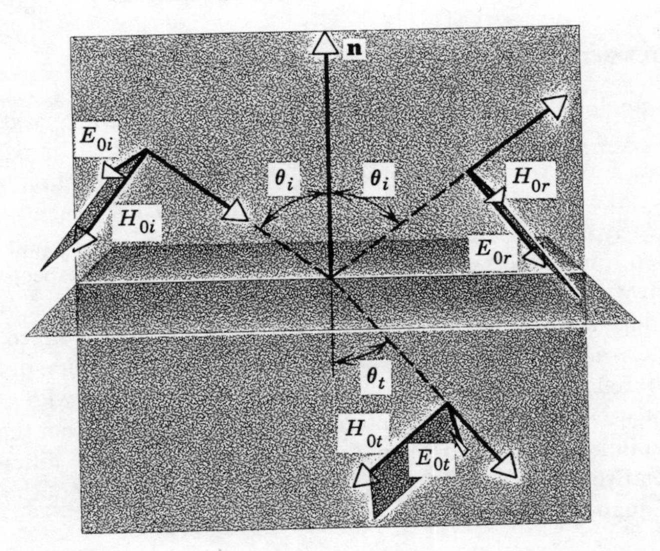

Figure 12. Reflection and refraction for normal polarization.

and, again with 9-3, the magnetic field amplitudes are

$$H_{0r} = H_{0i}\left(\frac{\dfrac{\eta_2}{\cos\theta_t} - \dfrac{\eta_1}{\cos\theta_i}}{\dfrac{\eta_2}{\cos\theta_t} + \dfrac{\eta_1}{\cos\theta_i}}\right) \tag{10-4a}$$

$$H_{0t} = H_{0i}\left[\frac{\dfrac{2\eta_1}{\cos\theta_t}}{\dfrac{\eta_2}{\cos\theta_t} + \dfrac{\eta_1}{\cos\theta_i}}\right] \tag{10-4b}$$

which also reduce to Equations 6-5 and 6-6 at normal incidence when proper account is taken of the convention employed to designate the positive sense of the field vectors.

Equations 9-5, 9-6, 10-3, and 10-4 were first derived for the reflection and refraction of light (with differences in notation) by Augustin Jean Fresnel in 1823 on the assumption that light is a vibration in an elastic ether. That they are a consequence of electromagnetic theory is still another indication of the electromagnetic nature of light.

The similarity in the forms of the relations for the reflected and transmitted amplitudes to the corresponding relations for transmission lines (Chapter 9, Problem 22) should be noted. In the next section these equations are applied to some interesting special cases.

11. Brewster's Angle and the Critical Angle

The analogy with transmission lines suggests that, with parallel polarization, the function $\eta_2 \cos\theta_t$ is in the nature of a load impedance on medium 1 for waves incident at an angle θ_i. The function $\eta_1 \cos\theta_i$ behaves like a characteristic impedance of medium 1 for such waves.

Similarly, for normal polarization the functions $\eta_2/\cos\theta_t$ and $\eta_1/\cos\theta_i$ are load impedances and characteristic impedances, respectively, for incident waves at an angle θ_i.

In transmission lines we saw that when the load impedance and the characteristic impedance matched, there was total transmission of the wave and no reflection. Let us look for analogous behavior with obliquely incident waves at a dielectric interface. We shall assume that the relative permeabilities of the two media are equal and that they differ only in their relative dielectric constants. For parallel polarization of the incident wave Equations 9-5*a* and 9-6*a* show that when

$$\eta_2 \cos\theta_t' = \eta_1 \cos\theta_i' \tag{11-1}$$

where θ_t' and θ_i' are the particular angles that satisfy 11-1, there is no reflected amplitude of either the electric or the magnetic field. With Snell's law, Equation 11-1 becomes

$$\sqrt{\mu_2/\epsilon_2}\sqrt{1 - (\mu_1\epsilon_1/\mu_2\epsilon_2)\sin^2\theta_i'} = \sqrt{\mu_1/\epsilon_1}\sqrt{1 - \sin^2\theta_i'} \tag{11-2}$$

from which

$$\sin\theta_i' = \sqrt{\epsilon_2/(\epsilon_1 + \epsilon_2)} \tag{11-3a}$$

or

$$\tan\theta_i' = \sqrt{\epsilon_2/\epsilon_1} \tag{11-3b}$$

Since the right side of 11-3*a* is less than unity, there is always an angle of incidence for which there is no reflected amplitude when the incident wave is polarized parallel to the plane of incidence.

For normal polarization, Equations 10-3*a* and 10-4*a* show that for no reflection

$$\frac{\eta_2}{\cos\theta_t'} = \frac{\eta_1}{\cos\theta_i'} \tag{11-4}$$

and substitutions from Snell's law into 11-4 show that no incident angle can be found to satisfy this relation. Therefore, in an incident wave of mixed polarization, incident at the angle defined by 11-3, the parallel component is completely transmitted while the normal component is partially transmitted and partially reflected. The reflected wave is thus completely polarized normal to the plane of incidence.

The angle θ_i' is known as Brewster's angle or the polarizing angle. The earliest method of producing and detecting polarized light employed a glass plate (or a stack of plates for greater sensitivity) oriented at this angle with respect to a beam of light. A second glass plate, or stack of plates, properly oriented relative to the reflected wave served as an analyzer. Partial polarization occurs at other angles, in that the two components are not reflected with equal amplitude. Use is made of this fact to reduce glare of light reflected from shiny objects by means of polaroid filters.

In transmission lines complete reflection occurs when the terminating impedance is zero, infinite, or a pure imaginary quantity. We look for circumstances in which the impedances of the second medium for obliquely incident waves takes on such extreme values. If the second medium is a good conductor, Equation 3-2 shows that its characteristic impedance is very close to zero, and Equations 9-5*a* and 10-4*a* show that there is then reflection at an equal amplitude with the incident wave, but with a reversal of the phase. For dielectrics, the characteristic impedance of the medium is a pure real number of finite magnitude, but $\cos\theta_t$ can be zero or imaginary. Since

$$\cos\theta_t = \sqrt{1 - (\epsilon_1/\epsilon_2)\sin^2\theta_i} \tag{11-5}$$

when

$$\sin\theta_i = \sqrt{\epsilon_2/\epsilon_1} \tag{11-6}$$

the impedance which the second medium presents to the incident beam is zero for parallel polarization and infinite for normal polarization. For angles of incidence greater than that given by Equation 11-6, these impedances are either positive or negative imaginary. Thus for an incident wave of either polarization where the angle of incidence is greater than $\arcsin\sqrt{\epsilon_2/\epsilon_1}$, there is complete reflection and no transmission. But 11-6 shows that such an angle has a real value only when the second medium has a smaller dielectric constant than the first. In optical terms, the light must be traveling in the denser and toward the less dense medium in order for total reflection to occur. The angle given by Equation 11-6 is called the critical angle.

Discussion questions

1. From your knowledge of transmission lines, discuss the method of handling the following problem: Determine the reflected amplitude of an obliquely incident wave on a dielectric layer of thickness *d* covering a semi-infinite second dielectric.

2. Can the velocity of energy propagation in an anisotropic medium be directly measured for a steady-state monochromatic wave? If not, how might the energy velocity be measured?

Problems

1. An electromagnetic wave is incident normally on the surface of a medium of relative dielectric constant 10 and conductivity 5 mhos per meter. The frequency of the wave is 10^{10} cycles per second. Find the amplitude and the phase of the reflected fields relative to the incident fields and the relative amplitude of the transmitted fields just at the surface. Find the skin depth.

2. Consider the questions of Problem 1 for a good conductor such as silver, with $\sigma = 6 \times 10^7$ mhos per meter. Find the Poynting vector in silver at the skin depth.

3. What is the skin resistance for electromagnetic waves of frequency 3×10^9 cycles per second on the inside of a silver pipe of rectangular cross-section with dimensions 4×7 cm.

4. The three relative dielectric constants of an anisotropic medium are 3.76, 4.84, and 6.30. The relative permeability is unity. Assign proper axes and determine the directions in which there is a single velocity of propagation. What is the angle between these two directions? What is this velocity?

5. In the medium of Problem 4, what are the two phase velocities associated with the direction that makes equal angles with the three coordinate axes? What are the velocities associated with the direction parallel to the y-axis?

6. For both of the directions in Problem 5, find $\mathbf{D}_1$ and $\mathbf{D}_2$ and find the directions of the ray velocities. Test whether $\mathbf{D}_1$, $\mathbf{D}_2$, and the phase velocity are mutually perpendicular. Illustrate these vectors in diagrams.

7. A thick piece of glass has a relative dielectric constant of 2.50. A light wave, incident on this glass at an angle of 30° with the normal, is polarized at 45° to the plane of incidence. What angle does the resultant electric field in the reflected wave make with the plane of incidence? What angle does the electric field in the refracted wave make with the plane of incidence?

8. Plot the reflected relative amplitude of each polarization against the angle of incidence for the glass in the previous problem. Make the graph for both air to glass and glass to air incidence.

9. The relative dielectric constant for a piece of flint glass is 3.50. A light wave reaches the glass air surface from the glass side, its direction making an angle of 60° with the normal to the surface. The light is plane polarized, with the plane of polarization making an angle of 45° with the plane of incidence. Find the magnitudes of the reflected amplitudes for each component of the incident wave and find the difference in phase between each reflected component and the incident wave. Discuss fully the nature of the polarization of the reflected beam.

10. An incident wave is polarized at 45° to the plane of incidence and the angle of incidence is Brewster's angle. What is the angle between the directions of the reflected and the transmitted waves?

11. Incident waves approach an interface from either side, each wave being incident at Brewster's angle. What is the angle between the directions of the two waves?

12. Show that the reflection and transmission ratios can be put into the forms:

$$\frac{E_{0rp}}{E_{0ip}} = \frac{H_{0rp}}{H_{0ip}} = \frac{\sin\theta_i\cos\theta_i - \sin\theta_t\cos\theta_t}{\sin\theta_i\cos\theta_i + \sin\theta_t\cos\theta_t}$$

$$\frac{E_{0rn}}{E_{0in}} = \frac{H_{0rn}}{H_{0in}} = -\frac{\sin(\theta_i - \theta_t)}{\sin(\theta_i + \theta_t)}$$

$$\frac{E_{0tp}}{E_{0ip}} = \frac{2\cos\theta_i\sin\theta_t}{\sin(\theta_i + \theta_t)\cos(\theta_i - \theta_t)} \qquad \frac{H_{0tp}}{H_{0ip}} = \frac{\sin 2\theta_i}{\sin(\theta_i + \theta_t)\cos(\theta_i - \theta_t)}$$

$$\frac{E_{0tn}}{E_{0in}} = \frac{2\cos\theta_i\sin\theta_t}{\sin(\theta_i + \theta_t)} \qquad \frac{H_{0tn}}{H_{0in}} = \frac{\sin 2\theta_i}{\sin(\theta_i + \theta_t)}$$

where p and n indicate "parallel" and "normal polarizations," respectively. These are called the "Fresnel" ratios.

13. A transparent medium of dielectric constant ϵ_2 is sandwiched between two semi-infinite media of constants ϵ_1 and ϵ_3, respectively. What is the impedance at surface 1-2 for a light wave incident normally from medium 1? What must be the thickness (in terms of the frequency of the incident light) and the relation of ϵ_2 to ϵ_1 and ϵ_3 in order that there be no reflection at the 1-2 surface? (cf: Chapter 9, Problem 15.)

REFERENCES

1. Born and Wolf, *Principles of Optics* (Pergamon Press), 1959.
2. Andrews, *Optics of the Electromagnetic Spectrum* (Prentice-Hall), 1960.
3. Ramo and Whinnery, *op. cit.* Chapter 6 and 7.
4. R. W. Ditchburn, *Light* (Blackie and Son, Limited) London and Glasgow, 1952. Chapter XVI.
5. Moon and Spencer, *Foundations of Electrodynamics* (Van Nostrand), 1960. Chapter 7.
6. Stratton, *Electromagnetic Theory* (McGraw-Hill), 1941. Chapter IX, Sections 9-4 to 9-9.
7. Smythe, *Static and Dynamic Electricity* (McGraw-Hill), 1950. Chapter XIII.

12

Guided electromagnetic waves

In this chapter we examine the influence of multiple boundaries which guide wave propagation in particular directions. Although we devote our attention to conducting boundaries, it should be noted for completeness that a dielectric boundary, such as, for example, the boundary between a plastic cylinder and air, can also guide the propagation of electromagnetic waves. The directivity of such a structure, however, is usually not so complete as with metallic boundaries, and they are in less common use.

One kind of waveguide, the transmission line, has already been considered in some detail in Chapter 9, although it was treated as a kind of a-c circuit that is a transition from lumped circuit theory to wave propagation. We shall consider it from the point of view of a waveguide later in this chapter. It is desirable, however, to begin the discussion of guides with a much simpler case, that of the propagation of waves between two parallel conducting planes.

1. Waves Between Parallel Conducting Planes

For the purpose of analyzing the guiding of waves by parallel planes it is instructive to reexamine the problem of reflection from a plane surface, treated in Chapter 11. That analysis was carried out with the use of vector notation, an economical method that puts much information into a small space. But certain features pertinent to guided waves are brought out more explicitly by an analysis in Cartesian coordinates.

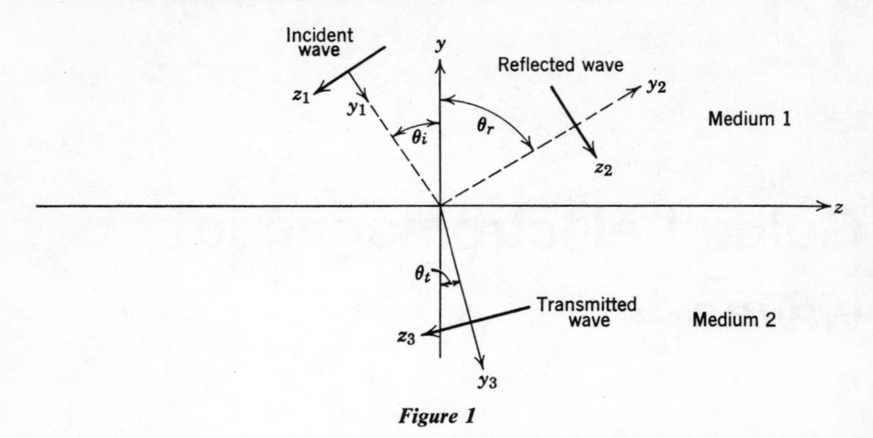

Figure 1

We start by expressing the incident, the reflected, and the refracted plane waves at a plane boundary in terms of Cartesian coordinates, assigning an individual coordinate system to each wave. These waves and their coordinate systems are identified by the subscripts 1, 2, and 3, respectively. Figure 1 shows the three coordinate systems and the directions of the coordinate axes for each wave. For each system the x-axis is directed into the paper. Although the axes for each system are indicated separately, it should be understood that the three systems have a common origin in the boundary surface. Finally, a fourth, unsubscripted, coordinate system is oriented in and normal to the boundary and will serve as a connecting link in relating the three plane waves.

Each plane wave is propagating in the positive direction of its own y-axis. The three electric and the three magnetic components of the fields are expressed as follows, each in terms of its own coordinate system:

(1-1)
$$E_i = E_{0i}e^{j(\omega t - \beta_1 y_1)} \qquad H_i = H_{0i}e^{j(\omega t - \beta_1 y_1)}$$
$$E_r = E_{0r}e^{j(\omega t - \beta_1 y_2)} \qquad H_r = H_{0r}e^{j(\omega t - \beta_1 y_2)}$$
$$E_t = E_{0t}e^{j(\omega t - \beta_2 y_3)} \qquad H_t = H_{0t}e^{j(\omega t - \beta_2 y_3)}$$

These functions are now expressed in terms of a common coordinate system by a method developed in elementary analytic geometry. When the coordinates of a point, P, are y_i, z_i in one system and y, z in another (Figure 2), these coordinates are related by

(1-2)
$$y_i = y \cos \varphi + z \sin \varphi$$
$$z_i = -y \sin \varphi + z \cos \varphi$$

where φ is the angle between the y_i and the y axes, measured in the positive sense from y_i to y. Applying the substitution in Equation 1-2 to the fields in 1-1, we get

$$(1\text{-}3)\qquad \begin{aligned} E_i &= E_{0i}e^{-j\beta_1 z \sin\theta_i}e^{j(\omega t+\beta_1 y \cos\theta_i)} \\ H_i &= H_{0i}e^{-j\beta_1 z \sin\theta_i}e^{j(\omega t+\beta_1 y \cos\theta_i)} \\ E_r &= E_{0r}e^{-j\beta_1 z \sin\theta_r}e^{j(\omega t-\beta_1 y \cos\theta_r)} \\ H_r &= H_{0r}e^{-j\beta_1 z \sin\theta_r}e^{j(\omega t-\beta_1 y \cos\theta_r)} \\ E_t &= E_{0t}e^{-j\beta_2 z \sin\theta_t j}e^{(\omega t+\beta_2 y \cos\theta_t)} \\ H_t &= H_{0t}e^{-j\beta_2 z \sin\theta_t}e^{j(\omega t+\beta_2 y \cos\theta_t)} \end{aligned}$$

The coordinate system for the incident wave has been rotated through an angle $\varphi = \pi - \theta_i$; for the reflected wave, $\varphi = \theta_r$; and for the refracted wave, $\varphi = \pi - \theta_t$.

Equations 1-1 and 1-3 are two different mathematical descriptions for the same physical system of fields. Whereas, for example, the first of 1-1 is an electric field of constant amplitude, E_0, over a plane parallel to the x_1z_1-plane, with its phase plane traveling in the y_1-direction with a velocity ω/β_1, the first of 1-3 is in the form of a nonuniform plane wave whose amplitude over a plane parallel to the xz-plane is a function of z (that is, $E_0e^{-j\beta_1 z \sin\theta_i}$), with its phase plane traveling in the negative y-direction with a velocity of $\omega/\beta_1 \cos\theta_i$. That this interpretation is not unique is seen by the fact that a rearrangement of the exponential terms results in an expression whose form is that of a nonuniform plane wave traveling in the positive z-direction with a velocity $\omega/\beta_1 \sin\theta_i$.

We assume that the incident, reflected, and refracted directions of propagation are coplanar. Using the subscript p to indicate the

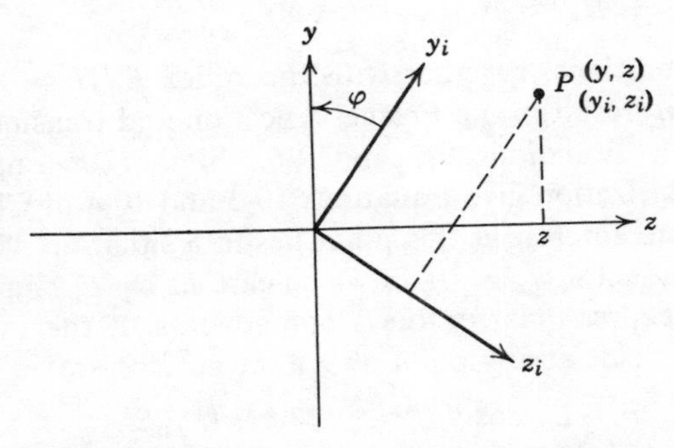

Figure 2. Rotation transformation of coordinate systems.

component of a field vector tangential to the boundary surface, we put the boundary conditions at $y = 0$ in the form:

(1-4)
$$E_{0ip}e^{-j\beta_1 z \sin\theta_i} + E_{0rp}e^{-j\beta_1 z \sin\theta_r} = E_{0tp}e^{-j\beta_2 z \sin\theta_t}$$
$$H_{0ip}e^{-j\beta_1 z \sin\theta_i} + H_{0rp}e^{-j\beta_1 z \sin\theta_r} = H_{0tp}e^{-j\beta_2 z \sin\theta_t}$$

where the common factor $e^{j\omega t}$ has been canceled out. The argument already employed in Chapter 11 that these boundary conditions cannot be a function of where on the boundary they are applied leads to equating the exponential terms, out of which come directly Snell's laws of reflection and refraction.

The boundary conditions are then very simply expressed as

(1-5)
$$E_{0ip} + E_{0rp} = E_{0tp}$$
$$H_{0ip} + H_{0rp} = H_{0tp}$$

Only now need we specify the polarization of the incident wave. As an example, we consider parallel polarization with the electric fields in the direction of the z_1, z_2, and z_3 axes, respectively, and the magnetic fields in the direction of the positive x_1, x_2, and x_3 axes, at a given instant of time. The tangential components of these field amplitudes are

(1-6)
$$E_{iz} = -E_{0i}\cos\theta_i \qquad H_{ix} = H_{0i}$$
$$E_{rz} = E_{0r}\cos\theta_i \qquad H_{rx} = H_{0r}$$
$$E_{tz} = -E_{0t}\cos\theta_t \qquad H_{tx} = H_{0t}$$

and thus Equations 1-5 become

(1-7)
$$-E_{0i}\cos\theta_i + E_{0r}\cos\theta_i = -E_{0t}\cos\theta_t$$
$$H_{0i} + H_{0r} = H_{0t}$$

These two conditions, together with the ratios $E_i/H_i = \eta_1$, $E_r/H_r = \eta_1$, and $E_t/H_t = \eta_2$ result in the Fresnel reflection and transmission relations of Chapter 11, Equations 9-5 and 9-6. Similar operations with perpendicular polarization give Equations 10-3 and 10-4 of Chapter 11.

Our particular interest at this point lies in a situation where medium 2 is a perfect conductor ($\eta_2 = 0$), in which case $E_{0r} = E_{0i}$, and consequently the complete expressions for the z components of the incident and reflected electric fields at any point above the surface are

(1-8)
$$E_{iz} = -E_{0i}\cos\theta_i e^{j(\omega t - \beta_1 z \sin\theta_i)} e^{j\beta_1 y \cos\theta_i}$$
$$E_{rz} = E_{0i}\cos\theta_i e^{j(\omega t - \beta_1 z \sin\theta_i)} e^{-j\beta_1 y \cos\theta_i}$$

The sum of the incident and reflected z components at a given point is the total z component,

$$E_{z\ \text{Total}} = E_{0i} \cos \theta_i[-2j \sin (\beta_1 y \cos \theta_i)]e^{j(\omega t - \beta_1 z \sin \theta_i)} \tag{1-9}$$

The total z component of the electric field is a traveling wave propagating in the z-direction with a velocity ($\omega/\beta_1 \sin \theta_i$) and with an amplitude that varies sinusoidally in the y-direction. Similar relations for the total y component of the electric field and the total magnetic field can easily be obtained:

$$E_{y\ \text{Total}} = -E_{0i} \sin \theta_i[2 \cos (\beta_1 y \cos \theta_i)]e^{j(\omega t - \beta_1 z \sin \theta_i)} \tag{1-10}$$

$$H_{x\ \text{Total}} = 2H_{0i} \cos (\beta_1 y \cos \theta_i)e^{j(\omega t - \beta_1 z \sin \theta_i)}$$

An examination of Equation 1-9 shows that $E_{z\ \text{Total}}$ is zero at all times and all z's when

$$\beta_1 y_n \cos \theta_i = n\pi \tag{1-11}$$

where n is an integer. It follows that we may place another perfectly conducting plane at the level y_n. With the boundary conditions at such a plane satisfied, there will be no change in the fields between the planes, although the fields above y_n will, of course, be eliminated. Physically, what has been done is to produce another reflection at the upper plane, so that the fields now progress in the z-direction in a zig-zag manner guided between the two infinite parallel planes.

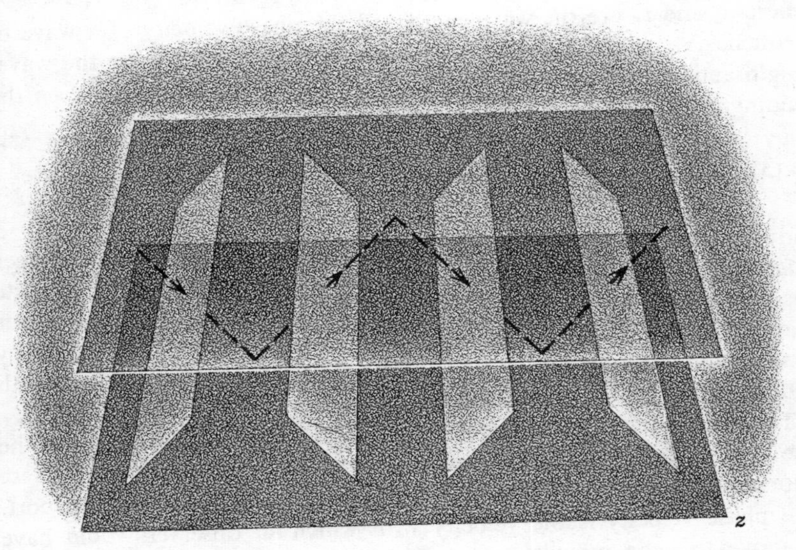

Figure 3. Propagation of a plane wave between two perfectly conducting planes.

By writing $\beta_1 = \omega/v_0$, where v_0 is the velocity of propagation of a wave of frequency ω in infinite space, we can convert Equation 1-11 to

$$\text{(1-12)} \qquad \cos\theta_i = \frac{n\pi v_0}{\omega y_n}$$

showing that for a fixed separation of the guiding planes, only certain angles of reflection enable a wave of a given frequency to propagate between the planes. There is, moreover, for given integer n a frequency, ω_c, below which the angle becomes imaginary and no propagation can take place. In free space propagation a wave of frequency ω_c would have a wavelength λ_c, giving the simple relationship

$$\text{(1-13)} \qquad \lambda_c = \frac{2y_n}{n}$$

Another interesting feature is evident from an examination of $\beta_1 \sin\theta_i = \beta_g$, the propagation constant for the guided wave in the z-direction. In any single frequency (monochromatic) traveling wave, the propagation constant can be written either as ω/v or as $2\pi/\lambda$, where v and λ are the phase velocity and the wavelength, respectively, of the traveling wave. For the wave propagating between the planes we may write

$$\text{(1-14)} \qquad \beta_g = \frac{2\pi}{\lambda_g} = \frac{\omega}{v_g} = \beta_1 \sin\theta_i = \frac{2\pi \sin\theta_i}{\lambda_0} = \frac{\omega \sin\theta_i}{v_0}$$

where λ_0 and v_0 are the wavelength and the phase velocity of a wave of frequency, ω, propagating in infinite space, and λ_g and v_g are the wavelength and phase velocity for propagation in the z-direction between the guiding planes. From Equation 1-14 we obtain

$$\text{(1-15)} \qquad \lambda_g = \frac{\lambda_0}{\sin\theta_i} \qquad v_g = \frac{v_0}{\sin\theta_i}$$

showing that the wavelength of the guided traveling wave is always longer than the wavelength of a plane wave of the same frequency in infinite space, and that the phase velocity of the guided wave is similarly greater than the phase velocity in infinite space. Note that when the frequency of the guided wave equals the cut-off frequency, then both the wavelength and the phase velocity are infinite.

The reader may feel an uneasiness about velocities greater than the free space velocity, but should remember that we are considering phase velocities of a wave, not velocities of masses or rates of energy transport. The phase velocity is the velocity with which an observer would have to move to keep up with some particular phase of the wave—say, a maximum. If we picture a uniform plane wave approaching a surface at an angle θ_i, we should need to move with a velocity v_0 in the direction of travel of the plane wave to stay with a given maximum, but should have to move with a velocity $v_0/\sin\theta_i$ to stay with this maximum if our motion were in the z-direction. The closer to normal incidence, the faster must the observer move parallel to the surface to keep up with a phase. But we shall now show that the rate at which energy progresses between the planes is less than the rate at which a uniform plane wave carries energy in infinite space, where, of course, the rate of energy transport and the phase velocity are the same.

2. Group or Energy Velocity Between Parallel Planes

Similar to the reasoning used in Section 4 of Chapter 11, we shall compute the average energy in the fields within a volume of unit width in the x-direction, of height y_n, and length dz in the direction of propagation, and equate this energy to the average energy which enters this volume in a time dt, where $dt = dz/v_e$, v_e being the average rate of propagation of energy between the planes.

The energy density in the fields is

$$\left(\frac{\epsilon}{2}\right)(E_y^2 + E_z^2) + \left(\frac{\mu}{2}\right)H_x^2 \tag{2-1}$$

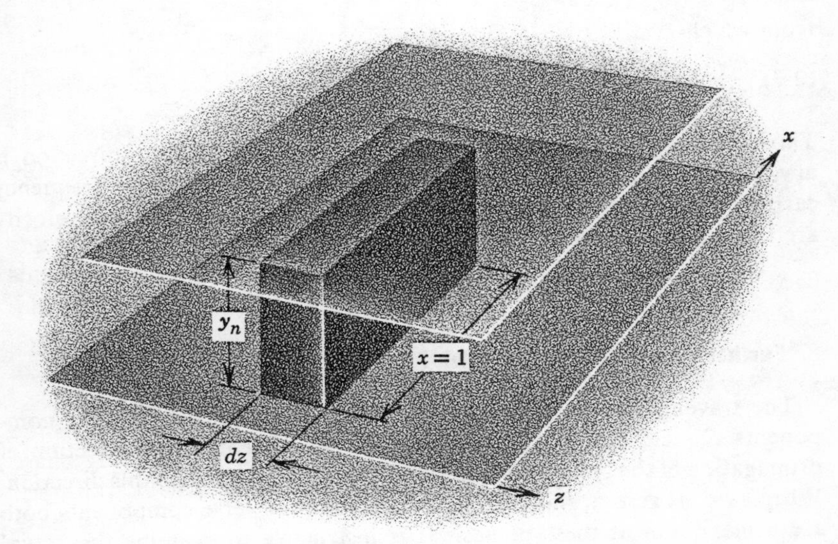

Figure 4. Determination of energy velocity.

The integral of this density over the volume $y_n\, dz$, using Equations 1-9, 1-10, and 1-11, and writing δ for $(\omega t - \beta_1 z \sin \theta_i)$, is

$$dW = y_n\, dz[\epsilon E_{0i}^2(\sin^2 \theta_i \cos^2 \delta + \cos^2 \theta_i \sin^2 \delta) + \mu H_{0i}^2 \cos^2 \delta] \tag{2-2}$$

The time average of dW over a whole number of cycles is $\frac{1}{2}$ of this quantity, since $\overline{\cos^2 \delta} = \overline{\sin^2 \delta} = \frac{1}{2}$. With the relation $E_{0i}/H_{0i} = \eta_1 = \sqrt{\mu/\epsilon}$, the time average energy in the volume becomes

$$\overline{dW} = y_n \mu H_{0i}^2\, dz \tag{2-3}$$

The instantaneous rate at which energy is entering the left face of this volume is the integral of the Poynting vector over this face:

$$\int_0^{y_n} S\, dy = -\int_0^{y_n} E_y H_x\, dy = 2E_{0i}H_{0i}y_n \sin \theta_i \cos^2 \delta \tag{2-4}$$

and the average rate is

$$E_{0i}H_{0i}y_n \sin \theta_i = \sqrt{\frac{\mu}{\epsilon}}\, H_{0i}^2 y_n \sin \theta_i \tag{2-5}$$

The energy entering in a time dt is now equated to the energy within the volume:

$$\sqrt{\frac{\mu}{\epsilon}}\, H_{0i}^2 y_n \sin \theta_i\, dt = \mu H_{0i}^2 y_n\, dz \tag{2-6}$$

from which

$$v_e = \frac{dz}{dt} = \frac{1}{\mu}\sqrt{\frac{\mu}{\epsilon}} \sin \theta_i = v_0 \sin \theta_i \tag{2-7}$$

Thus the average rate at which energy progresses in the z-direction is always less than the rate at which a plane wave of the same frequency carries energy in infinite space. The energy velocity and the phase velocity are related by

$$v_e v_g = v_0^2 \tag{2-8}$$

3. Further Discussion of Parallel Planes

The traveling wave of Equations 1-9 and 1-10 consists of field components E_z, E_y, and H_x, the first a field component in the direction of propagation of the guided wave and the last two transverse to this direction. Whereas E_z is zero at the guiding planes, the transverse components both are a maximum at these planes. It is instructive to examine the actual field patterns which are obtained by writing the real parts of the complex

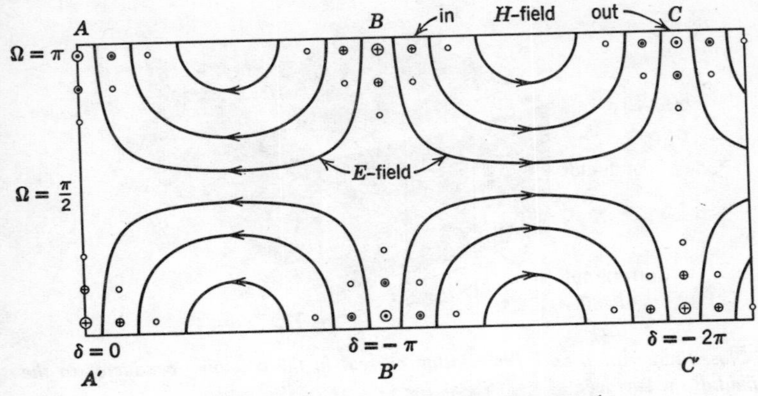

Figure 5. Field pattern between parallel planes.

expressions for the fields. For brevity we shall let $\beta_1 y \cos\theta_i = \Omega$ and again let $(\omega t - \beta_1 z \sin\theta_i) = \delta$.

$$E_z = \text{Real part of } [2E_{0i}\cos\theta_i \sin\Omega e^{-j(\pi/2)} e^{j\delta}] \tag{3-1}$$

$$= 2E_{0i}\cos\theta_i \sin\Omega \cos\left(\delta - \frac{\pi}{2}\right)$$

$$= 2E_{0i}\cos\theta_i \sin\Omega \sin\delta$$

$$E_y = \text{Real part of } [2E_{0i}\sin\theta_i \cos\Omega e^{j\pi} e^{j\delta}]$$

$$= -2E_{0i}\sin\theta_i \cos\Omega \cos\delta$$

$$H_x = \text{Real part of } [2H_{0i}\cos\Omega e^{j\delta}]$$

$$= 2H_{0i}\cos\Omega\cos\delta$$

For $n = 1$ the field pattern is shown in Figure 5. Note that δ decreases as z increases. This pattern moves in the direction of increasing z as time goes on.

To consider what happens within the conducting surfaces, we need a relation between the tangential magnetic field at a surface and the currents within the surface. Figure 6 shows a portion of the conductor forming one of the planes. Just at the surface the tangential magnetic field is H_x. An integration path extends many skin-depths into the conductor (for an ideal conductor the skin-depth is zero and any distance will serve). An integral of the magnetic field around this closed path is $H_x\,dx$ and is equal to the current enclosed. Thus the surface current density is H_x.

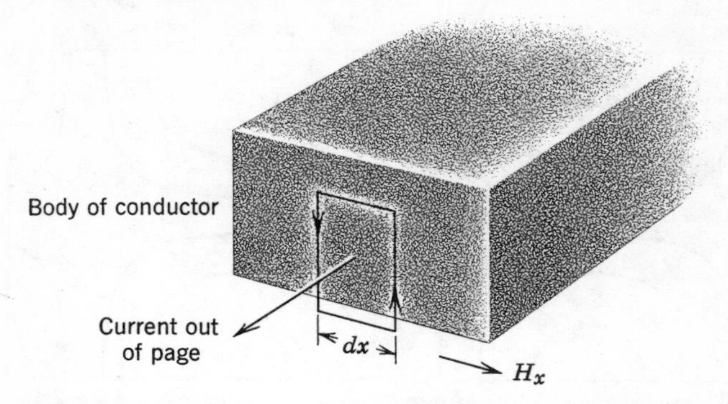

Figure 6. Closed integration path for relating current in the bounding conductor to the magnetic field at the surface.

Electric field lines start on positive charges and end on negative charges, the charge density being equal to the normal component of **D**, or ϵE_y. At A and A' there is an instantaneous negative charge density with a surface current directed toward the left. At B and B' there is a positive charge density with a surface current toward the right. Since the pattern moves toward the right as time goes on, the pattern of charges moves with the fields, and the currents are those consistent with the motion of the charges. The guiding of electromagnetic energy between parallel planes can be considered either in terms of the electromagnetic fields between the planes or in terms of the currents and charges in the conductors. The two descriptions are complementary.

We have discussed the interrelationship of fields with charges and currents in terms of the field pattern obtained when the integer n is set equal to unity. This number fixes the number of half-cycles of $\sin \Omega$ or $\cos \Omega$ across the separation of the planes. The reader will profit from drawing field patterns for other values of n.

In all the foregoing discussion the polarization of the wave is parallel to the plane of incidence. For polarization normal to the plane of incidence there is a somewhat different field pattern, but ω_c, v_g, and λ_g are the same.

With parallel polarization the magnetic field is entirely transverse to the direction of propagation, while with normal polarization the electric field is entirely transverse. These are two types of waves, identified as Transverse Magnetic (TM) and Transverse Electric (TE), respectively. The pattern illustrated in Figure 5 is the TM_1 mode, where the subscript is the value of n. In general, any system of propagating fields between parallel planes can be resolved into a linear combination of a TM and a TE mode, similar to the way in which an arbitrarily polarized plane wave in space can be resolved into a parallel and a perpendicularly polarized wave.

4. Rectangular Waveguides

Many of the properties of parallel plane waveguides are also found in cylindrical guides, of which the most commonly used cross sections are the rectangular and the circular. In practice, parallel plane guides find only very restricted application, but they do give a relatively simple introductory insight into guide behavior and, as we shall see, many of the concepts can be carried over almost intact to the understanding of pipe waveguides.

The basic principles of pipe waveguides are the same for any type of cross section, but the solutions for rectangular pipes require sines and cosines, with which the reader is familiar, whereas circular pipes require the usually less familiar Bessel functions. We shall, therefore, emphasize the treatment of rectangular waveguides, a choice which fits well with the more common use of rectangular rather than circular guides.

In Section 2 of Chapter 10 it was found that a consequence of the Maxwell field equations was that the electric and the magnetic fields satisfied the differential wave equation:

$$\nabla^2 \mathbf{E} = \mu\epsilon\ddot{\mathbf{E}} \qquad \nabla^2 \mathbf{H} = \mu\epsilon\ddot{\mathbf{H}} \tag{4-1}$$

and that, in Cartesian coordinates, the wave equation separates into relations of identical form for the individual components:

$$\begin{aligned} \nabla^2 E_x &= \mu\epsilon\ddot{E}_x \qquad & \nabla^2 H_x &= \mu\epsilon\ddot{H}_x \\ \nabla^2 E_y &= \mu\epsilon\ddot{E}_y \qquad & \nabla^2 H_y &= \mu\epsilon\ddot{H}_y \\ \nabla^2 E_z &= \mu\epsilon\ddot{E}_z \qquad & \nabla^2 H_z &= \mu\epsilon\ddot{H}_z \end{aligned} \tag{4-2}$$

These components of **E** and **H** are interrelated through the Maxwell field equations:

$$\nabla \times \mathbf{E} = -\mu\dot{\mathbf{H}} \qquad \nabla \times \mathbf{H} = \epsilon\dot{\mathbf{E}} \tag{4-3}$$

and must, in addition, satisfy the boundary conditions at the conducting walls of the pipe. We shall assume these walls to be perfect conductors and the space within the walls to be filled with a perfect, isotropic dielectric, so that the tangential components of the electric field at the walls vanishes. The normal components of **E**, on the other hand, can start or terminate on instantaneous charge concentrations on the walls. Since there can be no electromagnetic field within a perfect conductor, the normal components of **H** at the walls must vanish. The tangential components of **H** need not vanish, however, being associated with the surface currents in the walls.

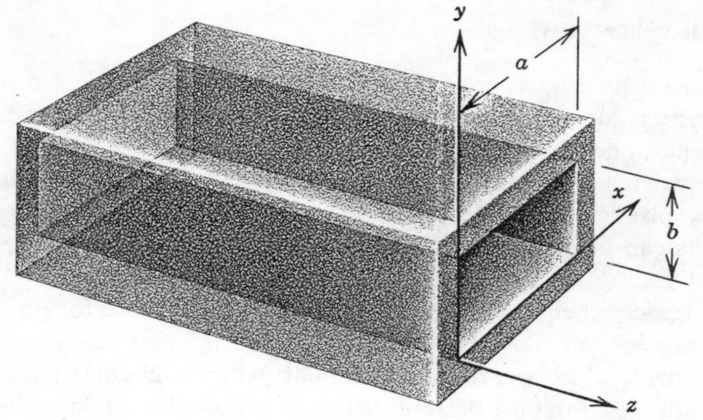

Figure 7. Reference axes on a rectangular waveguide.

We are particularly interested in investigating the type of field which has the form of a traveling wave propagating along the length of the pipe. Let us take the z-axis directed longitudinally and the x and y axes in the transverse plane, as in Figure 7. For propagation in the z-direction the dependence on time and on z for each field component is included in the function

$$e^{j(\omega t - \beta_g z)} \tag{4-4}$$

where β_g is a propagation constant for propagation in the z-direction, analogous to the β_g for parallel planes in Section 1. As before, β_g can be alternatively replaced by $2\pi/\lambda_g$, where λ_g is a wavelength measured in the z-direction, or ω/v_g, where v_g is the phase velocity of propagation in the z-direction. When the dependence on t and z is of the form of Equation 4-4, derivatives with respect to t or z are replaced by

$$\frac{\partial}{\partial t} = j\omega \qquad \frac{\partial}{\partial z} = -j\beta_g \tag{4-5}$$

Letting X be that part of any of the six field components that is a function of x and y, a typical equation of the group 4-2 becomes

$$\frac{\partial^2 X}{\partial x^2} + \frac{\partial^2 X}{\partial y^2} = -k^2 X \tag{4-6}$$

where

$$k^2 = \frac{\omega^2}{v_o^{\,2}} - \beta_g^{\,2} \tag{4-7}$$

Equation 4-6 is readily solved by separation of variables. Let

(4-8) $$X = F(x)G(y)$$

where F and G are functions of single variables only. Then 4-6 becomes, after rearranging:

(4-9) $$\left(\frac{1}{F}\right)\frac{\partial^2 F}{\partial x^2} + \left(\frac{1}{G}\right)\frac{\partial^2 G}{\partial y^2} = -k^2$$

Since each term on the left is a function of only one variable, the sum can be a constant for any combination of x and y only if

(4-10) $$\left(\frac{1}{F}\right)\frac{\partial^2 F}{\partial x^2} = -k_1^2 \quad \text{and} \quad \frac{1}{G}\frac{\partial^2 G}{\partial y^2} = -k_2^2$$

where k_1 and k_2 are constants which satisfy the relation

(4-11) $$k_1^2 + k_2^2 = k^2$$

Solutions of 4-10 that can satisfy the boundary conditions have the form

(4-12) $$F = \begin{pmatrix}\sin\\ \cos\end{pmatrix} k_1 x \qquad G = \begin{pmatrix}\sin\\ \cos\end{pmatrix} k_2 y$$

and a typical field component becomes

(4-13) $$X_i = C_i \begin{pmatrix}\sin\\ \cos\end{pmatrix} k_1 x \begin{pmatrix}\sin\\ \cos\end{pmatrix} k_2 y e^{j(\omega t - \beta_g z)} \qquad i = x, y, \text{ or } z$$

We find ourselves with six expressions for the six field components, in each of which there must be made two decisions between sine or cosine function forms, in addition to the evaluation of three constants, k_1, k_2, and C_i. These decisions are made in the light of the boundary conditions and the requirements of the Maxwell equations, 4-3.

The boundary conditions dictate some of the choices between the sine and cosine functions and also the nature of the constants k_1 and k_2. Thus the fact that E_x is tangential to the guide wall at $y = 0$, that E_y is tangential to the wall at $x = 0$, and that E_z is tangential at both $x = 0$ and $y = 0$ require that

(4-14)
$$E_x = C_x \begin{pmatrix}\sin\\ \cos\end{pmatrix} k_1 x \sin k_2 y e^{j(\omega t - \beta_g z)}$$

$$E_y = C_y \sin k_1 x \begin{pmatrix}\sin\\ \cos\end{pmatrix} k_2 y e^{j(\omega t - \beta_g z)}$$

$$E_z = C_z \sin k_1 x \sin k_2 y e^{j(\omega t - \beta_g z)}$$

Similar arguments concerning the normal components of **H** lead to

$$(4\text{-}15) \qquad H_x = D_x \sin k_1 x \begin{pmatrix} \sin \\ \cos \end{pmatrix} k_2 y e^{j(\omega t - \beta_g z)}$$

$$H_y = D_y \begin{pmatrix} \sin \\ \cos \end{pmatrix} k_1 x \sin k_2 y e^{j(\omega t - \beta_g z)}$$

$$H_z = D_z \begin{pmatrix} \sin \\ \cos \end{pmatrix} k_1 x \begin{pmatrix} \sin \\ \cos \end{pmatrix} k_2 y e^{j(\omega t - \beta_g z)}$$

There has been nothing in the argument so far that requires that all the constants k_1 and k_2 be the same in all the components, but they have been written so in anticipation of the results of introducing the Maxwell relations 4-3. Assuming that k_1 and k_2 are common to all the components, the boundary conditions yield further information. Since $E_x = 0$ at $y = b$ and $E_y = 0$ at $x = a$, it follows that

$$(4\text{-}16) \qquad \sin k_1 a = 0 \qquad \text{and} \qquad \sin k_2 b = 0$$

so that

$$(4\text{-}17) \qquad k_1 = \frac{m\pi}{a} \qquad \text{and} \qquad k_2 = \frac{n\pi}{b}$$

where m and n are integers.

There can thus be a theoretically infinite number of electromagnetic wave functions in the guide, each function characterized by a pair of integers, m and n, where these numbers identify the number of half-cycles of the transverse functions of the field components. Since the differential equations for the field components are linear, the general solution is a sum of the various individual wave functions. The transmission properties for given guide dimensions are determined by these characteristic numbers, m and n. We now develop these properties, leaving the wave functions to be completed later.

5. Transmission Properties of a Rectangular Waveguide

Using Equation 4-17, Equation 4-11 becomes, with the aid of 4-7

$$(5\text{-}1) \qquad k^2 = \left(\frac{\omega^2}{v_0^2}\right) - \beta_g^2 = \left(\frac{m\pi}{a}\right)^2 + \left(\frac{n\pi}{b}\right)^2$$

from which the propagation constant is obtained:

$$(5\text{-}2) \qquad \beta_g^2 = \left(\frac{\omega}{v_0}\right)^2 - \left(\frac{m\pi}{a}\right)^2 - \left(\frac{n\pi}{b}\right)^2$$

β_g becomes imaginary for angular frequencies below ω_c, where

$$\omega_c^2 = v_0^2\left[\left(\frac{m\pi}{a}\right)^2 + \left(\frac{n\pi}{b}\right)^2\right] \tag{5-3}$$

ω_c is called the "cut-off angular frequency." Thus, different from an ideal transmission line in which all frequencies have real propagation constants, ω/v_0, a waveguide acts like a high-pass filter, with cut-off frequencies depending on the characteristic numbers, m and n.

As with parallel plane guides, let λ_g be the wavelength in the guide; v_g, the phase velocity in the guide; λ_0 and v_0, the wavelength and phase velocity which plane waves of frequency ω have in infinite space. Then, using $\beta_g = \dfrac{\omega}{v_g} = \dfrac{2\pi}{\lambda_g}$, the reader can readily show that

$$v_g = \frac{v_0}{\sqrt{1 - (f_c/f)^2}} \tag{5-4}$$

$$\lambda_g = \frac{\lambda_0}{\sqrt{1 - (f_c/f)^2}} \tag{5-5}$$

where f_c and f are the linear frequencies corresponding to the angular frequencies ω_c and ω, respectively.

As with electromagnetic waves guided between parallel planes, the phase velocity and the guide wavelength increase as the frequency of the waves approaches the cut-off frequency, both being infinite at the cut-off frequency. For frequencies far from the cut-off frequency, on the other hand, both approach their free space values.

At this point the expressions for the field components as an infinite series of wave functions look somewhat formidable. In practice, rectangular waveguides usually have dimensions such that a is approximately double b. Equation 5-3 shows that if we take $a/b = 2$, then the lowest nonzero cut-off frequency occurs for $m = 1$, $n = 0$.

$$\omega_{c(1,0)} = v_0 \frac{\pi}{a} \tag{5-6}$$

The table of Figure 8 shows the ratios $[\omega_{c(m,n)}]/[\omega_{c(1,0)}]$ for $a/b = 2$.

No other than the (1, 0) wave can exist in the guide until the frequency reaches twice the (1, 0) cut-off frequency. From Equation 5-6 it can be shown that the wavelength in infinite space corresponding to this lowest cut-off frequency is $2a$. Most rectangular guides are designed to be used at frequencies which fall into the range between the lowest and the next lowest cut-off frequencies, so that only the lowest wave function, called

		n			
		0	1	2	3
m	0	0.00	2.00	4.00	6.00
	1	1.00	2.24	4.12	6.08
	2	2.00	2.83	4.47	6.33
	3	3.00	3.61	5.00	6.71

Figure 8. *Relative cut-off frequencies as a function of m and n for $a/b = 2$.*

the dominant mode, can propagate. For the dominant mode, the propagation constant, the velocity of transmission, and the guide wavelength are all a function only of the x dimension of the guide.

6. Completing the Waveguide Fields

We are now ready to resolve the remaining ambiguities and evaluate the constants in the field components by introducing the condition that these expressions must be consistent with the Maxwell field relations. When these relations are written out in component form, including the replacement for the differential operators of 4-5, they become

$$\text{(6-1)} \quad \left.\begin{aligned} &(a) \quad \frac{\partial E_z}{\partial y} + j\beta_g E_y = -j\omega\mu H_x \\ &(b) \quad -j\beta_g E_x - \frac{\partial E_z}{\partial x} = -j\omega\mu H_y \\ &(c) \quad \frac{\partial E_y}{\partial x} - \frac{\partial E_x}{\partial y} = -j\omega\mu H_z \end{aligned}\right\} \text{from } \nabla \times \mathbf{E} = -\mu\dot{\mathbf{H}}$$

$$\left.\begin{aligned} &(d) \quad \frac{\partial H_z}{\partial y} + j\beta_g H_y = j\omega\epsilon E_x \\ &(e) \quad -j\beta_g H_x - \frac{\partial H_z}{\partial x} = j\omega\epsilon E_y \\ &(f) \quad \frac{\partial H_y}{\partial x} - \frac{\partial H_x}{\partial y} = j\omega\epsilon E_z \end{aligned}\right\} \text{from } \nabla \times \mathbf{H} = \epsilon\dot{\mathbf{E}}$$

An examination of these equations shows that nowhere does E_z or its derivatives appear in the same equation with H_z or its derivatives. The reader may demonstrate for himself that it is possible to have one system of fields in which $E_z = 0$ and another in which $H_z = 0$, and that any general system of fields in a guide can be composed of a linear combination of these two sets. One set of components ($E_z = 0$) forms a traveling electromagnetic wave in which the electric field is entirely transverse to the guide (TE); whereas the other set ($H_z = 0$) is a wave in which the magnetic field is entirely transverse (TM). These two types play a role analogous to the parallel and normal polarization of plane waves in free space. Just as any general plane wave can be resolved into two polarizations, so an electromagnetic wave propagating in a guide can be resolved into a Transverse Electric and a Transverse Magnetic part. Moreover, just as the two polarizations of a plane wave behave differently when conditions, as at a boundary, are not symmetrical with respect to their two planes, so the two guide waves behave differently in a guide where $a \neq b$.

Although in practice waveguides are used only with the dominant mode, for completeness we shall substitute the general relations 4-14 and 4-15 into Equations 6-1 to obtain the field components for any combination of m and n. Later we reduce these to the special case of the dominant mode. This substitution resolves the remaining ambiguities of the sine and cosine functions, after which the same equations relate the constants C_x, C_y, C_z, D_x, D_y, and D_z. Although there are six equations in 6-1, only four are independent, so that two of the six constants remain arbitrary. This appears reasonable from either of two other considerations: (1) The differential equation we are solving is a second-order equation and hence has two arbitrary constants. (2) From physical considerations, the guide itself cannot completely determine the characteristics of the propagating waves; the source of the waves must also be considered in fixing their amplitude and orientation.

We shall let the two undetermined constants be C_z and D_z. By letting first $D_z = 0$ and then $C_z = 0$, we separate the TM and the TE types of waves as follows:

$$E_x = -j\left(\frac{k_1\beta_g}{k^2}\right)C_z \cos k_1x \sin k_2y \tag{6-2}$$

$$E_y = -j\left(\frac{k_2\beta_g}{k^2}\right)C_z \sin k_1x \cos k_2y$$

$$E_z = C_z \sin k_1x \sin k_2y \qquad \text{TM}$$

$$H_x = j\left(\frac{\omega\epsilon k_2}{k^2}\right)C_z \sin k_1x \cos k_2y$$

$$H_y = -j\left(\frac{\omega\epsilon k_1}{k^2}\right)C_z \cos k_1x \sin k_2y$$

$$\text{(6-3)} \qquad \begin{aligned} E_x &= j\left(\frac{\omega\mu k_2}{k^2}\right) D_z \cos k_1 x \sin k_2 y \\ E_y &= -j\left(\frac{\omega\mu k_1}{k^2}\right) D_z \sin k_1 x \cos k_2 y \\ H_x &= j\left(\frac{\beta_g k_1}{k^2}\right) D_z \sin k_1 x \cos k_2 y \\ H_y &= j\left(\frac{\beta_g k_2}{k^2}\right) D_z \cos k_1 x \sin k_2 y \\ H_z &= D_z \cos k_1 x \cos k_2 y \end{aligned} \qquad \text{TE}$$

Equations 6-2 and 6-3 show that the (1, 0) mode can exist only in the Transverse Electric (TE) form. Setting $n = 0$ makes $k_2 = 0$, which eliminates each of the five TM equations. It also eliminates two of the TE equations, leaving

$$\text{(6-4)} \qquad \begin{aligned} E_y &= -j\left(\frac{\omega\mu a}{\pi}\right) D_z \sin\left(\frac{\pi x}{a}\right) \\ H_x &= j\left(\frac{\beta_g a}{\pi}\right) D_z \sin\left(\frac{\pi x}{a}\right) \\ H_z &= D_z \cos\left(\frac{\pi x}{a}\right) \end{aligned} \qquad \text{TE}_{10}$$

where the fact that $k_1 = k = \pi/a$ has been used in the coefficients. Each of these expressions should, for completeness, be multiplied by the traveling wave factor, $e^{j(\omega t - \beta_g z)}$. The TE_{10} mode is independent of the y-coordinate; E_y and H_x vary with a half-period sine function over the x dimension of the guide, while the z component of H varies as a half-period cosine function.

It is instructive to plot the electric and the magnetic fields in the guide at an arbitrary time, say, $t = 0$. In Equations 6-4, the functions appear as complex numbers, but the physically real quantity to be associated with a complex number can be taken as either the real or the imaginary part, the choice being unimportant so long as it is made consistently. For plotting, we shall write the real parts of 6-4, remembering that $j = e^{j\pi/2}$ and $-j = e^{-j\pi/2}$.

$$\text{(6-5)} \qquad \begin{aligned} E_y &= \left(\frac{\omega\mu a D_z}{\pi}\right) \sin\left(\frac{\pi x}{a}\right) \cos\left(\omega t - \beta_g z - \frac{\pi}{2}\right) \\ H_x &= \left(\frac{\beta_g a D_z}{\pi}\right) \sin\left(\frac{\pi x}{a}\right) \cos\left(\omega t - \beta_g z + \frac{\pi}{2}\right) \\ H_z &= D_z \cos\left(\frac{\pi x}{a}\right) \cos(\omega t - \beta_g z) \end{aligned}$$

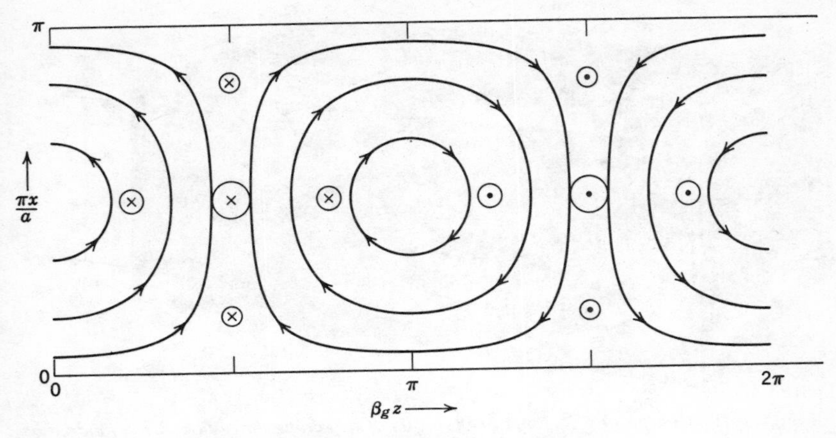

Figure 9. TE$_{10}$ *fields in a rectangular guide.*

H_z is out of phase with H_x by $-\pi/2$ and with E_y by $+\pi/2$, and has its maximum values at the walls, whereas E_y and H_x have their maximum values at the middle of the guide. In Figure 9 the E and the H fields are shown looking down on the surface $y = b$, with the lines showing the H-field and the circles the E-field. Dots in the circles indicate the E-field directed out of the paper and crosses, into the paper, the diameter of the circles proportional to the magnitude of the field. These magnitudes are independent of the y-coordinate. The field pattern moves in the positive z-direction as time increases.

As with parallel planes there can be associated with these fields a pattern of currents, charge densities, and voltages. Across the guide in the y-direction are voltages which are the line integrals of E; the voltage from top to bottom of the guide is $E_y b$. The E-field terminates on negative charge densities and starts on positive charge densities. The current in a wall is obtained from the tangential H-field at the wall as explained in Section 3. For the field pattern of Figure 9 the charges and currents are shown in Figure 10.

The reader should study this diagram carefully to see how the flow of charges in the walls is consistent with the movement down the guide of the charge concentrations associated with the electric field. Slots may be cut in a length of guide, as in a slotted line, in order to introduce probes with which to sample the fields. The pattern of currents suggests that a minimum of disturbance will be introduced by a slot cut longitudinally in the $y = 0$ or $y = b$ face, whereas a slot cut in the $x = 0$ or $x = a$ face would disrupt the transverse current pattern.

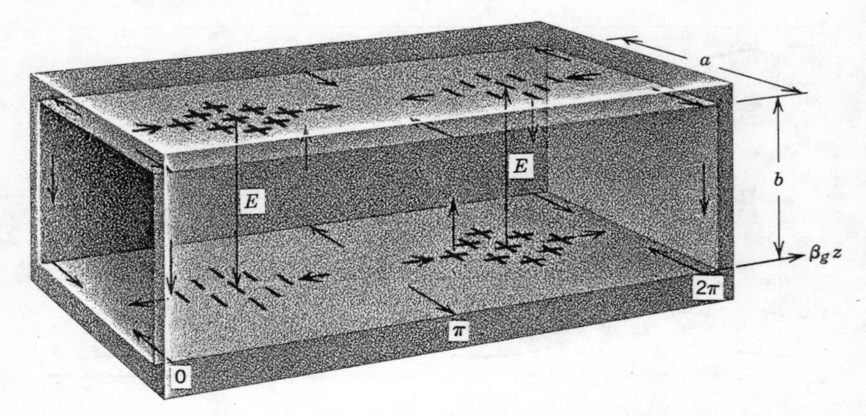

Figure 10. *Charge and currents in the walls of a rectangular guide,* TE_{10} *mode.*

7. Characteristic Impedance of a Waveguide

In both filters and transmission lines it was found profitable to define at a given distance from the generator (or load) the ratio of the voltage to the current as an impedance. The value of this ratio when there was no reflection of energy from the termination was called the characteristic impedance. It was similarly shown that in a propagating plane wave, the ratio of E to H had the nature of an impedance, characteristic of the medium.

In each of these examples the characteristic impedance could be computed equally well from a knowledge of the power, together with either the voltage or the current. That is,

$$Z_k = \frac{V}{I} = \frac{P}{I^2} = \frac{V^2}{P} \tag{7-1}$$

provided the system was ideal, so that the impedance was a pure real number.

It is tempting to investigate the possibility of defining such an impedance for an ideal waveguide, where we should also expect it to be a real number. But here we encounter some ambiguities. The longitudinal power flow, the transverse voltage, and the longitudinal current vary over a cross-sectional plane. What shall we say is *the* voltage, *the* current, or *the* power to be used in Equation 7-1?

A not unreasonable choice for the current is the total longitudinal current in the upper or lower wall. (All discussion of impedance is for the TE_{10} mode, for which there is no longitudinal current in the side walls.)

Like the current, the voltage varies from zero to a maximum across the face of the guide, and it is the practice to define *the* voltage for impedance purposes as the value at the center line, where it is a maximum. The power is determined by integrating the Poynting vector over the cross section. (Current, voltage, and power are all time averaged, or effective, values.) A difficulty that arises out of these choices—or any other that might be made—is that the three ways of obtaining impedance in Equation 7-1 give different answers. This ambiguity is not unduly serious in practice, although it is theoretically disturbing, because the three answers differ only by multiplying constants; they agree in their functional dependence on the dimensions and the medium of the guide.

The power, longitudinal current, and the maximum voltage are

$$\bar{P} = -\int_0^b \int_0^a \bar{E}_y \bar{H}_x \, dx \, dy \tag{7-2}$$

$$I_{\text{eff}} = \int_0^a \bar{H}_x \, dx \qquad \text{at } y = 0 \text{ or } y = b$$

$$V_{\text{eff}} = \bar{E}_{y\,\text{max}} b$$

These relations yield the following three characteristic impedances:

$$Z_{VI} = \left(\frac{\pi}{2}\right)\left(\frac{\omega\mu}{\beta_g}\right)\frac{b}{a} \tag{7-3}$$

$$Z_{PI} = \frac{\pi^2}{8}\left(\frac{\omega\mu}{\beta_g}\right)\frac{b}{a}$$

$$Z_{PV} = 2\left(\frac{\omega\mu}{\beta_g}\right)\frac{b}{a}$$

where the subscripts on each impedance indicate the quantities used in its evaluation.

It will be remembered that β_g for the dominant mode is a function only of the dimensions a. Equations 7-3 show that the characteristic impedance, however determined, can be controlled by a choice of the b dimension without affecting the other transmission properties.

For many cases where the concept of characteristic impedance enters into the design of waveguide systems (for example, the matching of impedances to eliminate reflections) only relative values of impedance are significant. It is immaterial which of the forms in 7-3 are used, so long as the choice is used consistently. Sometimes, however, the absolute value is significant. Then it is a matter of judgment on the part of the experimenter as to which form is most relevant to the problem, and usually the judgment must be checked experimentally. It is clear that the impedance

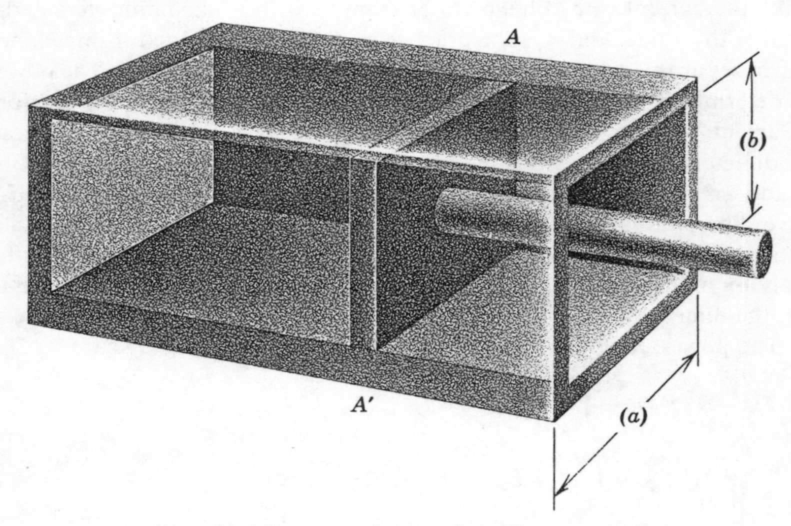

Figure 11. Elementary shorting plunger in a waveguide.

concept when applied to guides is somewhat arbitrary and must be used with care. It has very little significance for higher modes of propagation.

As an example of the way in which the information about the impedance of a waveguide can be used as a qualitative help in design, let us consider the construction of an adjustable shorting plunger at the end of a rectangular waveguide, a device which produces standing waves in a guide as does a shorted termination on a transmission line. First, consider the simplest way to make such a plunger, as illustrated in Figure 11. A metal block is fitted to the inside dimensions of the guide and makes sliding contact with the walls at its perimeter. Such a contact is less than perfect and grows worse with use. A better short across the plane A-A' is obtained by the construction shown in Figure 12, where the plunger is made in the form of a cylinder of rectangular cross section with walls parallel to the guide walls but separated from them by a small distance. This cylinder extends almost to a conducting plane that fills the cross section of the guide, the distance from the front of the cylinder to the rear plane being one-quarter of a guide wavelength. The cylinder is hollow, so that there is another one-quarter guide wavelength within it. The same reasoning concerning the reflection of impedances applies to guides as applied to transmission lines; impedances in a standing wave repeat every half wavelength and invert over a quarter wavelength. Thus zero impedance at the inside surface of the front plane is reflected at the points A and A' as a zero impedance, and a high impedance contact at B and B' is also reflected as a

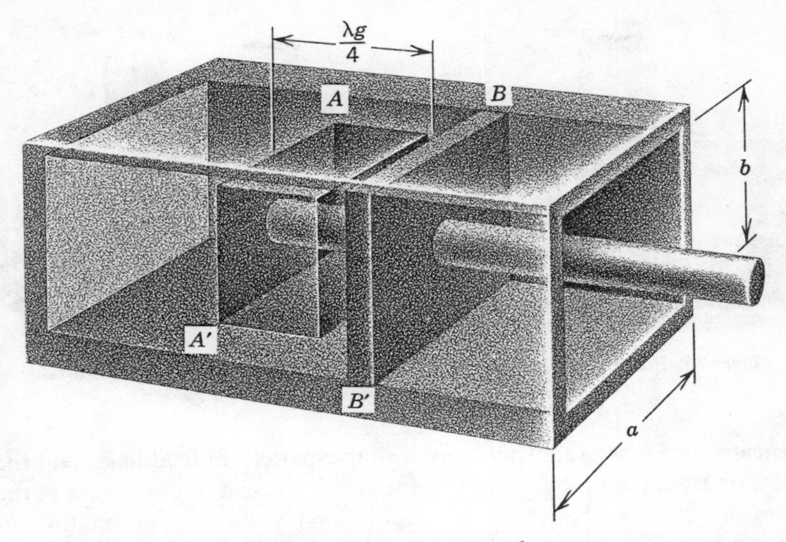

Figure 12. Improved shorting plunger.

low impedance at A and A'. Moreover, the space between the plunger and the walls is itself a guide with a small vertical spacing and thus, from 7-3, has a very low characteristic impedance. All of these factors combine to make the impedance over the plane A-A' a closer approximation to zero than can be obtained by an attempt to maintain a direct contact as in Figure 11. It is left to the reader to justify the fact that the gap between the plunger and the walls $x = 0$ and $x = a$ plays no part.

8. Sources of Fields in Guides

Fields are introduced into a waveguide in a number of ways: from apertures leading to other guides or cavities; by electronic generators mounted directly in the guide; by probes or loops mounted in the guide and connected by coaxial lines to electronic generators outside. We shall confine the discussion to the last of these.

One of the simplest methods for introducing power into a guide is to mount a small probe which penetrates into the volume and to which power is supplied by a coaxial line. Such a probe is illustrated in Figure 13. The outer conductor of the coaxial line is connected to the wall of the guide and the probe itself is an extension of the center conductor. Thus the probe acts as a small antenna radiating energy into the space of the guide. The electric field lines associated with the voltages on the probe are

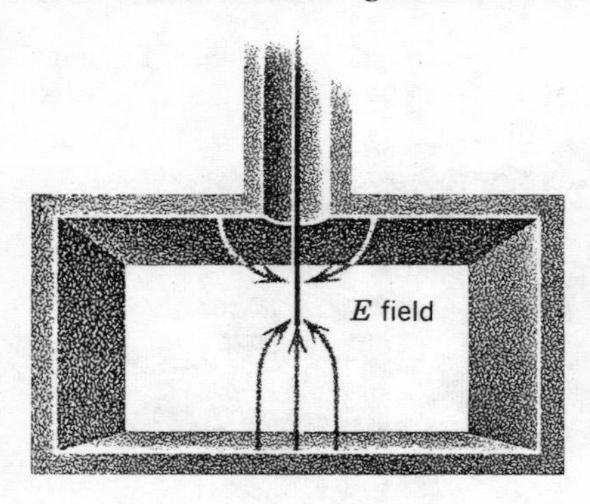

Figure 13. Probe in a waveguide.

Figure 14. Loop for introducing power in a guide.

illustrated in Figure 13. This is not a simple pattern of field lines, and the radiation from the probe might be expected to excite a large number of the waveguide modes. However, in a guide having the proper relation of dimensions to frequency, only the dominant mode is able to propagate, and this mode becomes well-developed in only a short distance from the probe.

Another way of introducing power is by means of a small loop antenna, as illustrated in Figure 14, where the currents in the loop at the end of the coaxial line produce magnetic fields within the guide. Such fields also do not conform to the magnetic field pattern of the dominant mode, but again, with the proper dimensions, it is the dominant mode alone that develops in the guide. Proper positioning of either probes or loops is important in the effectiveness with which energy can be introduced. Intuitively, it should be evident that if the fields introduced by the antenna element do not have major components which are parallel to the pattern of fields to be excited in the guide, the desired mode is only weakly excited.

Just as an antenna may be used reciprocally either for transmission or receiving, so probes and loops may also be used for the detection of fields within the guide or for the removal of energy from the guide. In the former case, the probe or loop will not give a correct picture of the conditions in the guide unless its penetration into the field space is very small, so as to produce a minimum of disturbance of the field pattern. For the latter case, on the other hand, it is the purpose of the probe or loop to alter the field pattern by abstracting energy and an appreciable penetration is employed.

The use of a probe as a detector has already been considered in connection with a slotted coaxial transmission line in Chapter 9, where the probe could be moved longitudinally to observe standing waves on the line. A comparable arrangement is a slotted wave guide with a small probe mounted on a carriage and penetrating into the field space. The uses of a slotted wave guide as a standing wave detector are analogous in all major details to a slotted transmission line, and most of the principles discussed in Chapter 9 can be carried over directly to the measurement of standing waves in guides. With transmission lines only impedance ratios appeared in the slotted line equations. The same is true for slotted wave guides, so that the ambiguities of Equations 7-3 do not cause difficulty unless it is necessary to obtain from the standing wave data a value for an absolute impedance. For many wave guide problems this is not necessary.

9. Resonant Cavities

When a rectangular wave guide of cross sectional dimensions a and b is closed off with a conducting plate at the plane $z = 0$ and $z = c$, the resulting cavity is resonant at frequencies which satisfy the boundary conditions within the cavity. By resonant is meant that standing electromagnetic waves are set up in the cavity by sources of energy such as, say, a probe that penetrates the field space, or possibly by atomic or molecular emissions within the cavity. For perfectly conducting walls, these waves, once started, continue indefinitely; for real walls, the wave amplitudes decay as the energy is dissipated as heat in the resistance of the walls unless the energy losses are replaced from an outside source.

We shall start by analyzing the simple case of a TE_{10} wave traveling in the z direction in a rectangular closed guide. Because of reflections at the ends there are two waves, one traveling in the positive and the other in the negative z direction. Consider in particular the electric field vector, E_y (from Equation 6-4):

$$\begin{aligned} E_{y+} &= -j\left(\frac{\omega\mu a}{\pi}\right) D_z \sin\left(\frac{\pi x}{a}\right) e^{j(\omega t-\beta_g z)} \\ E_{y-} &= -j\left(\frac{\omega\mu a}{\pi}\right) D_z' \sin\left(\frac{\pi x}{a}\right) e^{j(\omega t+\beta_g z)} \end{aligned} \tag{9-1}$$

where the second, traveling in the negative z direction is obtained from the first by a reversal of the sign of β_g, with the amplitude D_z' being fixed by the reflection conditions at the end.

For perfectly conducting walls the reflection conditions on a normally incident electric field wave requires a reflected wave of equal amplitude but reversed phase, in order that the net electric field at the $z = 0$ and $z = c$ surfaces may be zero. At any point the total electric field vector is the sum of the forward and the reflected fields:

$$E_{y\,\text{Total}} = -2\left(\frac{\omega\mu a}{\pi}\right) D_z \sin\left(\frac{\pi x}{a}\right) \sin(\beta_g z) e^{j\omega t} \tag{9-2}$$

The fact the $E_{y\,\text{Total}}$ is tangent also at the $z = c$ plane dictates that

$$\sin(\beta_g c) = 0 \qquad \text{so that } \beta_g = \frac{p\pi}{c} \tag{9-3}$$

where p is an integer, which together with $m = 1$ and $n = 0$, fixes the frequency for which such standing waves are possible. For, since

$$\beta_g^{\,2} = \left(\frac{p\pi}{c}\right)^2 = \omega^2\mu\epsilon - \left(\frac{m\pi}{a}\right)^2 - \left(\frac{n\pi}{b}\right)^2 \tag{9-4}$$

it follows that

$$\omega^2 = v_0^{\,2}\left[\left(\frac{m\pi}{a}\right)^2 + \left(\frac{n\pi}{b}\right)^2 + \left(\frac{p\pi}{c}\right)^2\right] \tag{9-5}$$

For the TE_{101} mode,

$$\omega^2 = v_0^{\,2}\left[\left(\frac{\pi}{a}\right)^2 + \left(\frac{\pi}{c}\right)^2\right] \tag{9-6}$$

The other components of the TE_{101} mode are readily found to be

$$\begin{aligned} H_{x\,\text{Total}} &= 2j\left(\frac{a}{c}\right) D_z \sin\left(\frac{\pi x}{a}\right) \cos\left(\frac{\pi z}{c}\right) e^{j\omega t} \\ H_{z\,\text{Total}} &= -2j D_z \cos\left(\frac{\pi x}{a}\right) \sin\left(\frac{\pi z}{c}\right) e^{j\omega t} \end{aligned} \tag{9-7}$$

By writing $j = e^{j\pi/2}$, the field expressions become

$$\begin{aligned} E_{y\,\text{Total}} &= -2\left(\frac{\omega\mu a}{\pi}\right) D_z \sin\left(\frac{\pi x}{a}\right) \sin\left(\frac{\pi z}{c}\right) e^{j\omega t} \\ H_{x\,\text{Total}} &= 2\left(\frac{a}{c}\right) D_z \sin\left(\frac{\pi x}{a}\right) \cos\left(\frac{\pi z}{c}\right) e^{j(\omega t+\pi/2)} \\ H_{z\,\text{Total}} &= 2 D_z \cos\left(\frac{\pi x}{a}\right) \sin\left(\frac{\pi z}{c}\right) e^{j(\omega t-\pi/2)} \end{aligned} \tag{9-8}$$

from which it appears that the electric field is a quarter cycle out of phase

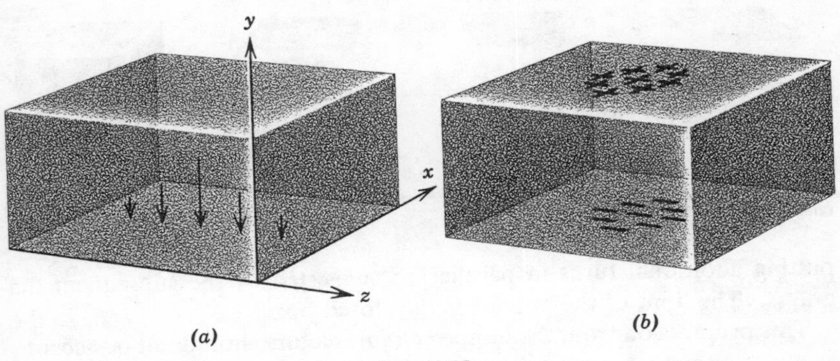

Figure 15

with the magnetic field. At no place within the resonant volume is there a net transport of energy.

It is instructive to diagram the distribution of the fields as functions of position and of time. Figure 15*a* shows the electric field and 15*b* the charge distribution on the walls at $\omega t = 0$, at which time the magnetic field components are zero. At $\omega t = \pi/2$ the electric field is zero, and the magnetic field appears as in Figure 16*a* and the wall currents as in 16*b*. The positive charges, which were on the upper surface in Figure 15*b*, are flowing toward the lower surface in Figure 16. At the time $\omega t = \pi$, the currents are again zero and the charge concentrations are reversed from those at $\omega t = 0$. The charges flow back and forth in the side walls to produce the concentrations in the top and bottom faces.

The following conceptual progression has been used to suggest a development from a low-frequency LC resonant circuit to a resonant cavity (Figure 17). To increase the resonant frequency, it is necessary to reduce the inductance, first by reducing the number of turns, and then by

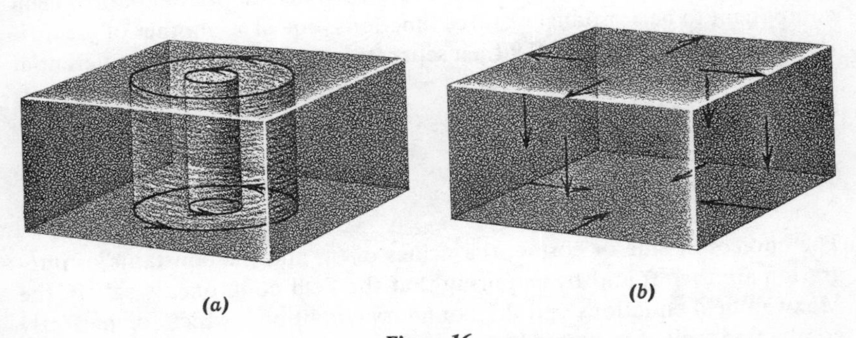

Figure 16

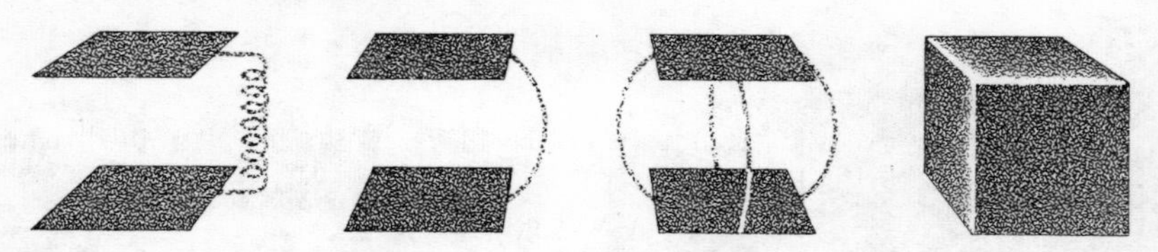

Figure 17. Successive reduction in L from an LC resonant circuit to a closed rectangular cavity.

putting additional turns in parallel, as suggested by the subsequent diagrams. The limit of the process is the closed box.

This progression from the lumped circuit picture should not be accepted as anything but a loose analogy, one that works pretty well for the special case of the TE_{101} mode. There are many more ways in which such a box can sustain standing waves, as many more ways as there are modes in the wave guide. A much more complete picture of the resonant cavity operation is obtained by developing the electromagnetic field from fundamental principles—much the same reasoning as was applied to the propagating wave in a guide.

We start with no assumptions about the field other than that it depends on time through the function $e^{j\omega t}$, and write the differential wave equation for each field component:

$$\nabla^2\mathbf{E} = -\frac{\omega^2}{v_0^2}\mathbf{E} \qquad (9\text{-}9)$$

$$\nabla^2\mathbf{H} = -\frac{\omega^2}{v_0^2}\mathbf{H}$$

These equations are solved by the same separation of variables technique as was used for the guide, except that we now do not make an assumption about the z dependence of the fields. We consider the spatial part of each component to be a product of three functions, one of x, another of y, and a third of z. The Equations 9-9 then separate into three ordinary differential equations whose solutions are sines or cosines, with each of which there is associated a constant k_i such that

$$k_1^2 + k_2^2 + k_3^2 = \frac{\omega^2}{v_0^2} \qquad (9\text{-}10)$$

The choices of sine or cosine, the values of k_i, and the constants of integration are then found by requiring that the field components satisfy the Maxwell field equations and the boundary conditions, which for perfectly conducting walls are particularly simple.

Before evaluating the constants of integration the solutions have the form

$$(9\text{-}11) \qquad E_x = C_x \cos\left(\frac{m\pi x}{a}\right) \sin\left(\frac{n\pi y}{b}\right) \sin\left(\frac{p\pi z}{c}\right) e^{j\omega t}$$

$$E_y = C_y \sin\left(\frac{m\pi x}{a}\right) \cos\left(\frac{n\pi y}{b}\right) \sin\left(\frac{p\pi z}{c}\right) e^{j\omega t}$$

$$E_z = C_z \sin\left(\frac{m\pi x}{a}\right) \sin\left(\frac{n\pi y}{b}\right) \cos\left(\frac{p\pi z}{c}\right) e^{j\omega t}$$

$$H_x = D_x \sin\left(\frac{m\pi x}{a}\right) \cos\left(\frac{n\pi y}{b}\right) \cos\left(\frac{p\pi z}{c}\right) e^{j\omega t}$$

$$H_y = D_y \cos\left(\frac{m\pi x}{a}\right) \sin\left(\frac{n\pi y}{b}\right) \cos\left(\frac{p\pi z}{c}\right) e^{j\omega t}$$

$$H_z = D_z \cos\left(\frac{m\pi x}{a}\right) \cos\left(\frac{n\pi y}{b}\right) \sin\left(\frac{p\pi z}{c}\right) e^{j\omega t}$$

It is left as an exercise for the reader to evaluate the constants. The four Maxwell field equations can be called upon to provide eight relations among the constants, but only four are independent relations. Two of the six constants, C_i and D_i, are arbitrary, as was the case with propagating fields in a guide. Choosing C_z and D_z as the arbitrary constants permits, as before, a division of the possible modes into two classes, TE_{mnp} and TM_{mnp}. m, n, and p form a set of characteristic integers which determine the frequency of resonance through Equation 9-10. Not more than one of the set can be zero.

10. Energy Losses in the Walls of a Resonant Cavity

The electromagnetic field within a cavity was found on the assumption that the walls are perfect conductors: For walls with finite conductivity, the tangential components of the electric field at the walls are no longer zero, since an electric field can exist within the walls. But the tangential electric field is small provided the conductivity is large. This tangential field is necessary for there to be current in the walls when the wall resistance is not zero, and the energy is taken from the fields to maintain these currents.

It is a good approximation, however, for the case of high wall conductivity, to assume that the fields within the cavity are negligibly different from those for perfectly conducting walls and that thus the currents in the walls are those that would exist in the ideal case. Then a computation of the electromagnetic energy converted to heat energy in the walls gives an upper limit to the heat losses.

It was found in Chapter 11 that the total heat loss in a semi-infinite conductor for currents parallel to the surface is the same as if a dc current equal to the total root-mean-square value of the actual current associated with the electromagnetic wave existed in a layer of thickness δ at the surface, where δ is the skin depth of penetration of the electromagnetic wave. Thus the power loss in the wall under an area $dx\,dz$ where the tangential magnetic field is H_x is

$$\text{(10-1)} \qquad \tfrac{1}{2}(H_x\,dx)^2\left(\frac{dz}{\sigma\delta\,dx}\right) = \frac{1}{2\sigma\delta}(H_x^{\,2}\,dx\,dz)$$

where H_x is to be evaluated at $y = 0$ or $y = b$. The factor $\frac{1}{2}$ comes from the time averaging of the field; σ is the conductivity of the wall. The heat generated per second in all the walls is then

$$\text{(10-2)} \qquad P_h = \frac{1}{2\sigma\delta}\left\{2\underset{(\text{at } x=0)}{\int_0^b\!\!\int_0^c} H_z^{\,2}\,dy\,dz + 2\underset{(\text{at } z=0)}{\int_0^a\!\!\int_0^b} H_x^{\,2}\,dx\,dy + 2\underset{(\text{at } y=0)}{\int_0^a\!\!\int_0^c}(H_y^{\,2} + H_z^{\,2})\,dx\,dz\right\}$$

the factor 2 at each integral coming from the symmetry of the currents in opposite walls. For the TE_{101} mode, Equation 10-2 gives for the wall losses

$$\text{(10-3)} \qquad P_h = \frac{D_z^{\,2}}{2\sigma\delta c^2}\,[2b(a^3 + c^3) + ac(a^2 + c^2)]$$

The quality of a resonant system can be expressed as

$$\text{(10-4)} \qquad Q = \frac{[\omega_r\ (\text{maximum energy stored in the fields})]}{(\text{rate of generation of heat})}$$

where ω_r is the frequency at which the system is resonant (cf. Chapter 8, Problem 13). Since the electric and the magnetic fields are not at maximum at the same time, the maximum stored energy can be found by integrating, say, over the energy density in the electric field at maximum. This maximum stored energy is

$$\text{(10-5)} \qquad U = \iiint\left[\frac{\epsilon E_{y\,\max}^2}{2}\right] dx\,dy\,dz = \frac{\epsilon\mu^2\omega^2 D_z^{\,2}a^3bc}{2\pi^2}$$

By using this with 10-3, we get

$$Q = \frac{2abc(a^2 + c^2)}{\delta[2b(a^3 + c^3) + ac(a^2 + c^2)]} \tag{10-6}$$

where use has been made of the relations

$$\omega^2 = \frac{\pi^2}{\mu\epsilon}\frac{a^2 + c^2}{a^2c^2} \quad \text{and} \quad \omega\mu\sigma = \frac{2}{\delta^2} \tag{10-7}$$

Q's of the order of 10,000 are not unusual for cavities which are resonant in the microwave region.

Let Δf be the frequency difference between the points on the resonant curve of a cavity where the power in the electromagnetic field is half of the power at the center frequency of the curve. Q is related to Δf by

$$Q = \frac{f_r}{\Delta f} \tag{10-8}$$

where f_r is the center, or resonant frequency. Q then is a measure of the sharpness of the resonant curve. In microwave cavities, the width of the resonance curve as defined in Equation 10-8 is of the order of 0.01% of the resonant frequency, which is to be compared to the situation in ordinary resonant LC circuits with Q's between 100 and 200, giving a width at resonance of the order of 1%. A microwave system can be made extremely frequency selective.

This method for finding the losses in a cavity can also be applied to finding the transmission losses in a wave guide, as well as to finding the losses in other modes in a cavity. More detailed discussions of this important aspect of the engineering applications can be found in references given at the end of this chapter.

11. Transmission-line Waves

It was noted at the end of Chapter 7 that the properties of ordinary transmission lines can be as well obtained from a consideration of the electric and magnetic fields in the space between the conductors as from a consideration of the voltages and currents on the conductors. An examination of Equations 6-2 and 6-3 serves as a beginning in such an analysis. In a two-conductor transmission line the electric and the magnetic fields are entirely transverse to the direction of propagation of the wave, whereas for hollow rectangular pipes we found that an E_z or an H_z or both were required. We must consider in the case of a transmission line what happens in 6-2 and 6-3 when *both* E_z and H_z are zero.

At first glance Equations 6-2 and 6-3 suggest that for $E_z = 0$ (and hence $C_z = 0$) all components in the TM waves are zero, since C_z appears in all; and similarly all components of the TE waves are zero if $H_z = 0$. But a closer examination shows that if at the same time $k^2 = 0$, then the x and y components are merely indeterminate, not necessarily zero. For these transverse components to exist, they must still obey the wave equation, 4-6, which now becomes:

$$\frac{\partial^2 X}{\partial x^2} + \frac{\partial^2 X}{\partial y^2} = 0 \qquad (11\text{-}1)$$

where X is the spatial part of any one of the four transverse components. In addition, with $k^2 = 0$, the propagation constant for these transverse fields becomes, from Equation 4-7:

$$\beta_g^{\,2} = \frac{\omega^2}{v_0^{\,2}} \qquad (11\text{-}2)$$

where v_0 is the propagation velocity in infinite space. This is the propagation velocity on a transmission line as found in Chapter 9.

But can such a wave pattern exist in a hollow guide? Equation 11-1, which is Laplace's equation requires that all points on a guide wall in a transverse plane be at the same potential. If there were differences of potential, there would be transverse currents, which in turn would be associated with longitudinal magnetic fields, contrary to the assumptions. One way in which a wall perimeter at a given z can be an equipotential is for the electric field lines to form a pattern of closed loops in the transverse plane, with the field magnitude zero at the walls. Since the electric and the magnetic fields must be mutually perpendicular, this requires that the magnetic field be radial, but nevertheless zero at the walls. But magnetic field lines must form closed loops, obviously impossible without a z component of H. (For $H_z \neq 0$, the closed loops of H could be in a plane parallel to the z axis, a condition which occurs in circular guides in certain modes.) It follows that waves which are both transverse magnetic and transverse electric (TEM waves) cannot exist in a guide which is simply a hollow pipe.

If, however, there is a second conductor, so that each conductor individually can be an equipotential in the transverse plane but with a difference of potential between them, evidently the conditions for TEM waves can be fulfilled. Such are the waves on a transmission line, and TEM waves are often called Transmission Line waves, sometimes also Principle waves.

Equation 11-1 is the same condition which is imposed upon electrostatic fields and steady state magnetic fields. Since the electric field in a capacitor and the magnetic field around an inductor obey Laplace's equation, it was therefore possible to analyse the behavior of a transmission line in terms of capacitative and inductive constants.

What was not evident in Chapter 9 was that transmission lines, in addition to propagating the TEM type of wave, can also propagate higher modes, providing the frequency is above the cut-off for these higher modes. Just as in hollow guides, such higher modes introduce complications, and transmission lines, usually coaxial, are used at lower frequencies where the cross-sectional dimensions of hollow guide become uncomfortably large, and hollow guide is used at very high frequencies where the proper choice of dimensions can prevent all but the TE_{10} mode from developing.

12. Some Waveguide Devices

Although it is not the purpose of this book to treat in detail the applications of electromagnetic theory, a brief discussion of some waveguide devices will provide a feeling for the way in which electromagnetic waves may be expected to behave. The comments in this section are qualitative in nature and the reader is referred to the references at the end of the chapter for more detailed discussion.

The fundamental purpose of a waveguide is to transmit electromagnetic energy along a prescribed path. This function in its simplest aspects has been discussed in detail in the previous sections. When it is necessary for the path to divide, junctions in wave guides are used (Figure 18). Such junctions are classed as either *E*-plane or *H*-plane junctions or as series or shunt Tees. The former labels are self-evident; the latter refer to the way

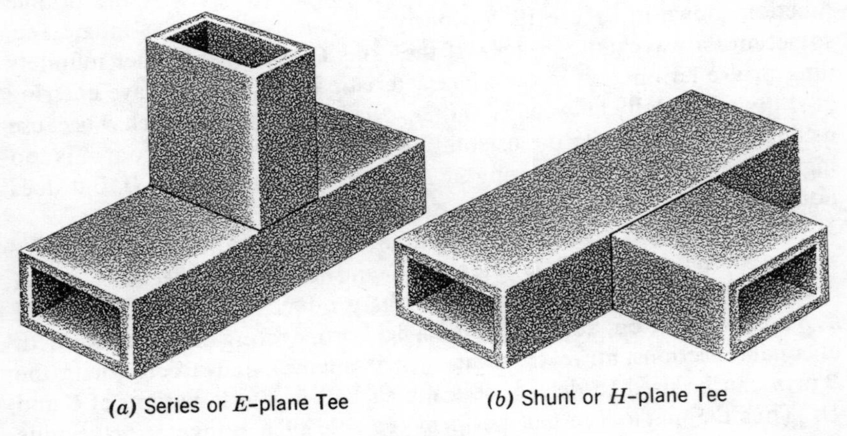

(*a*) Series or *E*-plane Tee (*b*) Shunt or *H*-plane Tee

Figure 18

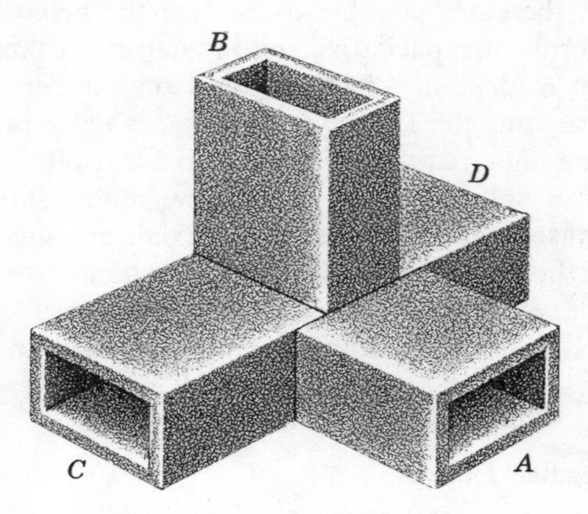

Figure 19. *Magic or hybrid Tee.*

in which the longitudinal currents behave. In the *E*-plane or series Tee junction, the currents in effect follow a series path; whereas in the *H*-plane or shunt Tee, the wall currents divide as in a parallel combination of elements. A junction is a discontinuity, and, in general, impedance matching is important if reflections are to be avoided. Various methods of creating an impedance match have been devised, such as introducing posts or vanes into the guide in the vicinity of a junction. Although knowledge of the theoretical behavior of waves at discontinuities is useful in the design of such constructions, the details are usually determined empirically.

An interesting combination of shunt and series Tees is the double junction shown in Figure 19, variously called a hybrid Tee, a magic Tee, sometimes a waveguide bridge. If the arms *C* and *D* are either infinitely long or are terminated in their characteristic impedance, a wave entering at *A* divides equally between *C* and *D*. No energy enters branch *B* because its cut-off frequency for the orientation of the fields coming from *A* is too high. Similarly a wave entering at *B* divides between *C* and *D*, but does not enter *A*.

With a wave entering at *A* and dividing between *C* and *D*, even if there are equal reflections at the ends of *C* and *D*, these reflections cancel upon returning to the junction as far as producing a field to propagate in *B* is concerned, but add to produce a field propagating back into *A*. With unequal reflections, however, some energy enters *B*, and a detector in the *B* arm can be used to determine a match between terminations at *C* and *D*. Thus the magic Tee can perform the role of a bridge at waveguide frequencies.

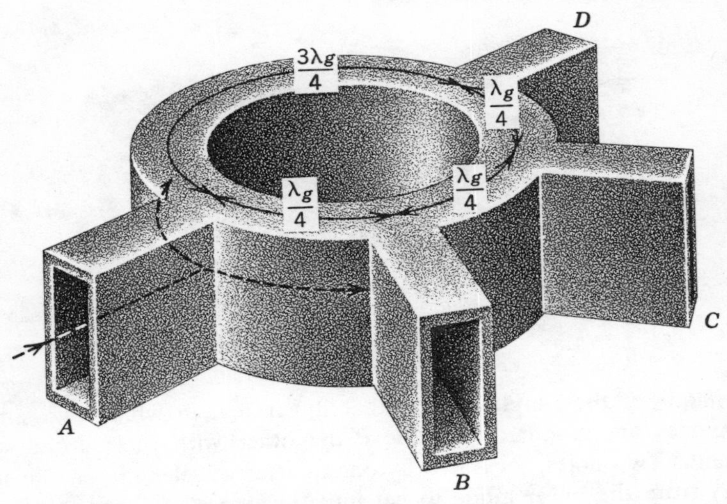

Figure 20. Ring waveguide junction.

The ring wave guide (Figure 20) acts in a way similar to that of the magic Tee. The ring is constructed to have an average circumference of one and one-half wavelengths. Consider a wave entering at *A*. It divides into two waves traveling around the ring in opposite directions. At *D* the two waves have traveled equal distances and hence combine constructively, with the result that energy appears in guide *D*. At *B* the path difference is one wavelength, again producing constructive interference, and energy appears in *B*. But at *C* the path difference is one half wavelength, there is destructive interference, and no energy appears in *C*. Similarly energy entering at *C* does not appear in *A*; etc.

A careful analysis shows that unbalanced impedances at *B* and *D* result in energy appearing in *C*, but that an impedance balance produces a null. The ring wave guide can serve as a bridge, like the magic Tee.

Another use for such wave guide bridges is to make possible the dual use of an antenna for both a transmitter and a receiver. Were both the transmitter and the receiver directly coupled to the antenna, the high power of the transmitter might burn out the receiver. But if the transmitter energy is fed into, say, *A*, on the ring, if the antenna is at *B*, if the receiver is at *C*, and a dummy matched load is at *D*, then the transmitter energy does not appear at the receiver, although signals from the antenna do. There is thus a 50% utilization of signal strength, both in transmitting and receiving.

Another device that makes use of the interference of waves is called a directional coupler (Figure 21). It is constructed in a number of different

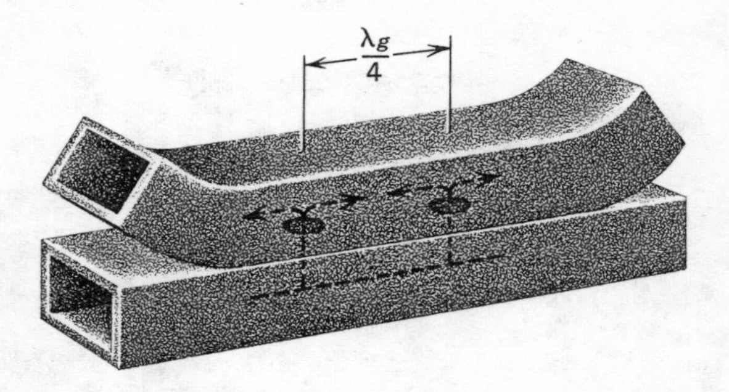

Figure 21. *A directional coupler.*

forms, one of the simplest being that illustrated, in which two similar waveguides are mounted one above the other with their broad walls adjacent. Two holes, separated by one-quarter wavelength, enable some energy from the lower guide to get into the upper. Suppose there is a source of energy at the left end and a load at the right end of the lower guide and consider first the forward wave from the generator. At hole *A* some energy enters the upper guide and divides right and left. A second pair of waves originates at hole *B*. The separation of the holes insures that the two waves traveling toward the left are one-half wavelength out of phase and destructively interfere, whereas the waves traveling toward the right are in phase and reinforce. For the reflected wave from the load the reverse is true. If detectors are located at each end of the upper guide with impedances adjusted to give no reflection, then the ratio of the signals at the detectors measures the standing wave ratio in the lower guide.

In an entirely different class are those waveguide devices in which there is an exchange of energy between an electromagnetic wave and an electron stream. One such device is the traveling wave tube amplifier. A wire in the shape of a helix has an electron beam traveling axially down its center.

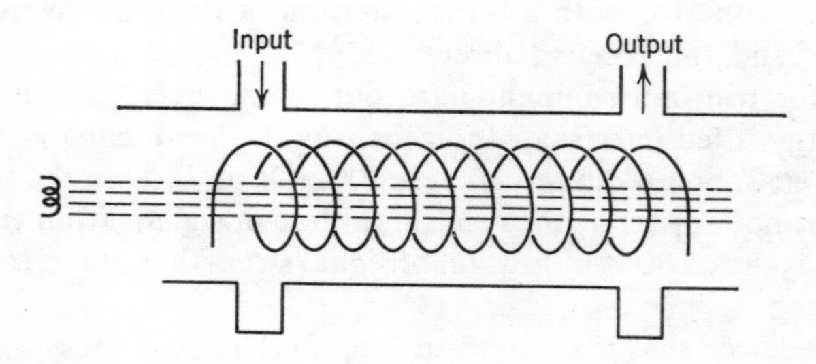

Figure 22. *Traveling wave amplifier.*

An electromagnetic wave is started on the helix and is guided in a helical path. The helix acts roughly like a parallel wire transmission line. Because of this guided wave there is an axially directed electric field at the electron beam. If the electromagnetic wave propagates on the helix at approximately the velocity of light, the axial field propagates along the axis at a much smaller velocity, determined by the pitch of the helix. It may, in fact, be made to propagate more slowly than the speed of the electrons in the beam. At those phases of the axial field where there is a retarding force on the electrons, the wave gains energy at the expense of the kinetic energy of the electrons. Where the field accelerates the electrons, they gain energy from the field. Because the effect of the field near the beginning of the helix is to produce a bunching of the electrons in the stream, a larger number are in those regions of a retarding field than in the regions of accelerating field, and there is a net gain of energy by the electromagnetic wave. Since the waveguide structure is similar to a transmission line, the velocity of the wave is nearly independent of the frequency over a wide range of frequencies, making the traveling wave tube a broad band amplifier.

The klystron oscillator or amplifier employs an interaction between an electron stream and the field in a resonant cavity. The cavities (Figure 23) are constructed in a more or less doughnut shape with parallel grids over the area of the hole. An electron beam passes through these grids.

First, consider what happens when an electron beam passes through the gap between the grids of the first cavity when there exists a standing wave oscillation in the cavity. The oscillation mode is such that the electric field at the gap is alternately parallel and antiparallel to the electron velocity in the gap. As successive electrons pass through the gap, they are either speeded up or slowed down, depending on the phase of the electric field at the time of their passing. In the region beyond the gap those

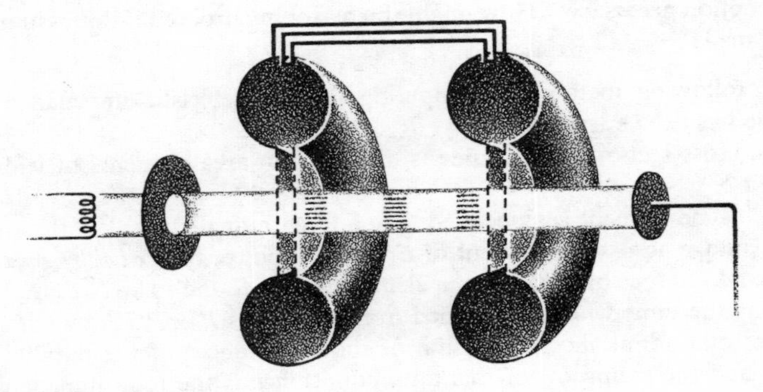

Figure 23. *A two-cavity klystron oscillator.*

electrons that were speeded up catch up with those ahead that were slowed down to form bunches of electrons.

A second cavity, identical with the first, is located at the point where these bunches of electrons are fully formed. As the bunches pass through the grids of this cavity they excite in it oscillations of the same frequency as in the first cavity. The power required to produce the bunches is small and it is the kinetic energy of the electron stream which is converted into electromagnetic energy of oscillations in the second cavity. Large amplifications are possible with this device. Different from the traveling wave tube, the klystron operates at a single frequency or at most over a very narrow band of frequencies.

When some of the energy in the second cavity is returned to the first in the proper phase, the system becomes self-exciting and is an oscillator. Another version of this oscillator is the reflex klystron, in which the electron beam is turned back on itself so that the bunches return through the original cavity gap, thus inducing oscillations in the same cavity that produced the bunches.

These brief descriptions of devices employing guided and standing electromagnetic waves are intended only to give an idea of the kind of uses to which confined waves can be put. The reader is urged to consult the references for a more complete analysis of their operation, for it is hoped that an interest may have been stimulated in what is a fascinating area of application of electromagnetic waves and also an area out of which important research tools develop.

Discussion Questions

1. The derivation of the Fresnel relations in Chapter 11 did not assume that the direction vectors of the incident, reflected, and refracted waves were coplanar, whereas the derivation of Section 1 in this chapter did. In the latter case, was this assumption necessary? How might the reasoning proceed if this assumption were not made?

2. The following method of computing the characteristic impedance of a waveguide has been suggested:

(*a*) The cross section of the guide is divided into area elements of width dx and height b.

(*b*) Each such element is associated with a longitudinal current in the wall of $H_x\,dx$, a voltage across the element of $E_y b$, and a power flow of $E_y H_x b\,dx$.

(*c*) The admittance of such an area element is then found to be $H_x dx/E_y b$, this result being the same whether obtained from the forms I/V, P/V^2, or I^2/P.

(*d*) The total admittance is the sum of the admittances of the parallel area elements, and the reciprocal of this total admittance is the total impedance of the guide.

There is thus one impedance by this method instead of the three obtained in Section 7. Criticize and comment on this argument.

Problems

1. Using the methods of Section 1, obtain the Fresnel transmission and reflection relations for both parallel and perpendicular polarization of the incident wave.

2. Let there be two traveling waves:

$$y_1 = A \sin [\omega t - \beta z]$$

$$y_2 = A \sin [(\omega + \Delta\omega)t - (\beta + \Delta\beta)z]$$

Show that the sum of these two waves is a wave of frequency $\omega + (\Delta\omega/2)$, of propagation constant $\beta + (\Delta\beta/2)$, and of amplitude $2A \cos [\frac{1}{2}(\Delta\omega t - \Delta\beta z)]$. Demonstrate that the amplitude pattern propagates with a velocity $\Delta\omega/\Delta\beta$. Assign numerical values and plot such a sum of two waves for several successive values of t. The velocity of the amplitude pattern is called the group velocity and is the same as the phase velocity only if all frequencies travel with the same velocity.

3. Show that the group velocity of the wave described in Section 1 for propagation between parallel planes is the same as the velocity of propagation of energy obtained in Section 2.

4. Using first the method of Section 2 and then the derivative $d\omega/d\beta$, find the group velocity or velocity of energy propagation for waves in a rectangular wave guide.

5. Show that the equations of the electric field lines in Figure 5 are $\sin \delta \cos \Omega = \sin \delta_0$, where δ_0 is the value of δ at the point where the line starts on a conducting plane.

6. For the TE_{10} mode in a rectangular wave guide, plot the ratio of the guide wavelength to the cut-off wavelength against the ratio of the frequency to the cut-off frequency.

7. Carry out the steps to obtain Equations 6-2 and 6-3.

8. At one cross section in a rectangular waveguide there is an abrupt change in dielectric. Show that there can be frequencies of propagation in the TE_{10} mode for which there is no transmission at this dielectric boundary. If the wave travels from dielectric ϵ_1 toward dielectric ϵ_2, determine what frequencies are not transmitted and any limitations on ϵ_1 and ϵ_2. Relate this effect to the critical angle in the reflection of light (Chapter 11, Section 11).

9. Obtain Equations 9-11 from Equations 9-9 and then evaluate the constants in terms of C_z and D_z. Reduce to the TE_{101} case and compare with Equations 9-2 and 9-7.

10. Find the heat losses per unit length in a waveguide which is supporting a propagating wave in the TE_{10} mode.

11. Analyze the phases of the waves circulating in a ring waveguide bridge and show that for input at A and equal reflections at B and D, there will be no transmission to C.

12. A rectangular box, $6 \times 8 \times 10$ cm, is silver-coated on the inside. List the lowest six resonant frequencies in this box. For the lowest frequency find the Q of the box. Assume the coating is appreciably thicker than δ (σ for silver $= 6 \times 10^7$ mhos per meter).

REFERENCES

1. Ramo and Whinnery, *op. cit*. Chapters 8, 9, 10.
2. Ginzton, *Microwave Measurements* (McGraw-Hill), 1957. An authoritative book.
3. Southworth, *Principles and Applications of Waveguide Transmission* (Van Nostrand), 1950. An older book with clear descriptions of many microwave devices.
4. Hemenway, Henry, Caulton, *Physical Electronics* (Wiley), 1962. Chapters 15 and 16 contain good qualitative descriptions of some microwave generators.
5. Atwater, *Introduction to Microwave Theory* (McGraw-Hill), 1962.

Many of the references cited earlier also contain sections on waveguide theory and devices.

13

Relativity and electromagnetism

The differential equations for the propagation of electromagnetic fields predict a velocity of propagation of $c = 1/\sqrt{\mu\epsilon}$, which in vacuum is 2.99792×10^8 meters per second, a value known to within about one kilometer per second.

In the early years of the electromagnetic theory it was assumed that the propagation of such waves could not take place without there being a medium, called the "ether," to support the propagation. Such a concept followed reasonably from an analogy with sound waves and water waves, where for waves to exist there had to be material medium to do the "waving." No properties of such an ether, however, could be found other than those inferred from the electromagnetic waves themselves, so that arguments for its existence went in a circle. It was often said merely that it was inconceivable for the waves to propagate without an ether. But an argument based on inconceivability may tell more about the inadequacy of the human intellect than about the natural processes which that intellect is attempting to investigate. The search for confirming evidence of the existence of the ether included a remarkable series of experiments which are now generally agreed to have been unsuccessful in verifying its existence.

If a medium was required for the propagation of light, it seemed reasonable that the wavelength should be affected by the motion of the observer relative to the medium, just as the wavelength of sound is affected by the motion of the observer relative to the air which carries the sound wave.

The Michelson-Morley interferometer experiment, in which interference between two coherent light beams in mutually perpendicular directions is observed, was expected to show a change in the interference pattern when the instrument was moving parallel to one of the beams and perpendicular to the other from the pattern when the instrument was rotated through 90° so that the relation of the motion to the light beams was reversed. This experiment has been repeated many times, with either no change in the interference pattern or with changes which were far smaller than could be justified by the ether theory, whether the source of the light was moving with the instrument or whether the source was starlight in which there was relative motion of the source and the instrument. A variety of other experiments have also been tried, with consistently negative results. Although ingenious theories were devised ascribing special properties to the ether to explain the results of each experiment, no set of properties were found to be consistent with all the experiments.

If the velocity of light were a function of the motion of the observer relative to the ether, such an ether could constitute a reference system for all motion. An absolute velocity, relative to this ether, could be assigned to a body, independent of its relation to other bodies. It should therefore be possible to be inside a closed room and, by measuring the velocity of light, to determine the absolute motion of the room without reference to bodies outside the room. The null results of the various attempts to observe a change in the velocity of light in vacuum finally led to the conclusion that such an idea was not tenable, and that motion has meaning only in a relative, and not in an absolute sense.

This conclusion, in itself, was hardly revolutionary. In actual practice neither the descriptions of position or of motion had ever been treated in any other way. A stated or implied reference system is always used in any equation describing physical events. When, as often happens, it is necessary to translate a description of an event relative to one reference system to a description relative to another, the *relative* motion of the two systems must always be taken into account in the translation. But it is at this point where the conceptual difficulties with the constant velocity of light arise, since apparently the velocity of light, measured relative to any reference—at least, any reference which has a uniform velocity—is the same. This puts the velocity of light in a rather special position in physical theory. Whereas the constancy of this velocity led to a rejection of the concept of absolute velocity, it at the same time seemed to be inconsistent with the classical ideas of transforming a mathematical description of events from one system to another which is in relative motion to the first. Surely if other velocities are affected by the relative motion, the velocity of light should be as well. The modern theory of relativity, which is discussed in this chapter, resolves this apparent dilemma by revising the method of transforming from one reference system to another.

This theory accepts as a postulate that absolute uniform motion has no meaning and that we can only deal with relative uniform motion. (Accelerated motion is another matter, dealt with in the general theory of relativity; the present discussion is confined to the uniform motion of reference systems.) It follows, then, that a valid statement of a physical law should have the same mathematical form relative to the coordinates of any system in uniform motion. If this were not true, we could conceivably infer information about the absolute motion of a laboratory from the form of the physical law observed in it without reference to the laboratory's motion relative to other bodies. Consequently a second postulate of the special theory of relativity is that a properly stated physical law has the same form when expressed in terms of the coordinates of any of a group of reference systems which are in uniform motion relative to one another. Such a statement of a physical law is said to be invariant to a coordinate transformation.

The criterion of invariance is also not new to theoretical physics. It has long been recognized as an important property of a properly stated physical law in terms of the classical, or Galilean, transformation, which is discussed in Section 1. The advent of relativity theory has, in a sense, given it new significance, by emphasizing that the invariance must be in terms of a transformation that retains the constancy of the speed of light, which the Galilean transformation does not do.

The electromagnetic theory, in which the velocity of light plays a special role, was peculiarly important in the development of the theory of relativity. In this chapter we develop some of the more fundamental parts of this theory, with special emphasis on its relation to electromagnetism. Space does not permit a complete discussion, and many of the applications to mechanics will have to be omitted.

The theory of relativity is the result of the contributions of many men—Lorentz, FitzGerald, Poincaré, Minkowski, Tolman, Lewis—but foremost among them is Albert Einstein, whose publication in 1905 of the Special Theory of Relativity is credited with producing the fundamental clarification of the conceptual difficulties that have been described above.

1. The Galilean Transformation

In order to appreciate the new ideas introduced by the special theory of relativity, we review briefly the classical concepts of the transformation of a mathematical description from one system of coordinates to another.

Figure 1 shows a system of coordinates S relative to which the position

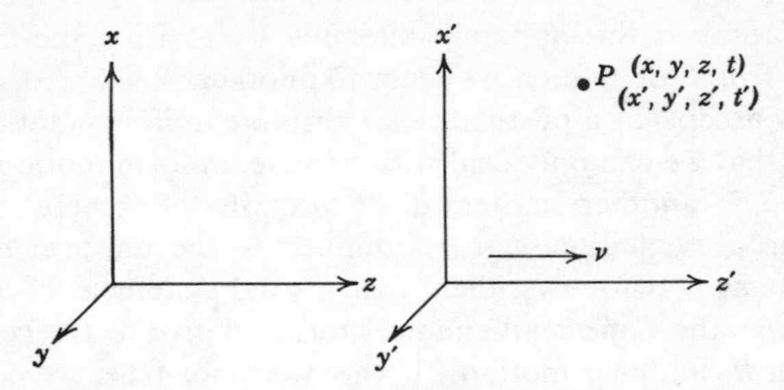

Figure 1. Coordinate system S′ is moving with velocity v parallel to the z-axis of system S. The location of point P may be referred to either axis.

of, say, a particle is $P(x, y, z)$ at a time t. A second coordinate system S' has, for simplicity, its axes respectively parallel to those of S and is moving relative to S with a velocity v in the z-direction. In S' the position is $P(x', y', z')$ at the time t'. We wish to relate the two sets of coordinates of the point. It is a usually unstated assumption of the classical theory that it is possible for observers, one located in each system, to have time measuring instruments which are synchronized so that a measurement of time is the same for both. We shall assume that time in both systems is measured from the instant when the two origins are coincident. The classical, or Galilean, transformation of coordinates that follows from these assumptions is

$$x' = x \qquad y' = y \qquad z' = z - vt \qquad t' = t \tag{1-1}$$

where the last relation is included for comparison with later theory. Differentiation of 1-1 with respect to time yields the transformation of the velocity components of the particle:

$$u_x' = u_x \qquad u_y' = u_y \qquad u_z' = u_z - v \tag{1-2}$$

The extension to the transformation of accelerations is obvious.

It turns out that the results of such a Galilean transformation agree well with experiment within the accuracy of laboratory measurements for the ordinary velocities of material bodies encountered on the earth in other than experiments in the atomic realm. Such velocities are many orders of magnitude smaller than the velocity of light. But when the Galilean transformation is applied where there are velocities within a few magnitudes of the velocity of light, the results are no longer satisfactory, a fact which turns out to be directly related to the constancy of the speed of light.

Let us imagine the following thought experiment. Suppose that at the instant that the two origins are coincident a flash of light is produced at the common origin, and then the wavefront of this light pulse is followed as seen in the two systems. The equation of the spherical wavefront seen in system S is

$$x^2 + y^2 + z^2 = c^2t^2 \tag{1-3}$$

and the transformation of this relation to the S' system results in

$$x'^2 + y'^2 + (z' + vt')^2 = c^2t'^2 \tag{1-4}$$

This latter should also be a spherical wavefront with the center at the origin of S', if an observer in S' is to measure the speed of light independent of his motion. Equation 1-4 is at each instant of time the equation of a sphere, but it is a sphere whose center moves with time along the z' axis. S' therefore measures different speeds of light depending on the direction from the origin at which his instruments are located. The Galilean transformation is evidently inconsistent with the evidence of the constancy of the speed of light, independent of the uniform motion of the observer.

2. The Lorentz Transformation

Since the classical transformation breaks down in predicting the behavior of a light wavefront, perhaps the solution of this difficulty will provide us with the correct transformation applicable also in the general case of the motion of any bodies. Certain demands may reasonably be made of such a transformation of coordinates when it is found. It must, of course, leave the velocity of light unchanged; but we also expect it to reduce to the Galilean transformation for motions in which the speeds are small compared to the speed of light, for there is ample evidence that the Galilean transformation is the proper one for such speeds. Moreover, since it should reduce to the Galilean transformation, we may hope, pending tests, that it will also be a linear transformation, involving only first powers of the coordinates.

It is reasonable to expect that the coordinate x of a point referred to the S system will be x' in the S' system, where $x = x'$; and similarly for $y = y'$. (What would be the consequences, if this were not true?) In other words, position coordinates on axes perpendicular to the relative motion of the reference systems can reasonably be expected to be unaffected by the motion. In the classical transformation the z' coordinate is a function of both z and t, and we shall try a linear relationship of this type for the new transformation. An examination of Equations 1-3 and 1-4 suggests that a defect in the classical transformation may lie in assuming that the time is identical in the two systems. As a trial we shall assume that t' is also a function of z and t. Thus we try

(2-1) $$x' = x \qquad y' = y \qquad z' = az + bt \qquad t' = ez + ft$$

where the constants a, b, e, and f are to be found.

Before discussing other conditions whose satisfaction leads to determining these constants, we can immediately note that the origin of the S' system, for which $z' = 0$, is at the point $z = vt$, as seen from S. The third relation in 2-1, using this fact, leads to $v = -(b/a)$.

We wish to find a transformation for which the speed of light is not affected by the relative uniform motion. This implies that a wavefront which starts from a flash of light at the coincident origins shall propagate as a spherical wavefront with center at the origin in either system. That is, the equation of the wavefront in the S system shall be

(2-2) $$x^2 + y^2 + z^2 = (ct)^2$$

and in the S' system shall be

(2-3) $$x'^2 + y'^2 + z'^2 = (ct')^2$$

The constants a, b, e, and f in Equations 2-1 are now to be determined so that 2-1 transforms 2-3 into 2-2. When we substitute the Equations 2-1 into 2-3, getting

(2-4) $$x^2 + y^2 + (az + bt)^2 = c^2(ez + ft)^2$$

we find that 2-4 will be identical with 2-2, provided

(2-5) $$a^2 - c^2e^2 = 1 \qquad ab - c^2ef = 0 \qquad f^2 - (b/c)^2 = 1$$

which, together with the condition that $b/a = -v$, results in

(2-6) $$a = f = \pm\frac{1}{\sqrt{1-\beta^2}} \;\Bigg|\; e = \mp\frac{\beta/c}{\sqrt{1-\beta^2}} \;\Bigg|\; b = \mp\frac{v}{\sqrt{1-\beta^2}}$$

where $\beta = v/c$. Thus the transformation equations become

(2-7)
$$x' = x$$
$$y' = y$$
$$z' = \frac{\pm z \mp vt}{\sqrt{1-\beta^2}}$$
$$t' = \frac{\pm t \mp \beta z/c}{\sqrt{1-\beta^2}}$$

The sign ambiguities are resolved by requiring that Equations 2-7 reduce to the classical transformation, 1-1, for small values of v compared to c. Thus finally,

$$\begin{aligned} x' &= x \\ y' &= y \\ z' &= \gamma(z - vt) \\ t' &= \gamma(t - \beta z/c) \end{aligned} \tag{2-8}$$

where $\gamma = 1/\sqrt{1 - \beta^2}$.

Either by solving Equations 2-7 for x, y, z, and t or by noting that the relative velocity of S as seen from S' is $-v$ and hence that the inverse transformation should have the same form except for the change of sign of v, we get

$$\begin{aligned} x &= x' \\ y &= y' \\ z &= \gamma(z' + vt') \\ t &= \gamma(t' + \beta z'/c) \end{aligned} \tag{2-9}$$

These equations were first obtained by Lorentz for the purpose of reconciling the null result of the Michelson–Morley experiment with the ether theory. Einstein showed that the retention of the ether theory was both unnecessary and misleading; he developed the general implications of the transformation equations. Their acceptance does not, of course, rest solely on the fact that they maintain the constancy of the speed of light and that they reduce to the classical form for small relative velocities of the reference frames. Predictions based on them must be, and have been, tested by experiment, especially in the realm of atomic and nuclear physics. One of the more spectacular predictions was that the measured mass of a body is a function of its motion relative to an observer, and that this mass is a form of energy. It also was shown that no body can have a speed greater than the speed of light.

In what follows we shall examine some of the implications of this transformation to obtain a better appreciation of its beauty and its logic and to see how the equations of electromagnetism remain invariant under such a transformation.

3. The FitzGerald-Lorentz Contraction and the Time Dilatation

Let us imagine an observer, whose frame of reference is S, making measurements of the positions of the front and back ends of a rod which is moving past him with a velocity v, the direction of the velocity and the length of the rod being parallel. The observer is located at the point z, and reads on his clock a time t_1 as the front end of the rod passes him and a time t_2 as the back end passes him. He thus concludes that the length of the rod is

$$L = v(t_2 - t_1) \tag{3-1}$$

In a frame of reference S', fixed to the rod, the front end is at z_1' and the back end at z_2' so that in the S' frame the length of the rod is

$$L' = z_1' - z_2' \tag{3-2}$$

Transforming the right side of Equation 3-1 into the primed coordinates, we get

$$L = v\gamma\left(t_2' + \frac{\beta}{c} z_2' - t_1' - \frac{\beta}{c} z_1'\right) \tag{3-3}$$

where t_2' and t_1' are the times in S' corresponding to t_2 and t_1 in S. But also

$$z = \gamma(z_1' + vt_1') = \gamma(z_2' + vt_2') \tag{3-4}$$

from which

$$L' = (z_1' - z_2') = v(t_2' - t_1') \tag{3-5}$$

Using Equation 3-5 in 3-3, we find that

$$L = L'\sqrt{1 - \beta^2} \tag{3-6}$$

Evidently the length of a rod as measured by an observer with respect to whom the rod is moving is reduced by a factor $\sqrt{1 - \beta^2}$ over the length of the rod as measured in a frame of reference with respect to which the rod is at rest. This effect is known as the FitzGerald-Lorentz contraction.

Now let us consider a source of light in the S' frame which flashes at regular intervals, so that the time between two flashes is

$$\Delta t' = t_2' - t_1' \tag{3-7}$$

The light source is located at the point z' in S'. By trial the observer in S sets up a detector at point z_1 which is at the location of the light source when the first of a pair of successive flashes occurs (at t_1') and another detector at z_2 where the next light flash occurs (at t_2'). These detectors are coupled with identical clocks which S has checked for synchronism (see Discussion Question 1). Hence S measures the first flash at z_1 at a time t_1 and the second at the point z_2 at the time t_2. He concludes that the interval between flashes is

$$\Delta t = t_2 - t_1 \tag{3-8}$$

Using the transformation equation for the time coordinate, we find that

$$\begin{aligned} \Delta t &= \gamma\left(t_2' + \frac{\beta}{c} z'\right) - \gamma\left(t_1' + \frac{\beta}{c} z'\right) \\ &= \frac{\Delta t'}{\sqrt{1 - \beta^2}} \end{aligned} \tag{3-9}$$

and hence S measures an interval between flashes which is longer by his clocks than the interval measured in a frame of reference in which the light source is at rest.

As an example, suppose S' were moving relative to S with a velocity $v = (3/5)c$, and the light flashes were produced once per second as S' measures them. S would then record the time interval between flashes as 1.25 seconds on his clocks and would conclude that the S' clocks were running slow. Since the effect involves the square of the relative velocity, it is independent of the sign of v. Thus S', observing a time interval between events in S, measures a longer interval than S measures.

4. Transformations as Rotations

The transformation equations of 2-8 may be written out in the following ordered form.

$$\begin{aligned} x' &= x + 0 + 0 + 0 \\ y' &= 0 + y + 0 + 0 \\ z' &= 0 + 0 + \gamma z - \gamma\beta(ct) \\ (ct') &= 0 + 0 - \gamma\beta z + \gamma(ct) \end{aligned} \tag{4-1}$$

The fourth equation has been multiplied by c so that all terms have the dimensions of length. Still another type of representation is:

$$\begin{pmatrix} x' \\ y' \\ z' \\ ct' \end{pmatrix} = \begin{pmatrix} 1 & 0 & 0 & 0 \\ 0 & 1 & 0 & 0 \\ 0 & 0 & \gamma & -\gamma\beta \\ 0 & 0 & -\gamma\beta & \gamma \end{pmatrix} \begin{pmatrix} x \\ y \\ z \\ ct \end{pmatrix} \tag{4-2}$$

Here the groups of terms in large parentheses are called matrices and the symbols are to be interpreted in the following way: the first term in the matrix on the left is equal to the sum of the respective products of the first row in the 4 × 4 matrix with the single-column matrix on the right; the second term in the row matrix on the left is equal to the sum of the respective products of the second row of the 4 × 4 matrix with the column matrix; etc. The 4 × 4 matrix of coefficients is called the transformation matrix from the unprimed to the primed systems.

The reverse transformation from the primed to the unprimed system is evidently

$$\begin{pmatrix} x \\ y \\ z \\ ct \end{pmatrix} = \begin{pmatrix} 1 & 0 & 0 & 0 \\ 0 & 1 & 0 & 0 \\ 0 & 0 & \gamma & +\gamma\beta \\ 0 & 0 & +\gamma\beta & \gamma \end{pmatrix} \begin{pmatrix} x' \\ y' \\ z' \\ ct' \end{pmatrix} \tag{4-3}$$

The reversed transformation matrix is obtained from 4-2 by reversing the signs of the off-diagonal terms.

In Equations 4-2 and 4-3 the quantity ct has the dimensions of length, suggesting that it can be treated on an equal basis with the other three dimensions, so that an event can be identified completely in a four-dimensional coordinate system. The simplification of choosing the relative motion of the systems parallel to the z-axes means that in many simple cases we can represent the coordinate systems by showing only the z- and the (ct)-axes, leaving the other two to the imagination.

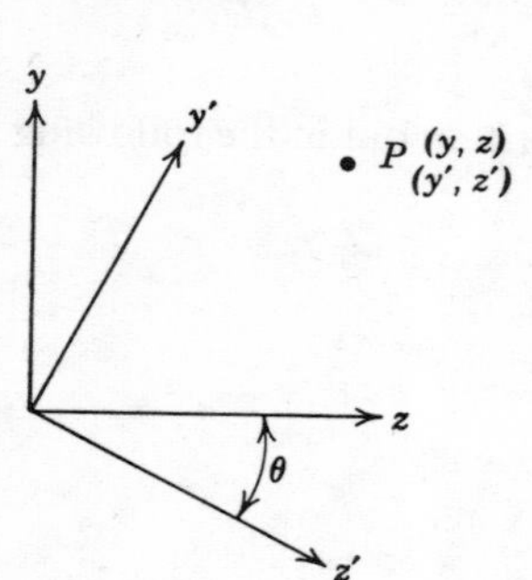

Figure 2

An analogy with three-dimensional space will illustrate some of the ideas and expressions to be used with reference to these transformations. We encountered the operation of turning one three-dimensional system of coordinates into another in Section 1 Chapter 12. A special case will be outlined here. If the axis of rotation is the x-axis, the transformation equations are:

$$\begin{aligned} x' &= x \\ y' &= y \cos \theta + z \sin \theta \\ z' &= -y \sin \theta + z \cos \theta \end{aligned} \tag{4-4}$$

and the inverse transformation:

$$
\begin{aligned}
x &= x' \\
y &= y' \cos\theta - z' \sin\theta \\
z &= y' \sin\theta + z' \cos\theta
\end{aligned} \tag{4-5}
$$

is obtained by reversing the signs of the off-diagonal terms.

These transformations may be expressed in 3 × 3 matrix notation as

$$
\begin{pmatrix} x' \\ y' \\ z' \end{pmatrix} = \begin{pmatrix} 1 & 0 & 0 \\ 0 & \cos\theta & +\sin\theta \\ 0 & -\sin\theta & \cos\theta \end{pmatrix} \begin{pmatrix} x \\ y \\ z \end{pmatrix} \tag{4-6}
$$

corresponding to 4-4, and

$$
\begin{pmatrix} x \\ y \\ z \end{pmatrix} = \begin{pmatrix} 1 & 0 & 0 \\ 0 & \cos\theta & -\sin\theta \\ 0 & +\sin\theta & \cos\theta \end{pmatrix} \begin{pmatrix} x' \\ y' \\ z' \end{pmatrix} \tag{4-7}
$$

corresponding to 4-5. The 3 × 3 matrix in either Equation 4-6 or 4-7 is called a rotation matrix, and is characterized by the fact that those terms located symmetrically with respect to the diagonal have the same magnitude but differ in sign. The expression

$$
\begin{vmatrix} 1 & 0 & 0 \\ 0 & \cos\theta & +\sin\theta \\ 0 & -\sin\theta & \cos\theta \end{vmatrix} \tag{4-8}
$$

is called the determinant of the matrix in 4-6. The value of the determinant, obtained by the standard rules, is always unity if its matrix is a matrix of rotation of one 3-space coordinate system into another. (A determinant may be evaluated, but a matrix is only an array of functions.) A determinant is distinguished from the matrix from which it is derived by the use of straight lines instead of parentheses.

Not only does the matrix of 4-6 (or 4-7) change the coordinates of one system into those of another system which is rotated with respect to the first, but the matrix may be applied to the transformation of the components of any vector in one system to the corresponding vector components in the other.

By analogy with the three dimensional rotation discussed above, the matrices of 4-2 and 4-3 can be looked upon as rotation matrices in the four-dimensional space of which x, y, z, and ct are the components. The analogy is not exact in every respect. Although the determinant of the 4-space matrix does have the value unity, the off-diagonal terms are not opposite in sign, and we shall quickly see an interesting geometrical interpretation resulting from this fact.

We can make the analogy between 3-space rotations and 4-space rotations closer by a modification in the time coordinate. From Equations 2-2 or 2-3, if x, y, z, and ct are the four-dimensional coordinates of a point on a spherical light wave that started at the origin, these coordinates are related by

$$x^2 + y^2 + z^2 - (ct)^2 = 0 \tag{4-9}$$

If, however, $xyz(ct)$ describe an event that occurred at a time less than the time required for a light wave from the origin to reach the point xyz, the left side of 4-9 is nonzero and positive. If the time of occurrence is a longer time, the left side is negative. From the nature of the transformation 2-8, we see that $x^2 + y^2 + z^2 - (ct)^2$ is an invariant under change of coordinate system.

Except for the negative sign before the time variable, this quantity is similar to the expression for the square of the distance from the origin to a point in 3-space. In order to make this analogy closer, the transformation equations, such as 4-1, are written by multiplying the last one by $j = \sqrt{-1}$:

$$\begin{aligned} x' &= x + 0 + 0 + 0 \\ y' &= 0 + y + 0 + 0 \\ z' &= 0 + 0 + \gamma z + j\beta\gamma(jct) \\ (jct') &= 0 + 0 + (-j\beta\gamma z) + \gamma(jct) \end{aligned} \tag{4-10}$$

If we consider the four coordinates in 4-space as being $xyz(jct)$, and write them as x_1, x_2, x_3, and x_4, then the position vector in 4-space takes the form

$$x_1{}^2 + x_2{}^2 + x_3{}^2 + x_4{}^2 = R^2 \tag{4-11}$$

which has a formal parallelism to the 3-space form and also emphasizes the essentially symmetry among the four coordinates. A 4-vector from the origin to the point $x_1x_2x_3x_4$ can thus by analogy be written as

$$\mathbf{R} = \mathbf{i}x_1 + \mathbf{j}x_2 + \mathbf{k}x_3 + \mathbf{l}x_4 \tag{4-12}$$

where **i**, **j**, **k**, and **l** are unit vectors along the respective axes. By the nature of the transformation in 4-space, the form of the vector **R** remains invariant with a shift to a second coordinate system moving with constant velocity relative to the first. As in the case of three dimensions the transformation matrix may also be used to express any four-dimensional vector in a new coordinate system. When such a vector is involved in a properly expressed law of nature, then the form of the law is left unchanged when expressed in the new coordinate system. A number of familiar laws—Newton's law of motion, for example—when expressed in terms of coordinates of 4-space, do not have this invariance, and hence are unsatisfactory as expressions of a truly general physical law. They are acceptable as approximations only when the velocities involved are negligible with respect to the speed of light. Some of the subsequent discussion will be concerned with finding expressions for the familiar electromagnetic quantities in such a form that they will have this invariance in 4-space.

5. Geometrical Representation of a Rotation in 4-Space

Before we consider a relativistic formulation of electromagnetic relations, however, let us examine a geometrical representation of the transformation in 4-space. It is fortunate that only two of the four axes are involved in this transformation for the simplified case we have been discussing. We measure the quantity ct along one axis and z along another, leaving the x- and y-axes to be imagined. To determine how to draw the ct'- and the z'-axes for a system moving with a velocity v relative to the unprimed system, we compare Equation 4-10 with 4-4 to find that the following are analogous:

$$\text{(5-1)} \qquad \sin\theta \leftrightarrow j\beta\gamma \qquad \cos\theta \leftrightarrow \gamma \qquad \tan\theta \leftrightarrow j\beta$$

Let ψ be the angle of rotation of the axes for the 4-space case so that

$$\sin\psi = j\beta\gamma \qquad \cos\psi = \gamma \qquad \tan\psi = j\beta$$

Then, using the exponential form of the tangent of an angle,

$$\text{(5-2)} \qquad \tan\psi = \frac{e^{j\psi} - e^{-j\psi}}{j(e^{j\psi} + e^{-j\psi})} = j\beta$$

we find that

$$\text{(5-3)} \qquad \psi = -j\ln\left(\frac{1-\beta}{1+\beta}\right)^{1/2}$$

and thus in 4-space the rotation angle is imaginary. It is still possible, however, to draw a geometrical analogue for this case by noting first of all that, in the plane of Figure 3 where $x = y = 0$, the case of $z^2 = (ct)^2$, representing the points on a light wavefront that originated at the origin, appear in this diagram as the points on the 45° line and that these points are symmetrical for any system. This suggests that the axes of the second system be represented as in Figure 4. In this representation the new axes are no longer mutually perpendicular. Figure 4 also shows the coordinates in the two systems of a general point P, which in the first system are z and (ct) and in the second system, z' and (ct').

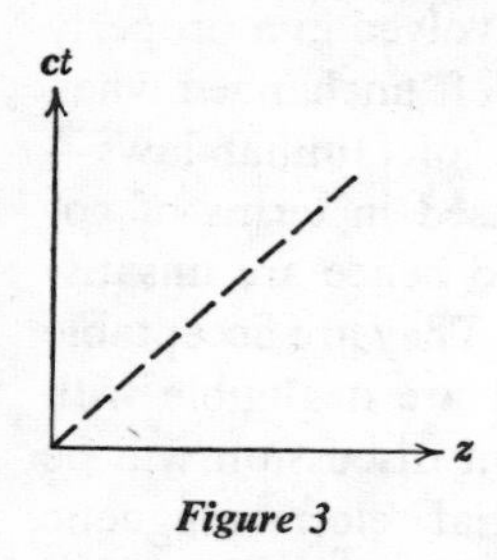

Figure 3

As was seen in the discussion of the FitzGerald-Lorentz contraction and the time dilatation, the scale of measurement along the unprimed axes is not the same as along the primed axes. In what follows we shall need to distinguish between what shall be called geometrical lengths and scaled lengths, the former being lengths on the primed axes measured on the graph using the same scale as for the unprimed

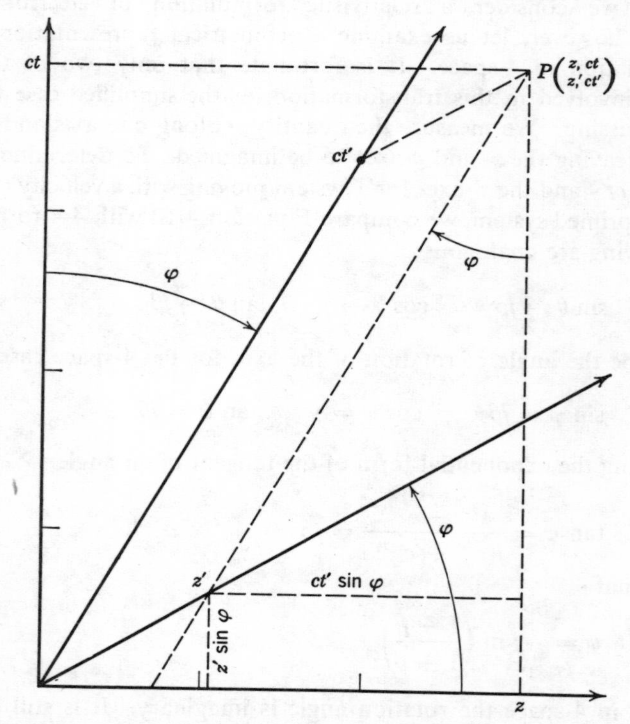

Figure 4. Geometrical representation of a relativistic transformation.

axes, whereas the latter are the corrected lengths after the appropriate scale factor has been applied. The geometrical lengths are identified with the subscript g. The construction lines shown on the graph of Figure 4 can be used to show that the geometrical lengths along the primed axes are related to lengths along the unprimed by

$$z_g' = \frac{[z \cos \varphi - (ct) \sin \varphi]}{[\cos^2 \varphi - \sin^2 \varphi]} \tag{5-4}$$

$$(ct')_g = \frac{[-z \sin \varphi + (ct) \cos \varphi]}{[\cos^2 \varphi - \sin^2 \varphi]}$$

where the angle φ, shown on the graph, will shortly be related to the constants β and γ of the transformation.

To see how the scale factor is to be established, let us first return to the corresponding 3-space case of Figure 2 and review how the scale is determined in ordinary geometry, where, in spite of familiarity with the use of such a graph, the explicit procedure in creating a measurement scale on the axes may have been forgotten. We shall discuss the procedure for a two-dimensional system of axes; the extension to three-dimensions follows readily. A property of the point $P(yz)$ that is invariant with a rotation of axes is its distance from the origin. We create a scale for the axes by drawing circles given by

$$y^2 + z^2 = (\text{constant})^2 \tag{5-5}$$

for successive values of the constant differing by equal increments as shown in Figure 5. The measurement scale on the axes is fixed by the intersections of these circles with the axes.

By an analogous procedure we create scales on the axes in relativistic space. For a point in such space, a property invariant to a Lorentz transformation is

$$x^2 + y^2 + z^2 + (jct)^2 = (\text{constant})^2 \tag{5-6}$$

which, in the plane of Figure 4, reduces to

$$z^2 - (ct)^2 = (\text{constant})^2 \tag{5-7}$$

This is a family of hyperbolas for which the constant can be either real or imaginary. Figure 6 shows such a family of hyperbolas imposed on the axes of Figure 4, the family being obtained by taking unit increments in the magnitude of the constant for the successive hyperbolas. It is left as a problem for the reader to show that

$$z' = z_g' \sqrt{\cos^2 \varphi - \sin^2 \varphi} \tag{5-8}$$

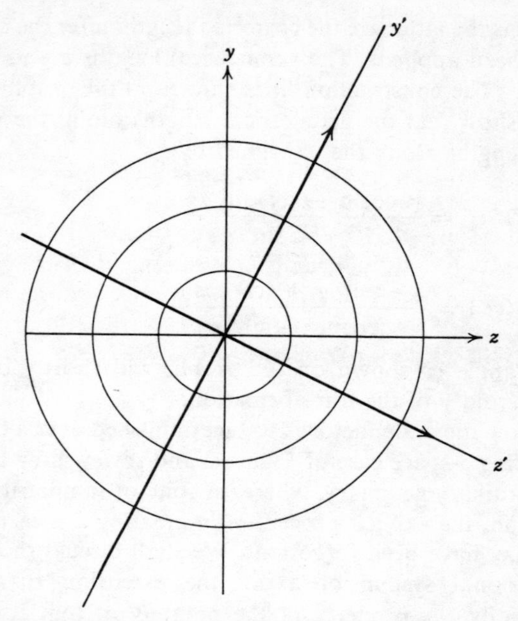

Figure 5. The scaling circles of ordinary geometry.

Using 5-8, Equations 5-4 become

(5-9)
$$z' = \frac{[z \cos \varphi - (ct) \sin \varphi]}{\sqrt{\cos 2\varphi}}$$

$$(ct') = \frac{[-z \sin \varphi + (ct) \cos \varphi]}{\sqrt{\cos 2\varphi}}$$

where the substitution $\cos 2\varphi = \cos^2 \varphi - \sin^2 \varphi$ has been used. Comparing these expressions with 4-10, it is seen that

(5-10)
$$\gamma = \frac{\cos \varphi}{\sqrt{\cos 2\varphi}} \qquad \gamma\beta = \frac{\sin \varphi}{\sqrt{\cos 2\varphi}}$$

so that

(5-11)
$$\tan \varphi = \beta$$

Equations 5-10 and 5-11 should be compared with 5-2 to see the relation between φ and ψ.

We note that the higher the relative velocity of the primed system the greater the angle φ, but it is the property of the scaling hyperbolas that they cut the primed axes at equal intervals if they cut the unprimed axes at equal intervals. Thus both systems use a uniform scale.

Let us examine the measurements of a length in the two systems. Suppose that a rod of unit length L' is at rest in the primed system with, at the instant of zero time, its end points represented by P_1' and P_2'. In 4-space such representative points are not at rest, but move parallel to the time axis with increasing time. Thus the point P_1' follows the line *a-a'* and the point P_2' follows the line *b-b'*. Suppose the observer in the unprimed system makes what he considers simultaneous measurements of the end points of the rod; that is, he measures the separation at, say, the two points, P_1 and P_2. Even a rough measurement on Figure 6 shows that these two points are less than unit length apart on the unprimed axis.

More precisely, the geometrical length of the rod, $P_2' - P_1' = L_g'$, is related to the length $P_2 - P_1 = L$, by

$$\frac{L_g'}{L} = \frac{\cos \varphi}{\cos 2\varphi} \tag{5-12}$$

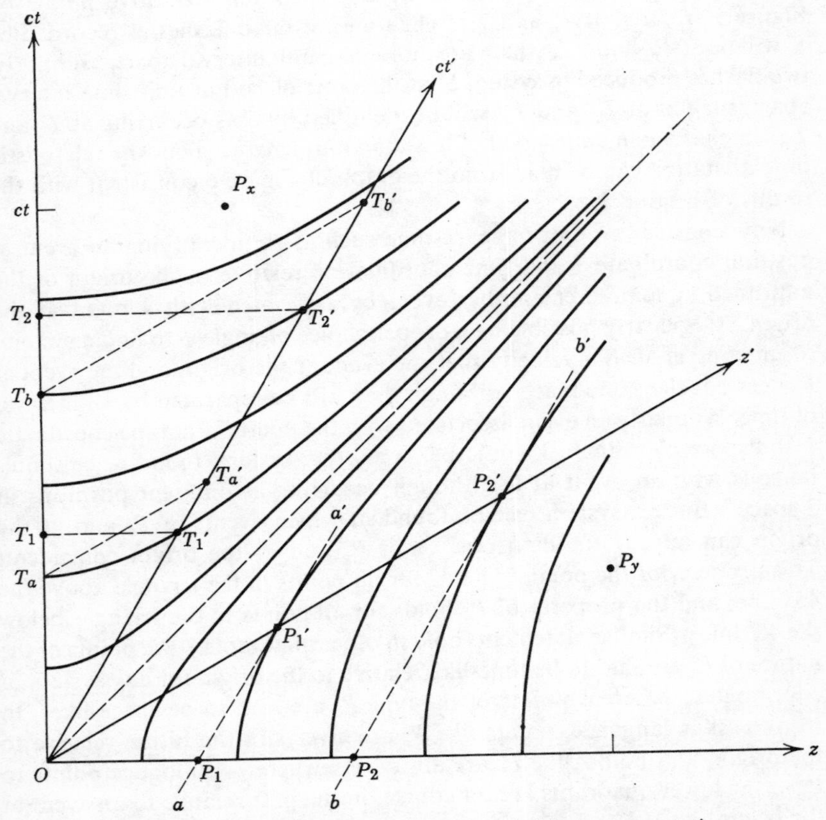

Figure 6. The scaling hyperbolas of the relativistic representation.

from which, using 5-8, we get

$$(5\text{-}13)\qquad L = \frac{[L_g'\sqrt{\cos 2\varphi}]\sqrt{\cos 2\varphi}}{\cos \varphi} = \frac{L'\sqrt{\cos 2\varphi}}{\cos \varphi}$$

where L' is the length of the rod on the scale of S'. Inserting Equation 5-10, we find that 5-13 becomes

$$(5\text{-}14)\qquad L = L'\sqrt{1 - \beta^2}$$

Similarly, if this rod of unit length were at rest in the unprimed system, the representative points of the ends would follow lines parallel to the (ct)-axis, from which it can be seen that observer in the primed system, making what he considers simultaneous measurements, would also observe a length reduce by the same factor.

Now let two successive light flashes be produced, say, in the unprimed system, both being produced at the same place. Let them be separated by a unit time interval as measured by S. Their representative points on Figure 6 are, say, at T_a and T_b. The times of these flashes as recorded by S' will be at T_a' and T_b', which are more than unit interval apart. Similarly, two flashes produced in system S' at the same place but unit time interval apart, such as at T_1' and T_2' will be identified by S as occurring at T_1 and T_2, which is again an interval of less than unit length. Thus the relativistic time dilatation can be read from the graph of Figure 6 consistent with the results of Section 3.

Now consider a point in space-time such as P_x, identifying an event at position coordinate 1 and time coordinate 3 relative to the origin of the unprimed system. Let another event be associated with a point at the origin. It is clearly possible by a proper choice of angle ϕ to find a system, S', moving at such a velocity that the event at the origin and the event at P_x occur at the same place, although they will be separated by an interval of time. Similarly an event associated with the point P_y, at time coordinate 1 and space coordinate 3, could be, in another system, found to be simultaneous with an event at 0, although occurring at different positions in 3-space. But no system can be found in which events at P_x and at the origin can appear simultaneous, or at P_y and at the origin coincident. This property of the point P_x holds for all points in the triangle above the 45° line, and the property of P_y holds for all points in the triangle below the 45° line. (Similar statements hold in other quadrants.) All points of the nature of P_x are said to be time-like relative to the origin for any system of coordinates; whereas points of the type P_y are said to be space-like. In Minkowski's language, points like P_x are always in the future relative to the origin, and points like P_y are always elsewhere. Analogous points to P_x in the lower quadrants are, of course, in the past relative to an event at the origin.

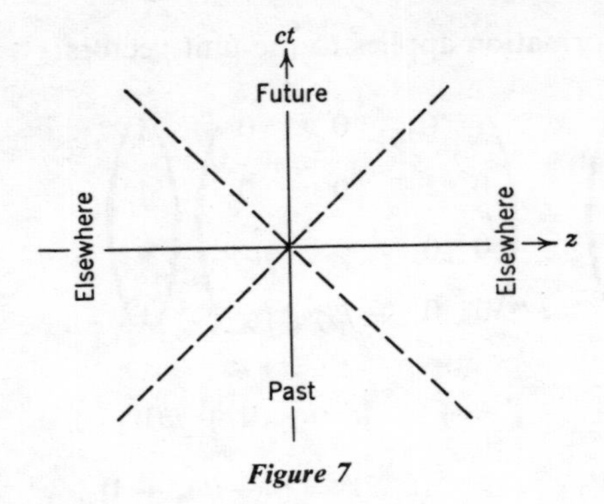

Figure 7

Many people find a geometric analogy such as the foregoing helpful to their thinking about problems in relativistic terms. Others do not. It is not a necessary step in the development of the theory. But every student should find the following-through and the understanding of the analysis of the geometrical interpretation will make him more at ease, more facile with the concepts of relativity, and more able to deal intelligently with the manipulations that are now to be discussed.

6. 4-Vectors and Operators

We have expressed the position vector of an event in space-time by

$$\mathbf{R} = \mathbf{i}x_1 + \mathbf{j}x_2 + \mathbf{k}x_3 + \mathbf{l}x_4 \tag{6-1}$$

where **i**, **j**, **k**, and **l** are unit vectors associated with the four directions in 4-space and x_1, x_2, x_3, and x_4 correspond to x, y, z, and jct. We can readily show that this vector retains its form with a change of coordinate system, becoming

$$\mathbf{R}' = \mathbf{i}'x_1' + \mathbf{j}'x_2' + \mathbf{k}'x_3' + \mathbf{l}'x_4' \tag{6-2}$$

The components of **R** are transformed by

$$\begin{pmatrix} x_1' \\ x_2' \\ x_3' \\ x_4' \end{pmatrix} = \begin{pmatrix} 1 & 0 & 0 & 0 \\ 0 & 1 & 0 & 0 \\ 0 & 0 & \gamma & j\beta\gamma \\ 0 & 0 & -j\beta\gamma & \gamma \end{pmatrix} \begin{pmatrix} x_1 \\ x_2 \\ x_3 \\ x_4 \end{pmatrix} \tag{6-3}$$

yielding the by now familiar relations

$$x_1' = x_1 \qquad x_2' = x_2 \qquad x_3' = \gamma(x_3 + j\beta x_4) \tag{6-4}$$

$$x_4' = \gamma(-j\beta x_3 + x_4)$$

A similar transformation applies to the unit vectors:

$$\begin{pmatrix} \mathbf{i}' \\ \mathbf{j}' \\ \mathbf{k}' \\ \mathbf{l}' \end{pmatrix} = \begin{pmatrix} 1 & 0 & 0 & 0 \\ 0 & 1 & 0 & 0 \\ 0 & 0 & \gamma & j\beta\gamma \\ 0 & 0 & -j\beta\gamma & \gamma \end{pmatrix} \begin{pmatrix} \mathbf{i} \\ \mathbf{j} \\ \mathbf{k} \\ \mathbf{l} \end{pmatrix} \tag{6-5}$$

and yields

$$\mathbf{i}' = \mathbf{i} \qquad \mathbf{j}' = \mathbf{j} \qquad \mathbf{k}' = \gamma(\mathbf{k} + j\beta\mathbf{l}) \tag{6-6}$$

$$\mathbf{l}' = \gamma(-j\beta\mathbf{k} + \mathbf{l})$$

Combining Equations 6-4, 6-6, and 6-2 gives

$$\begin{aligned} \mathbf{R} &= \mathbf{i}x_1 + \mathbf{j}x_2 + \gamma^2(\mathbf{k} + j\beta\mathbf{l})(x_3 + j\beta x_4) \\ &\quad + \gamma^2(-j\beta\mathbf{k} + \mathbf{l})(-j\beta x_3 + x_4) \\ &= \mathbf{i}x_1 + \mathbf{j}x_2 + \mathbf{k}\gamma^2(x_3 + j\beta\, x_4 - \beta^2 x_3 - j\beta\, x_4) \\ &\quad + \mathbf{l}\gamma^2(j\beta\, x_3 - \beta^2 x_4 - j\beta\, x_3 + x_4) \end{aligned} \tag{6-7}$$

which reduces to 6-1.

Another expression we shall encounter is the generalization of the differential operator del to 4-space. Let us indicate it by the symbol $\square$.

$$\square = \mathbf{i}\frac{\partial}{\partial x_1} + \mathbf{j}\frac{\partial}{\partial x_2} + \mathbf{k}\frac{\partial}{\partial x_3} + \mathbf{l}\frac{\partial}{\partial x_4} \tag{6-8}$$

It, too, is invariant to a change of coordinate system, and hence the 4-space generalization of the Laplacian, called the D'Alembertian,

$$\square^2 = \frac{\partial^2}{\partial {x_1}^2} + \frac{\partial^2}{\partial {x_2}^2} + \frac{\partial^2}{\partial {x_3}^2} + \frac{\partial^2}{\partial {x_4}^2} \tag{6-9}$$

is also invariant.

7. The 4-Space Current Density

The 3-space current density has the components J_x, J_y, and J_z. The quantity $c\rho$, where ρ is the charge density, has the same dimensions as current density. Let us therefore construct a 4-space current density vector in the following way:

$$J_1 = J_x \qquad J_2 = J_y \qquad J_3 = J_z \qquad J_4 = jc\rho \tag{7-1}$$

The 4-space divergence of this current density is

$$\text{(7-2)} \qquad \square \cdot \mathbf{J} = \frac{\partial J_1}{\partial x_1} + \frac{\partial J_2}{\partial x_2} + \frac{\partial J_3}{\partial x_3} + \frac{\partial J_4}{\partial x_4}$$
$$= \frac{\partial J_x}{\partial x} + \frac{\partial J_y}{\partial y} + \frac{\partial J_z}{\partial z} + \frac{\partial (jc\rho)}{\partial (jct)}$$
$$= \nabla \cdot \mathbf{J}_t + \frac{\partial \rho}{\partial t}$$

where J_t stands for the 3-space current density. Comparison with the relation showing the continuity of charge (Chapter 10, Section 1) shows that

$$\text{(7-3)} \qquad \square \cdot \mathbf{J} = 0$$

which is a concise and elegant form of the law of conservation of charge.

8. The 4-Space Vector Potential

The vector potential was defined in Section 17 of Chapter 5 as

$$\text{(8-1)} \qquad \mathbf{A} = \frac{\mu_0 I}{4\pi} \oint \frac{d\mathbf{l}}{r}$$

and is thus seen to have the dimensions of volt-sec/meter. Let us formulate a 4-space vector potential by using the three components of the ordinary vector potential and a fourth formed from the ratio of the scalar potential to the velocity of light:

$$\text{(8-2)} \qquad A_1 = A_x \qquad A_2 = A_y \qquad A_3 = A_z \qquad A_4 = \frac{j\phi}{c}$$

The four components are thus dimensionally consistent.

In Chapter 10, Section 5, the Lorentz gauge condition

$$\text{(8-3)} \qquad \nabla \cdot \mathbf{A} = -\mu\epsilon\dot{\phi}$$

was used to obtain the separated differential equations for the vector potential, **A**, and the scalar potential, ϕ. Setting the divergence of the 4-space vector **A** equal to zero results in a concise statement of this gage condition, 8-3:

$$\text{(8-4)} \qquad \square \cdot \mathbf{A} = \frac{\partial A_1}{\partial x_1} + \frac{\partial A_2}{\partial x_2} + \frac{\partial A_3}{\partial x_3} + \frac{\partial A_4}{\partial x_4}$$
$$= \frac{\partial A_x}{\partial x} + \frac{\partial A_y}{\partial y} + \frac{\partial A_z}{\partial z} + \frac{\partial (j\phi/c)}{\partial (jct)}$$
$$= \nabla \cdot \mathbf{A}_t + \left(\frac{1}{c^2}\right)\frac{\partial \phi}{\partial t} = 0$$

where $\mathbf{A}_t$ stands for the 3-space vector potential.

Now consider the equation:

$$\square^2 \mathbf{A} = -\mu \mathbf{J} \qquad \text{(4-space)} \tag{8-5}$$

This separates into the four equations:

$$\frac{\partial^2 A_x}{\partial x^2} + \frac{\partial^2 A_x}{\partial y^2} + \frac{\partial^2 A_x}{\partial z^2} + \frac{\partial^2 A_x}{\partial (jct)^2} = -\mu J_x \tag{8-6a}$$

$$\frac{\partial^2 A_y}{\partial x^2} + \frac{\partial^2 A_y}{\partial y^2} + \frac{\partial^2 A_y}{\partial z^2} + \frac{\partial^2 A_y}{\partial (jct)^2} = -\mu J_y \tag{8-6b}$$

$$\frac{\partial^2 A_z}{\partial x^2} + \frac{\partial^2 A_z}{\partial y^2} + \frac{\partial^2 A_z}{\partial z^2} + \frac{\partial^2 A_z}{\partial (jct)^2} = -\mu J_z \tag{8-6c}$$

$$\left(\frac{j}{c}\right)\left[\frac{\partial^2 \phi}{\partial x^2} + \frac{\partial^2 \phi}{\partial y^2} + \frac{\partial^2 \phi}{\partial z^2} + \frac{\partial^2 \phi}{\partial (jct)^2}\right] = -\mu (jc\rho) \tag{8-6d}$$

which reduce to

$$\nabla^2 A_x - \left(\frac{1}{c^2}\right)\ddot{A}_x = -\mu J_x \tag{8-7a}$$

$$\nabla^2 A_y - \left(\frac{1}{c^2}\right)\ddot{A}_y = -\mu J_y \tag{8-7b}$$

$$\nabla^2 A_z - \left(\frac{1}{c^2}\right)\ddot{A}_z = -\mu J_z \tag{8-7c}$$

$$\nabla^2 \phi - \left(\frac{1}{c^2}\right)\ddot{\phi} = -\frac{\rho}{\epsilon} \tag{8-7d}$$

Thus the single 4-space equation of 8-5 is the equivalent of the four equations of 5-6 of Chapter 10. Since these were obtained from the introduction of the vector and scalar potentials into the four Maxwell field equations, 8-5 can be considered as one way of writing the field equations in relativistic 4-space notation.

9. The Field Equations in 4-Space

Let us use the expressions for the electric field and the magnetic flux density in terms of the potentials (Chapter 6, Section 10):

$$\mathbf{E} = -\dot{\mathbf{A}} - \nabla\phi \qquad \text{and} \qquad \mathbf{B} = \nabla \times \mathbf{A} \tag{9-1}$$

as a guide in forming expressions for **E** and **B** in terms of 4-space vectors.

For example:

$$E_x = \frac{-\partial A_x}{\partial t} - \frac{\partial \phi}{\partial x} = \frac{-jc\,\partial A_1}{\partial x_4} - \left(\frac{c}{j}\right)\frac{\partial A_4}{\partial x_1} \tag{9-2}$$

$$= jc\left[\frac{\partial A_4}{\partial x_1} - \frac{\partial A_1}{\partial x_4}\right]$$

The quantity in brackets in the last expression is abbreviated as F_{41}, where the subscripts correspond to the subscripts of the A's in the order in which they appear. Thus the three components of **E** are written:

$$E_x = jcF_{41} \tag{9-3}$$

$$E_y = jcF_{42}$$

$$E_z = jcF_{43}$$

or in general, $E_i = jcF_{4i}$. Similarly

$$B_x = \frac{\partial A_3}{\partial x_2} - \frac{\partial A_2}{\partial x_3} = F_{32} \tag{9-4}$$

$$B_y = F_{13}$$

$$B_z = F_{21}$$

Evidently, from the way in which the functions F_{ij} are formed it follows that $F_{ij} = -F_{ji}$ for $j \neq i$ and $F_{ij} = 0$ for $j = i$.

These functions may be used to express Ampère's law, if we write this law in the form:

$$\nabla \times \mathbf{B} = \mu\mathbf{J} + \mu\epsilon\dot{\mathbf{E}} \tag{9-5}$$

The x component of 9-5 may be written successively as

$$\frac{\partial B_z}{\partial y} - \frac{\partial B_y}{\partial z} = \mu J_x + \frac{\mu\epsilon\,\partial E_x}{\partial t} \tag{9-6}$$

$$\frac{\partial F_{21}}{\partial x_2} - \frac{\partial F_{13}}{\partial x_3} = \mu J_1 + \frac{jc\mu\epsilon\partial(jcF_{41})}{\partial x_4}$$

$$\frac{\partial F_{21}}{\partial x_2} + \frac{\partial F_{31}}{\partial x_3} + \frac{\partial F_{41}}{\partial x_4} = \mu J_1$$

To which we formally add $\partial F_{11}/\partial x_1 = 0$, thereby making it possible to write Equation 9-6 in the concise form:

(9-7) $\square \cdot \mathbf{F}_1 = \mu J_1$

where $\mathbf{F}_1 = \mathbf{i}F_{11} + \mathbf{j}F_{21} + \mathbf{k}F_{31} + \mathbf{l}F_{41}$
Following the same pattern, three other component equations can be written:

(9-8) $\square \cdot \mathbf{F}_2 = \mu J_2$ where $\mathbf{F}_2 = \mathbf{i}F_{12} + \mathbf{j}F_{22} + \mathbf{k}F_{32} + \mathbf{l}F_{42}$

$\square \cdot \mathbf{F}_3 = \mu J_3$ where $\mathbf{F}_3 = \mathbf{i}F_{13} + \mathbf{j}F_{23} + \mathbf{k}F_{33} + \mathbf{l}F_{43}$

$\square \cdot \mathbf{F}_4 = \mu J_4$ where $\mathbf{F}_4 = \mathbf{i}F_{14} + \mathbf{j}F_{24} + \mathbf{k}F_{34} + \mathbf{l}F_{44}$

The last equation is readily shown to be equivalent to Gauss's law, $\nabla \cdot \mathbf{E} = \rho/\epsilon$. Thus the general form:

(9-9) $\square \cdot \mathbf{F}_i = \mu J_i \qquad i = 1, 2, 3, 4$

includes two of the four 3-space Maxwell field equations.

The vectors $\mathbf{F}_1$, $\mathbf{F}_2$, $\mathbf{F}_3$, and $\mathbf{F}_4$ can be arranged in columns to form a matrix

(9-10) $$(F) = \begin{pmatrix} 0 & F_{12} & F_{13} & F_{14} \\ F_{21} & 0 & F_{23} & F_{24} \\ F_{31} & F_{32} & 0 & F_{34} \\ F_{41} & F_{42} & F_{43} & 0 \end{pmatrix} = \begin{pmatrix} 0 & -B_z & B_y & -E_x/jc \\ B_z & 0 & -B_x & -E_y/jc \\ -B_y & B_x & 0 & -E_z/jc \\ E_x/jc & E_y/jc & E_z/jc & 0 \end{pmatrix}$$

When such a matrix is multiplied by a row matrix from the left (multiplying by a column matrix on the left has no meaning), rows are combined with columns in a way similar to the case encountered previously where a matrix multiplied a column on the right. If (F) of 9-10 is multiplied by a row matrix formed from the differential operator $\square$, the four equations of 9-7 and 9-8 are succinctly expressed as

(9-11) $$\left(\frac{\partial}{\partial x_1}\frac{\partial}{\partial x_2}\frac{\partial}{\partial x_3}\frac{\partial}{\partial x_4}\right)\begin{pmatrix} 0 & F_{12} & F_{13} & F_{14} \\ F_{21} & 0 & F_{23} & F_{24} \\ F_{31} & F_{32} & 0 & F_{34} \\ F_{41} & F_{42} & F_{43} & 0 \end{pmatrix} = \mu(J_1\, J_2\, J_3\, J_4)$$

or, more concisely:

(9-12) $(\square)(F) = \mu(J)$

thus expressing both Gauss's law and Ampère's law in a few brief symbols.

The formulation of Faraday's law in these terms goes as follows, using the x component as an illustration:

$$(9\text{-}13) \qquad (\nabla \times \mathbf{E})_x = \frac{\partial E_z}{\partial y} - \frac{\partial E_y}{\partial z} = \frac{-\partial B_x}{\partial t}$$

$$jc\left[\frac{\partial F_{43}}{\partial x_2} - \frac{\partial F_{42}}{\partial x_3}\right] = -(jc)\left[\frac{\partial F_{32}}{\partial x_4}\right]$$

$$\frac{\partial F_{34}}{\partial x_2} + \frac{\partial F_{42}}{\partial x_3} + \frac{\partial F_{23}}{\partial x_4} = 0$$

Three additional equations may be written on this pattern:

$$(9\text{-}14) \qquad \frac{\partial F_{13}}{\partial x_4} + \frac{\partial F_{41}}{\partial x_3} + \frac{\partial F_{34}}{\partial x_1} = 0$$

$$\frac{\partial F_{12}}{\partial x_4} + \frac{\partial F_{41}}{\partial x_2} + \frac{\partial F_{24}}{\partial x_1} = 0$$

$$\frac{\partial F_{12}}{\partial x_3} + \frac{\partial F_{23}}{\partial x_1} + \frac{\partial F_{31}}{\partial x_2} = 0$$

where the indices appear in cyclic order in the clockwise sense in each differentiation. The last equation of this group is found to correspond to $\nabla \cdot \mathbf{B} = 0$ and the first three to Faraday's law.

These equations can be given the conciseness of 9-12 by a somewhat artificial maneuver. A function called the "dual" of F is defined in the following manner:

$$(9\text{-}15) \qquad F_{jk} = F_{mn}{}^{*}$$

where the formal rule for choosing the indices is that the pattern, *jkmn*, is to be obtained from the sequence, 1234, by an even number of interchanges. A somewhat simpler method will be clear from an illustration: suppose F_{jk} to be, say, B_x; the indices mn are then chosen to be those associated with E_x. A matrix of F^* similar to 9-10 is obtained by simply interchanging the B's and the E's terms of 9-10. Remembering that $F_{mn}{}^{*} = -F_{nm}{}^{*}$ for $n \neq m$ and $F_{mn}{}^{*} = 0$ for $n = m$, Equations 9-14 take the form:

$$(9\text{-}16) \qquad (\square)(F^{*}) = 0$$

Though this latter procedure may seem contrived, its virtue is that all the Maxwell field equations are expressed in a matrix form, so that they may be manipulated by the rules of matrix algebra.

10. Summary of the Relativistic Electromagnetic Relations

For ready reference, the components of the various 4-space vectors are listed below, together with their 3-space equivalents:

J_1	J_x	$F_{41} = F_{32}{}^*$	E_x/jc
J_2	J_y	$F_{42} = F_{13}{}^*$	E_y/jc
J_3	J_z	$F_{43} = F_{21}{}^*$	E_z/jc
J_4	$jc\rho$	$F_{32} = F_{41}{}^*$	B_x
A_1	A_x	$F_{13} = F_{42}{}^*$	B_y
A_2	A_y	$F_{21} = F_{43}{}^*$	B_z
A_3	A_z		
A_4	$j\phi/c$		

and we also list the electromagnetic relations:

4-space	3-space
$\square \cdot \mathbf{J} = 0$	$\nabla \cdot \mathbf{J}_t = -\dot{\rho}$
$\square \cdot \mathbf{A} = 0$	$\nabla \cdot \mathbf{A}_t = -\mu\epsilon\dot{\phi}$
$\square^2\mathbf{A} = -\mu\mathbf{J}$	$\begin{cases} \nabla^2\mathbf{A}_t - \mu\epsilon\ddot{\mathbf{A}}_t = -\mu\mathbf{J}_t \\ \nabla^2\phi - \mu\epsilon\ddot{\phi} = -\dfrac{\rho}{\epsilon} \end{cases}$
$(\square)(F) = \mu(\mathbf{J})$	$\begin{cases} \nabla \times \mathbf{B} = \mu\mathbf{J}_t + \mu\epsilon\dot{\mathbf{E}} \\ \nabla \cdot \mathbf{E} = \dfrac{\rho}{\epsilon} \end{cases}$
$(\square)(F^*) = 0$	$\begin{cases} \nabla \times \mathbf{E} = -\dot{\mathbf{B}} \\ \nabla \cdot \mathbf{B} = 0 \end{cases}$

11. Transformations of the Electromagnetic Field Functions

We shall transform the functions $\mathbf{J}$, ρ, $\mathbf{A}$, ϕ, $\mathbf{E}$, and $\mathbf{B}$, measured in a system S, to similar functions as observed from another coordinate system, S', which is moving with a velocity v in the z direction relative to S.

The transformation is most easily symbolized and the operations performed, by means of the transformation matrix operating on each vector.

For the current density we write

$$\begin{pmatrix} J_1' \\ J_2' \\ J_3' \\ J_4' \end{pmatrix} = \begin{pmatrix} 1 & 0 & 0 & 0 \\ 0 & 1 & 0 & 0 \\ 0 & 0 & \gamma & j\beta\gamma \\ 0 & 0 & -j\beta\gamma & \gamma \end{pmatrix} \begin{pmatrix} J_1 \\ J_2 \\ J_3 \\ J_4 \end{pmatrix} \tag{11-1}$$

By combining rows and columns we obtain

$$J_1' = J_1 \qquad J_2' = J_2 \qquad J_3' = \gamma(J_3 + j\beta J_4) \tag{11-2}$$
$$J_4' = \gamma(-j\beta J_3 + J_4)$$

which, translated into 3-space terms, is

$$J_x' = J_x \qquad J_y' = J_y \qquad J_z' = \gamma(J_z - v\rho) \tag{11-3}$$
$$\rho' = \gamma(\rho - \beta J_z/c)$$

Thus, if in the S system there is a stationary charge density, ρ, with $J_x = J_y = J_z = 0$, an observer in the S' system measures not only a charge density, $\gamma\rho$, but also a current density, $-\gamma v\rho$. This is hardly an unexpected result, inasmuch as for S' the charges in S are moving in the negative z'-direction with a velocity v. On the other hand if a stationary wire in S lies parallel to the z-axis and carries a current of density J_z, with $\rho = 0$ (equal positive and negative charges in the wire), the observer in S' sees a current density of γJ_z and also a charge density of $-\gamma\beta J_z/c$. Why?

By a similar process the transformation of the vector potential yields

$$A_x' = A_x \qquad A_y' = A_y \qquad A_z' = \gamma(A_z - \beta\phi/c) \tag{11-4}$$
$$\phi' = \gamma(\phi - vA_z)$$

To transform the electric and magnetic field components, we shall first need the transformations of the differential operators:

$$\frac{\partial}{\partial x_1'} = \frac{\partial}{\partial x_1} \qquad \frac{\partial}{\partial x_2'} = \frac{\partial}{\partial x_2} \tag{11-5}$$
$$\frac{\partial}{\partial x_3'} = \gamma\left[\frac{\partial}{\partial x_3} + \frac{j\beta\partial}{\partial x_4}\right] \qquad \frac{\partial}{\partial x_4'} = \gamma\left[\frac{-j\beta\partial}{\partial x_3} + \frac{\partial}{\partial x_4}\right]$$

With these, the x component of $\mathbf{E}$, for example, in the moving system becomes

(11-6)
$$E_x' = jcF_{41}' = jc\left[\frac{\partial A_4'}{\partial x_1'} - \frac{\partial A_1'}{\partial x_4'}\right]$$
$$= jc\left[\gamma\left(\frac{\partial}{\partial x_1}\right)(-j\beta A_3 + A_4) - \gamma\left(\frac{-j\beta\partial}{\partial x_3} + \frac{\partial}{\partial x_4}\right)A_1\right]$$
$$= jc\gamma\left[\frac{\partial A_4}{\partial x_1} - \frac{\partial A_1}{\partial x_4} + j\beta\left(\frac{\partial A_1}{\partial x_3} - \frac{\partial A_3}{\partial x_1}\right)\right]$$
$$= jc\gamma[F_{41} + j\beta F_{13}]$$
$$= \gamma(E_x - vB_y)$$

Similar operations on the other components result in:

(11-7)
$$E_y' = \gamma(E_y + vB_x) \qquad E_z' = E_z$$
$$B_x' = \gamma[B_x + (\beta/c)E_y] \qquad B_y' = \gamma[B_y - (\beta/c)E_x] \qquad B_z' = B_z$$

Those components of the electric and the magnetic fields which are transverse to the motion of the S' system are modified whereas the longitudinal components are unaffected. A few simple illustrations will serve to make plausible these field transformations. As a first example, suppose there is a long straight wire parallel to the z-axis at rest in S, and that this wire carries a static charge. Figure 8 shows a view looking in the direction of the negative z-axis, with the moving S' system coming out from the paper with a velocity v. To an observer in S' the wire looks not only like a line charge whose volume density is γ times the volume density as measured in S, but also like a current in the $-z'$-direction, whose density is v times the observed charge density, as, of course, it should since to S' the charges are moving toward negative z. Whereas in S there is a static field with x and y components, in S' this field has the same components with magnitudes γ times their magnitude in S, and in addition S' sees, associated with the current density, a B-field whose x component is $\gamma(\beta/c)E_y$ and whose y component is $-\gamma(\beta/c)E_x$.

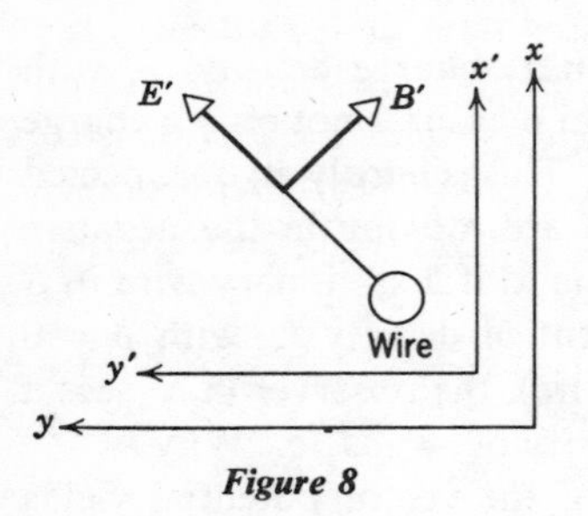

Figure 8

If, instead of a statically charged wire in S, the wire carries a current of density J_z, an observer in S measures only a B-field with components transverse to the direction of the current. The observer in S' also sees a current density of magnitude γ times the current density in S, associated with which are B' components which are γ times the B components in S; but in addition he sees a charge density, $-\gamma(\beta/c)J_z$ and associated with it E' field components, $E_y' = \gamma vB_x$ and $E_x' = -\gamma vB_y$. (Again, why?)

Since the measurement of currents which are transverse to the motion of the observer are not affected by that motion, it is consistent that the longitudinal component of the B-field is also unaffected. For imagine a solenoid in S whose axis is parallel to the z-axis. Since the current is entirely transverse, both S and S' see the same current and hence the same B-field along the z-axis.

The independence of E_z of the motion of the observer is consistent with the transformation of charge densities, as is illustrated by the following thought situation. Suppose a parallel plate capacitor, with surface charge densities σ and $-\sigma$, is oriented with the planes of the plates normal to the z-axis and at rest in S. Neglecting the fringing field, the electric field between the plates has only a z component, which is related to the surface charge density by $E_z = \sigma/\varepsilon$. Let the effective volume charge density in the plates be ρ. This is observed by S' as $\gamma\rho$ but the change is the result of the FitzGerald-Lorentz contraction, which acts only in the z-direction. Hence, the surface charge density is unaffected and the electric field remains the same.

We see that the electric and magnetic fields are not independent entities, but are complementary functions that depend on the relative motion of the observer with respect to the system observed. An apparent motion of an electric field is associated with the appearance of a magnetic flux density, whereas an apparent motion of a B-field is associated with the appearance of an E-field. A few additional examples in the following sections will illustrate this interrelationship further.

12. Fields of a Moving Point Charge

Let a point charge Q be at rest at the origin in system S. The field of this charge, as measured by an observer in S, is a static field of magnitude

$$E = \frac{Q}{4\pi\epsilon r^2} \tag{12-1}$$

where r is the distance from the origin. At a point $P(xyz)$ the components of E will be

$$E_x = \frac{Qx}{4\pi\epsilon r^3} \qquad E_y = \frac{Qy}{4\pi\epsilon r^3} \qquad E_z = \frac{Qz}{4\pi\epsilon r^3} \tag{12-2}$$

and since it is a static charge, $B_x = B_y = B_z = 0$.

Let an observer be moving with a system S' with a velocity v in the positive z-direction. For convenience, we shall consider the situation at the instant when the origins of the two systems coincide. The components of the fields as observed in S' are, by Equations 11-6 and 11-7:

$$E_x' = \gamma E_x \qquad E_y' = \gamma E_y \qquad E_z' = E_z \tag{12-3}$$

$$B_x' = \left(\frac{\gamma\beta}{c}\right) E_y \qquad B_y' = -\left(\frac{\gamma\beta}{c}\right) E_x \qquad B_z' = 0$$

and, since we count $t' = 0$ at the instant of coincidence of the origins, the coordinates of P are

$$x = x' \qquad y = y' \qquad z = \gamma(z' + vt') = \gamma z' \tag{12-4}$$

so that the distance of the point P from the origin is

$$r = \sqrt{x^2 + y^2 + z^2} = \sqrt{x'^2 + y'^2 + \gamma^2 z'^2} \tag{12-5}$$

and 12-3 becomes

$$E_x' = \frac{\gamma Q}{4\pi\epsilon} \frac{x'}{(x'^2 + y'^2 + \gamma^2 z'^2)^{3/2}} \tag{12-6}$$

$$E_y' = \frac{\gamma Q}{4\pi\epsilon} \frac{y'}{(x'^2 + y'^2 + \gamma^2 z'^2)^{3/2}}$$

$$E_z' = \frac{\gamma Q}{4\pi\epsilon} \frac{z'}{(x'^2 + y'^2 + \gamma^2 z'^2)^{3/2}}$$

$$B_x' = \frac{\gamma Q\beta}{4\pi\epsilon c} \frac{y'}{(x'^2 + y'^2 + \gamma^2 z'^2)^{3/2}}$$

$$B_y' = \frac{-\gamma Q\beta}{4\pi\epsilon c} \frac{x'}{(x'^2 + y'^2 + \gamma^2 z'^2)^{3/2}}$$

$$B_z' = 0$$

Let us examine the electric field in the plane $y' = 0$. The magnitude of this field is

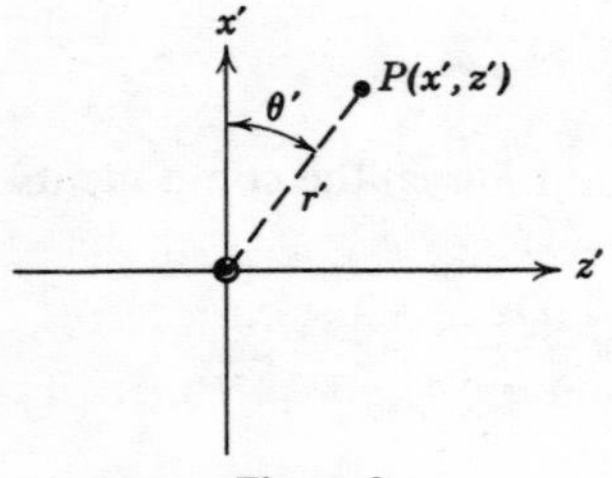

Figure 9

$$E' = \sqrt{E_x'^2 + E_z'^2} = \frac{\gamma Q}{4\pi\epsilon r'^2} \frac{1}{[1 + \gamma^2\beta^2 \cos^2\theta']^{3/2}} \tag{12-7}$$

A plot of E' vs θ' for several values of β is shown in Figure 10. The faster the observer moves with respect to the charge, the more concentrated is the electric field in the vicinity of the $z' = 0$ plane, whereas at speeds which are negligible with respect to the speed of light, the field is isotropic.

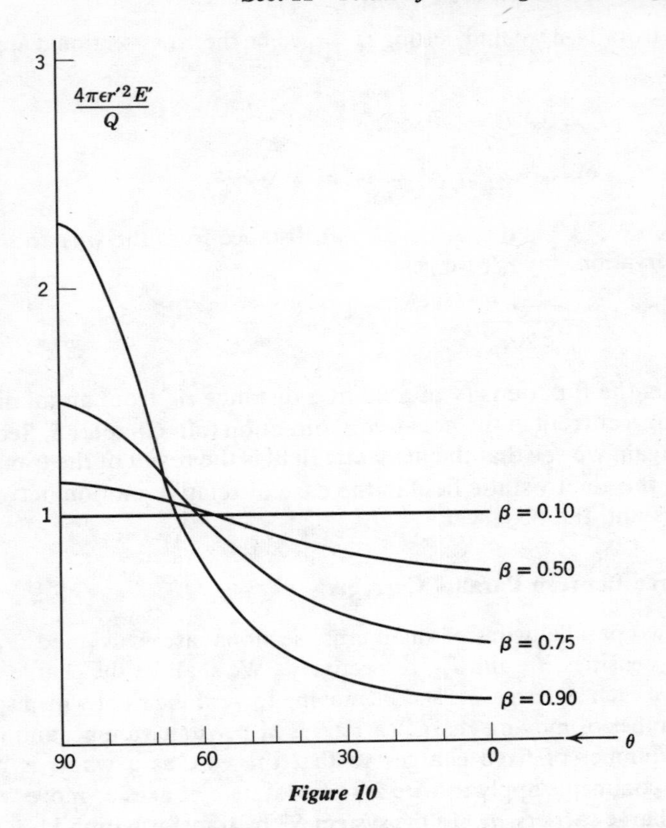

Figure 10

Let us now imagine an infinitely long wire along the z-axis of S, the wire carrying a static charge whose linear density is λ per unit length. The charge on an element of length of the wire, dz, is then λdz, which from Equation 11-3, is $(\lambda'/\gamma)(\gamma dz')$ or $\lambda' dz'$ in S_2' since $J_z = 0$. Thus the magnitude of the B-field from this element of charge is, for the moving observer:

$$dB' = \sqrt{dB_x'^2 + dB_y'^2} = \frac{\gamma v \mu \lambda'}{4\pi} \frac{\sqrt{x'^2 + y'^2}\, dz'}{(x'^2 + y'^2 + \gamma^2 z'^2)^{3/2}} \qquad \text{(12-8)}$$

where $\beta/\epsilon c = \mu v$. When 12-8 is integrated from $z' = -\infty$ to $z' = +\infty$, the total magnetic flux density seen by an observer moving with a velocity v parallel to the wire with a static charge is

$$B' = \frac{(\gamma \mu v \lambda')}{2\pi\sqrt{x'^2 + y'^2}} \qquad \text{(12-9)}$$

Again from 11-3 we find, letting $\mathscr{A} = \mathscr{A}'$ be the cross-sectional area of the wire,

$$(12\text{-}10) \qquad I' = \mathscr{A}'J' = -\gamma\mathscr{A}'v\rho$$

$$= -\gamma\mathscr{A}'v\lambda/\mathscr{A} = -\gamma v\lambda = -v\lambda'$$

Since $\sqrt{x'^2 + y'^2}$ is the perpendicular distance from the wire to the point of observation, say r_0', we get

$$(12\text{-}11) \qquad B' = \frac{-\mu I'}{2\pi r_0'}$$

which is the flux density at a point a distance r_0' from an infinite wire carrying a current in the negative z' direction (cf: Chapter 5, Section 6). Thus again we see that the magnetic field is the result of the transformation of the electrostatic field in the case of relative motion between the charges and the observer.

13. Force Between Parallel Currents*

Let two parallel wires of small cross-sectional areas, $\mathscr{A}_1$ and $\mathscr{A}_2$, carry current densities, J_{z1} and J_{z2}, respectively. We shall let the charge carriers in 1 have each a charge, q_1, and be moving toward the right with a speed v_1. The number of moving charge carriers is n_1 per unit volume, and is equal to the number of fixed charges so that the wire as a whole is neutral. Similar comments apply to wire 2. Let a system of axes S' move with one of the charge carriers, q_1. In the system S', by transformation 11-3, there is observed a charge density ρ' in wire 2 of

$$(13\text{-}1) \qquad \rho' = \frac{-\gamma(\mathbf{v}_1 \cdot \mathbf{J}_{z2})}{c^2}$$

where the dot product allows for $\mathbf{v}$ being either parallel or antiparallel to $\mathbf{J}$. This charge density multiplied by $\mathscr{A}_2$ (whose transverse dimensions are

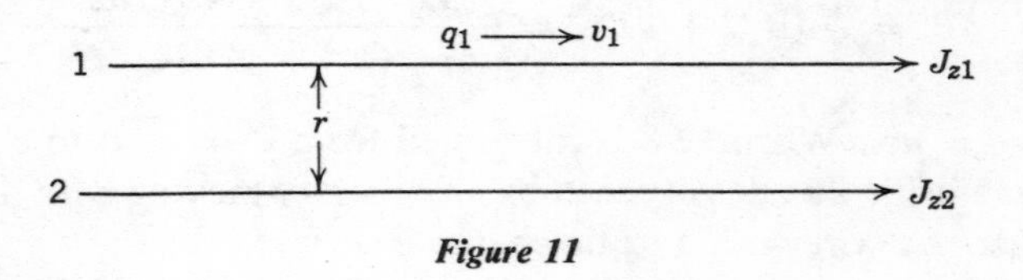

Figure 11

* For an elaboration of the topic of this section see also D. L. Webster, *Am. Jour. Phys.*, **29,** 841 (1961).

not altered by the motion) is the linear charge density in wire 2 as seen by q_1. The electric field at q_1 associated with this linear charge density is (see Chapter 1, Problem 4)

$$E' = \frac{-[\gamma(\mathbf{v}_1 \cdot \mathbf{J}_{z2})\mathscr{A}_2]}{2\pi\epsilon c^2 r} \tag{13-2}$$

Since the charge per unit length in wire 1 is $n_1 q_1 \mathscr{A}_1$, the force per unit length on wire 1 is

$$f' = \frac{-[\gamma(\mathbf{v}_1 \cdot \mathbf{J}_{z2})\mathscr{A}_2](n_1 q_1 \mathscr{A}_1)}{2\pi\epsilon c^2 r} \tag{13-3}$$

The currents in the wires can be expressed as

$$I_1 = n_1 q_1 v_1 \mathscr{A}_1 \tag{13-4}$$

$$I_2 = J_{z2}\mathscr{A}_2$$

so that 13-3 becomes

$$f' = \frac{-(\gamma I_1 I_2)}{2\pi\epsilon c^2 r} = \frac{-(\gamma\mu I_1 I_2)}{2\pi r} \tag{13-5}$$

a relation which reduces to the force per unit length on one of two parallel conductors, as found in Chapter 5, for velocities of charge carriers small compared to the velocity of light.

This result can also be inferred from the arguments of Section 12. Note that the force between the wires is one of attraction for charges of like sign flowing in the same direction in the two wires, regardless of whether the carriers are positively or negatively charged.

14. The Lorentz Force and the 4-Space Force

We now return to the transformations of the field components, repeating the results found in Section 11:

$$E_x' = \gamma(E_x - vB_y) \qquad E_y' = \gamma(E_y + vB_x) \qquad E_z' = E_z \tag{14-1}$$

$$B_x' = \gamma\left[B_x + \left(\frac{\beta}{c}\right)E_y\right] \qquad B_y' = \gamma\left[B_y - \left(\frac{\beta}{c}\right)E_x\right] \qquad B_z' = B_z$$

where the velocity v is taken parallel to the z-axis. These are special cases of more general forms, in which the fields are grouped into components perpendicular and parallel to the motion:

$$(14\text{-}2) \qquad \mathbf{E}_{\parallel}' = (\mathbf{E} + \mathbf{v} \times \mathbf{B})_{\parallel} \qquad \mathbf{E}_{\perp}' = \gamma(\mathbf{E} + \mathbf{v} \times \mathbf{B})_{\perp}$$

$$\mathbf{B}_{\parallel}' = \left(\mathbf{B} - \frac{\mathbf{v} \times \mathbf{E}}{c^2}\right)_{\parallel} \qquad \mathbf{B}_{\perp}' = \gamma\left(\mathbf{B} - \frac{\mathbf{v} \times \mathbf{E}}{c^2}\right)_{\perp}$$

where $(\mathbf{v} \times \mathbf{B})_{\parallel} = (\mathbf{v} \times \mathbf{E})_{\parallel} = 0$ have been added for the sake of symmetry.

At low velocities the force acting on a charge q is

$$(14\text{-}3) \qquad K = q(\mathbf{E} + \mathbf{v} \times \mathbf{B})$$

This quantity is called the Lorentz force, and gives the complete story of forces acting on charges in non-relativistic electromagnetic fields.

Now let us investigate the result of forming the product of the 4-vector J with the 4×4 matrix of Equation 9-10:

$$(14\text{-}4) \qquad (J_1\, J_2\, J_3\, J_4) \begin{pmatrix} 0 & -B_z & B_y & -E_x/jc \\ B_z & 0 & -B_x & -E_y/jc \\ -B_y & B_x & 0 & -E_z/jc \\ E_x/jc & E_y/jc & E_z/jc & 0 \end{pmatrix} = (k)$$

By writing the first three components of J as $\rho\mathbf{v}$ the four components of k can be generalized as

$$(14\text{-}5) \qquad k_1 = \rho[E_x + (\mathbf{v} \times \mathbf{B})_x]$$

$$k_2 = \rho[E_y + (\mathbf{v} \times \mathbf{B})_y]$$

$$k_3 = \rho[E_z + (\mathbf{v} \times \mathbf{B})_z]$$

$$k_4 = (j\rho/c)(\mathbf{v} \cdot \mathbf{E})$$

which reduces to the results of the operation in 14-4 for $v_x = v_y = 0$. The first three components of k are the components of the force per unit volume in a region where the charge density is ρ, whereas the last component is a measure of the power expended per unit volume on the moving charges.

An interesting manipulation becomes possible if we define a quantity

$$(14\text{-}6) \qquad T_{ab} = \frac{1}{\mu}\left[\sum_{i=1}^{4} F_{ai}F_{ib} + \frac{1}{4}\sum_{i=1}^{4}\sum_{j=1}^{4} \delta_{ab}F_{ij}^{\,2}\right]$$

where the F's are defined in 9-10 and $\delta_{ab} = 1$ when $a = b$ and $\delta_{ab} = 0$

when $a \neq b$. The subscripts a and b take on values from 1 to 4. A somewhat tedious but straightforward expansion of 14-6 gives

$$(14\text{-}7)\qquad T_{11} = \frac{1}{2\mu}\left[(B_x^2 - B_y^2 - B_z^2) + \frac{1}{c^2}(E_x^2 - E_y^2 - E_z^2)\right]$$

$$T_{22} = \frac{1}{2\mu}\left[(B_y^2 - B_z^2 - B_x^2) + \frac{1}{c^2}(E_y^2 - E_z^2 - E_x^2)\right]$$

$$T_{33} = \frac{1}{2\mu}\left[(B_z^2 - B_x^2 - B_y^2) + \frac{1}{c^2}(E_z^2 - E_x^2 - E_y^2)\right]$$

$$T_{44} = \frac{1}{2\mu}\left[(B_x^2 + B_y^2 + B_z^2) + \frac{1}{c^2}(E_x^2 + E_y^2 + E_z^2)\right]$$

$$T_{12} = T_{21} = \frac{1}{\mu}\left[B_x B_y + \frac{1}{c^2} E_x E_y\right]$$

$$T_{13} = T_{31} = \frac{1}{\mu}\left[B_x B_z + \frac{1}{c^2} E_x E_z\right]$$

$$T_{23} = T_{32} = \frac{1}{\mu}\left[B_y B_z + \frac{1}{c^2} E_y E_z\right]$$

$$T_{14} = T_{41} = \frac{1}{jc\mu}[E \times B]_x$$

$$T_{24} = T_{42} = \frac{1}{jc\mu}[E \times B]_y$$

$$T_{34} = T_{43} = \frac{1}{jc\mu}[E \times B]_z$$

and it can be verified that the force density of 14-5 is given by

$$(14\text{-}8)\qquad k_a = \sum_{i=1}^{4} \frac{\partial T_{ia}}{\partial x_i}$$

The terms T_{ab} are the elements of a matrix

$$(14\text{-}9)\qquad (T) = \begin{pmatrix} T_{11} & T_{12} & T_{13} & T_{14} \\ T_{21} & T_{22} & T_{23} & T_{24} \\ T_{31} & T_{32} & T_{33} & T_{34} \\ T_{41} & T_{42} & T_{43} & T_{44} \end{pmatrix}$$

whose terms may also be grouped into 4-vectors, analogous to the 4-vectors $\mathbf{F}_i$ of 9-8:

$$(14\text{-}10)\qquad \begin{aligned} \mathbf{T}_1 &= \mathbf{i}T_{11} + \mathbf{j}T_{21} + \mathbf{k}T_{31} + \mathbf{l}T_{41} \\ \mathbf{T}_2 &= \mathbf{i}T_{12} + \mathbf{j}T_{22} + \mathbf{k}T_{32} + \mathbf{l}T_{42} \\ \mathbf{T}_3 &= \mathbf{i}T_{13} + \mathbf{j}T_{23} + \mathbf{k}T_{33} + \mathbf{l}T_{43} \\ \mathbf{T}_4 &= \mathbf{i}T_{14} + \mathbf{j}T_{24} + \mathbf{k}T_{34} + \mathbf{l}T_{44} \end{aligned}$$

Then, analogous to Equations 9-8, we write

$$(14\text{-}11)\qquad k_1 = \Box \cdot \mathbf{T}_1 = (\nabla \cdot \mathbf{T}_{t1}) + \frac{\partial T_{41}}{\partial x_4}$$

Here T_{t1} represents the first three components of the 4-vector, $\mathbf{T}_1$. Expressions for k_2, k_3, and k_4 are similar. Since k_1 is the x component of the force per unit volume in a region containing charges, the integral of k_1 over a finite volume in 3-space is the total force acting on charges within that volume. Performing this integration on Equation 14-11, we get

$$(14\text{-}12)\qquad \begin{aligned} \int k_1\, d\mathscr{V} &= \int (\nabla \cdot \mathbf{T}_{t1})\, d\mathscr{V} - \frac{j}{c}\int \left[\frac{\partial T_{41}}{\partial t}\right] d\mathscr{V} \\ &= \int \mathbf{T}_{t1} \cdot d\mathscr{A} - \frac{\partial}{\partial t}\int \frac{S_x}{c^2}\, d\mathscr{V} \end{aligned}$$

where S_x is the x-component of the Poynting vector.

Since the total force is the rate of change of momentum, 14-12 can be rearranged as

$$(14\text{-}13)\qquad \left(\frac{\partial}{\partial t}\right)\int \left[p_x + \left(\frac{S_x}{c^2}\right)\right] d\mathscr{V} = \int T_{t1} \cdot d\mathscr{A}$$

where p_x is the x component of momentum per unit volume of the charged bodies acted upon by the fields. The term S_x/c^2 is also a momentum per unit volume, being the x component of the momentum density of the electromagnetic field. Since the right side of Equation 14-13 is an x component of force obtained by integrating T_{t1} over a closed area, it follows that T_{t1} has the nature of an x component of force per unit area. It may be either a pressure or a shearing stress.

As an illustration, let us consider a plane wave incident normally on a surface in which the wave is completely absorbed. Let this surface be normal to the x-direction, and the electric and magnetic field vectors be E_y and H_z. As shown in Figure 12, the absorbing surface area vector is in the negative x-direction. From Equations 14-10 and 14-7 the space

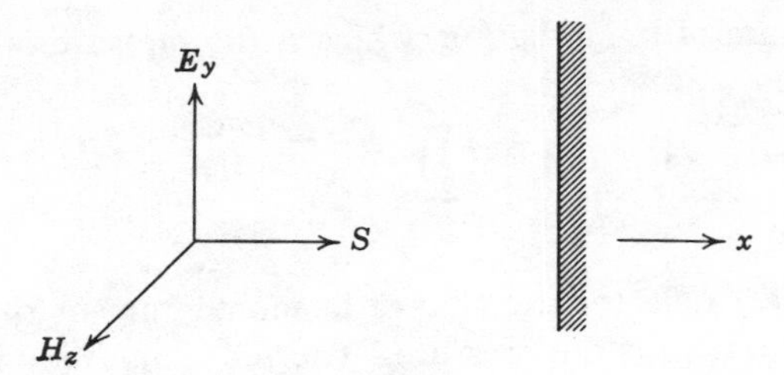

Figure 12

components of T_{t1} are

$$\mathbf{T}_{t1} = \left(\frac{\mathbf{i}}{2}\right)[\mu(H_x^2 - H_y^2 - H_z^2) + \epsilon(E_x^2 - E_y^2 - E_z^2)] + \mathbf{j}[\mu H_x H_y + \epsilon E_x E_y] + \mathbf{k}[\mu H_x H_z + \epsilon E_x E_z] \tag{14-14}$$

Since $H_x = H_y = E_x = E_z = 0$, there is left in 14-14:

$$\mathbf{T}_{t1} = -\left(\frac{\mathbf{i}}{2}\right)[\mu H_z^2 + \epsilon E_y^2] \tag{14-15}$$

which is the energy density in the incident fields at the surface. When Equation 14-15 is integrated over a unit area of the absorbing surface, we find that the pressure of the radiation is

$$P_r = (1/2)(\mu H_z^2 + \varepsilon E_y^2) = \mu H_z^2 \tag{14-16}$$

since $E_y = \sqrt{\mu/\varepsilon} H_z$. This is the maximum instantaneous pressure in terms of the field amplitudes. The average pressure, assuming sinusoidal time variations is

$$\bar{P}_r = \frac{\mu H_z^2}{2} \tag{14-17}$$

If, instead of a perfect absorber, the surface is a perfect reflector, then the sum of the incident and reflected electric field at the surface is zero and the sum of the incident and reflected magnetic field is $2H_z$. The same reasoning as before leads to an average pressure of μH_z^2, twice that found for the absorber.

Equation 14-13 suggests another approach to this analysis. The Poynting vector in the incident wave has the value $E_y H_z$, and hence there is a momentum density of radiation in this wave of $E_y H_z/c^2$. In a cylindrical column of unit cross-sectional area based on the perfect absorber and extending toward the left a distance equal to the value of c, the total momentum of the radiation has a time average of $E_y H_z/2c$. All of this radiation disappears at the surface in 1 second, and thus the rate of change of the momentum of the radiation, which is the pressure of the radiation on the surface, is

$$(14\text{-}18) \qquad \bar{P}_r = \left(\frac{1}{2c}\right)(E_y H_z) = \left(\frac{1}{2}\right)\sqrt{\mu\epsilon}\sqrt{\mu/\epsilon}\, H_z^{\ 2} = \left(\frac{\mu}{2}\right) H_z^{\ 2}$$

as before.

For the perfect reflector the change in momentum of the radiation is twice as great, and so is the pressure. The reflecting body is assumed to have a large enough mass that a negligible momentum is transferred to it.

The fourth component of k has a somewhat unexpected interpretation. From Equation 14-8 it has the form

$$(14\text{-}19) \qquad k_4 = \nabla \cdot \mathbf{T}_{t4} + \frac{\partial T_{44}}{\partial x_4}$$

Equations 14-7 show that $\mathbf{T}_{t4} = (1/jc)(E \times H) = \mathbf{S}/jc$ where $\mathbf{S}$ is the Poynting vector, while $T_{44} = U$, where U is the total energy density in the electromagnetic field. Thus, including k_4 from Equation 14-5, 14-19 becomes

$$(14\text{-}20) \qquad \left(\frac{j\rho}{c}\right)(\mathbf{v} \cdot \mathbf{E}) = -\left(\frac{j}{c}\right)(\nabla \cdot \mathbf{S}) - \left(\frac{j}{c}\right)\left(\frac{\partial}{\partial t}\right) U$$

whose integral over a closed volume,

$$(14\text{-}21) \qquad \int \rho \mathbf{v} \cdot \mathbf{E}\, d\mathscr{V} = -\int \mathbf{S} \cdot d\mathscr{A} - \frac{\partial}{\partial t}\int U\, d\mathscr{V}$$

is just the Poynting equation for the conservation of energy in the electromagnetic field. Thus, both the conservation of momentum and the conservation of energy are associated in the single 4-vector for the force per unit volume.

It is the 4-vector force density, k, rather than its Newtonian counterpart, which is invariant with a change to another coordinate system in relative motion. It is a generalization of the Newtonian force that includes not only the familiar rate of change of momentum, but the energy conservation relation as well and shows the intimate interrelation between momentum and energy. Although it was developed here out of the electromagnetic relations, the nature of force should be the same for mechanical interactions as for electromagnetic interactions, and the foregoing suggests a way of writing the relativistic expression for the mechanical force. The left side of 14-13 contains a term which is the sum of the 3-space momentum and a momentum term formed from an energy function. The relativistic mechanical force turns out to have three components related to the ordinary 3-space momentum and a fourth formed from the energy function.

But it is not the intention of this chapter to treat of the mechanical implications of relativistic theory. The interested reader is urged to consult some of the references given at the end of this chapter.

15. Conclusion

The fact that the speed of light is observed to have the same value as measured in different coordinate systems in relative uniform motion was shown to lead to a revision of the method of translating physical descriptions from one reference frame to another. Such a transformation can be looked upon as a kind of rotation of coordinate systems by analogy with coordinate rotations in 3-space. Four-space generalizations of the potential, the current density, the electric and the magnetic field vectors, and the force density were then formed, each of which can be shown to be invariant with a transformation to another reference system in relative motion to the first. These generalizations are elegant, concise formulations of the electromagnetic field relations that had been developed in previous chapters.

Such generalizations, moreover, emphasize the relative nature of $\mathbf{E}$ and $\mathbf{B}$, of $\mathbf{A}$ and ϕ, of $\mathbf{J}$ and ρ. What may be a purely static electric field to one observer is both an electric and a magnetic field to another. A pure scalar potential to one observer is both a vector and a scalar potential to another. A static charge density transforms to a charge density and a current density. But one quantity, the total charge in a closed system, is always conserved, irrespective of the motion of the observer.

This chapter has been only an introduction to the ideas of relativity in electromagnetism. No attempt has been made to introduce explicitly the refinements that tensor algebra brings to the subject. And certainly no treatment of relativistic electromagnetism is complete without also a full discussion of relativistic mechanics. The dividing line between electromagnetism and mechanics is not sharp, and the reader should feel stimulated to bridge this dividing line, to see the unity of the whole, to see similarities and analogies in the structures and the results in the different conventional subdivisions of physics.

It is hoped that the reader will be able to find pleasure in the beauty of the logical structure of electromagnetism, to feel a dissatisfaction where that logical structure is incomplete, and to be stimulated by that dissatisfaction to an effort to achieve a clarification. Physics has benefited often from a search for the practical application, but more often its enrichment has come from the desire for a clearer, more self-consistent explanation, from the need for a deeper insight, a broader, more coherent picture of the world. Such an insight has a beauty of its own, and the physicist who seeks it is engaged in the creation of the beautiful, just as much as is the artist. The author hopes that some of the beauty of the theory of electromagnetism has been communicated to the reader in this book.

Discussion Questions

1. Devise a thought experiment whereby in principle an observer could check several clocks at different places at rest in his own system of reference for synchronism.

2. Discuss the postulate of invariance. Is it a necessary feature of a physical law? A highly important feature? A desirable feature? If necessary, important, or desirable, from what point of view? For what purpose?

3. An observer in S' moves parallel to a conductor in S which carries a current of density J_z. Relativistic reasoning leads to the conclusion that S' also sees a charge density of $-\gamma(\beta/c)J_z$. Can this be made plausible by arguments of the type used in Section 11 to make other relativistic conclusions plausible?

4. (To be considered after Problems 15-17). Discuss the idea of assuming the electron mass to be entirely intrinsic to its electromagnetic field. Just where do the discrepancies lie? Suggest possible physical reasons why this concept is not adequate. To what extent would the results of Problems 15-17 be affected if, instead of a spherical shell charge, the electron were taken to be a sphere with the charge Q uniformly distributed throughout its volume?

Problems

1. Fill in the details in the derivation that progresses from Equations 2-2 and 2-3 to Equation 2-8.

2. By differentiating the relations in the Lorentz transformation, obtain the relations between the velocity components of a point as observed in the S' frame to the components as observed in the S frame.

3. Show that the components of the operator $\Box$ become

$$\begin{aligned}\partial/\partial x_1' &= \partial/\partial x_1\\ \partial/\partial x_2' &= \partial/\partial x_2\\ \partial/\partial x_3' &= \gamma(\partial/\partial x_3 + j\beta\,\partial/\partial x_4)\\ \partial/\partial x_4' &= \gamma(\partial/\partial x_4 - j\beta\,\partial/\partial x_3)\end{aligned}$$

when transformed from the S to the S' reference frame.

4. With the help of construction lines added to Figure 2, derive Equation 4-4 as the transformation between two 3-space coordinate systems, one of which has been rotated through an angle θ about the x-axis relative to the other.

5. Show the reasoning that leads to Equation 5-4 and then derive 5-8 and 5-9.

6. In Section 3 there was described a method by which an observer in frame S

determines the times when the two ends of a rod, moving with a velocity v, pass him in order to determine the length of the rod. Show how these measurements appear on a diagram such as Figure 6 and show that the FitzGerald-Lorentz contraction is obtained from reasoning with this diagram.

7. Show that relation between $T_b' - T_a'$ to $T_b - T_a$ in Figure 6 is that of the time dilatation discussed in Section 3.

8. Show that the total charge within a closed volume is conserved in a transformation between frames of reference in relative uniform motion.

9. In a frame S what is the 4-space potential function A due to a point charge Q at rest at the origin of S? Transform this potential to the potential A' in a reference frame S' moving with a velocity v in the z-direction relative to S. Show that the field components of Equations 12-6 follow from A' through Equations 9-3 and 9-4.

10. A point charge Q is moving with a velocity v along the z-axis relative to an observer at rest. Determine the angle of a cone about the z-axis enveloping a fraction, f, of the total flux of the electric field of this charge as the observer sees it. Also show that the total flux of the electric field is Q/ϵ.

The following four problems refer to a spherical shell of radius r_0 carrying a total charge Q distributed uniformly over its surface. The shell is moving with a velocity v along the z-axis.

11. Express the energy density in the electric field, the energy density in the magnetic flux field, and the Poynting vector, as functions of the distance r from the charge and the angle θ which r makes with the z-axis.

12. Find the total energy in the electric field over all space outside the sphere. Show that for $v = 0$, this energy reduces to $U_{E_0} = Q^2/8\pi\epsilon r_0$.

13. Find the total energy in the magnetic flux field outside the sphere, and show that for $\beta^2 \ll 1$ this energy reduces to $U_B = (\frac{2}{3})U_{E_0}\beta^2$ (Note that for small x, $\arctan x = x - x^3/3$.)

14. Find the total momentum of the electromagnetic field outside the sphere, and show that for $\beta^2 \ll 1$ this momentum reduces to $G = (\frac{4}{3})(U_{E_0}v/c^2)$.

15. It is a consequence of the theory of relativity that mass and energy are related by $E = mc^2$. An attempt has been made to develop a theory of the electron by assuming that its entire mass consists of the energy in its surrounding fields. Assuming that the charge on the electron (1.6×10^{-19} coulomb) is uniformly distributed over the surface of a sphere of radius r_0 and using the known mass of the electron (9.1×10^{-31} kilogram), find the value of r_0 for the electron at rest. Is this a plausible value?

16. If the energy of the electron is entirely in its associated fields, then the energy in the magnetic flux field should be its energy of motion, its kinetic energy. If at low velocities, the kinetic energy is $mv^2/2$, what is the mass of the electron in motion, assuming the spherical shell picture of Problem 15? Does this reduce to the rest mass of Problem 15?

17. The mass of the moving electron can also be found from the momentum of the electromagnetic field. At low velocities the momentum of the electron is mv. What mass is thus implied for the spherical shell electron of Problem 15? Compare with the previous values.

18. Show that Equation 14-8 produces the four equations for the force density of 14-5.

19. Show that the matrix T defined in 14-6 is also given by

$$T = (\tfrac{1}{2}\mu)[(F)(F) - (F^*)(F^*)]$$

where F is the matrix shown in 9-10 and F^* is the matrix formed by the rule of 9-15. (In a product of two matrices, the term at the intersection of the mth row and the nth column is the sum of the products of the terms in the mth row of the first matrix with the nth column of the second matrix.)

20. In the matrix equation

$$(F') = (Tr)(F)(\tilde{Tr})$$

(F) is the matrix of Equation 9-10, (Tr) is the transformation matrix of Equation 6-3, and $(\tilde{Tr})$ is obtained from (Tr) by exchanging symmetrically located terms on either side of the diagonal. Perform the operations to obtain the matrix (F') corresponding to (F) in a system S' moving at a velocity v along the z-axis relative to S. Show that the components of (F') are consistent with the electric and magnetic field components after transformation as found in Equations 11-6 and 11-7.

21. What is the radiation pressure exerted by a plane wave incident normally on a pure dielectric of dielectric constant ϵ?

REFERENCES

1. Jackson, *Classical Electrodynamics* (Wiley). Chapter 11.
2. Moon and Spencer, *Foundations of Electrodynamics* (Van Nostrand). Chapter 12.
3. Sommerfeld, *Electrodynamics*, Part III (Academic Press).
4. Born, *Einstein's Theory of Relativity* (Dutton). Chapter VI.
5. Joos, *Theoretical Physics* (Stechert). Chapter X.
6. Leighton, *Principles of Modern Physics* (McGraw-Hill). Chapter I.
7. Brehme, "A Geometric Representation of Galilean and Lorentz Transformations," *Am. Jour. Phys.* **30**, 489 (1962).
8. Sears, "Some Applications of the Brehme Diagram," *Am. Jour. Phys.*, **31**, 269 (1963).
9. Daubin, "A Geometrical Introduction to Special Relativity," *Am. Jour. Phys.*, **30**, 818 (1962).
10. Lass, "Accelerating Frames of Reference and the Clock Paradox," *Am. Jour. Phys.*, **31**, 274 (1963).

Appendix A
Summary of vector algebra and calculus

For convenient reference, the sections in the text where vector algebra and calculus are presented are here listed:

	Chapter	Section
The Scalar Product	1	9
The Operator Del	1	13
The Divergence	1	13
Gauss's Theorem or Divergence Theorem	1	14
The Gradient	2	6
The Laplacian	2	13
The Vector or Cross Product	5	2
The Curl	5	14
Stokes's Theorem	5	14

Relations involving the operator del, which are useful in transforming vector relations, are listed below. (Here **A** and **B** are vector functions of position and ψ is a scalar function of position.)

(A-1) $\nabla \cdot (\psi \mathbf{A}) = (\nabla \psi) \cdot \mathbf{A} + \psi(\nabla \cdot \mathbf{A})$

(A-2) $\nabla \times (\psi \mathbf{A}) = (\nabla \psi) \times \mathbf{A} + \psi(\nabla \times \mathbf{A})$

(A-3) $\nabla \cdot (\mathbf{A} \times \mathbf{B}) = \mathbf{B} \cdot (\nabla \times \mathbf{A}) - \mathbf{A} \cdot (\nabla \times \mathbf{B})$

(A-4) $\nabla \times (\mathbf{A} \times \mathbf{B}) = (\mathbf{B} \cdot \nabla)\mathbf{A} - \mathbf{B}(\nabla \cdot \mathbf{A}) - (\mathbf{A} \cdot \nabla)\mathbf{B} + \mathbf{A}(\nabla \cdot \mathbf{B})$

(A-5) $$\nabla(\mathbf{A}\cdot\mathbf{B}) = (\mathbf{B}\cdot\nabla)\mathbf{A} + (\mathbf{A}\cdot\nabla)\mathbf{B} + \mathbf{B}\times(\nabla\times\mathbf{A}) + \mathbf{A}\times(\nabla\times\mathbf{B})$$

(A-6) $$\nabla\times\nabla\psi = 0$$

(A-7) $$\nabla\cdot(\nabla\times\mathbf{A}) = 0$$

(A-8) $$\nabla\times(\nabla\times\mathbf{A}) = \nabla(\nabla\cdot\mathbf{A}) - \nabla^2\mathbf{A}$$

(A-9) (Gauss's Theorem) $$\int \nabla\cdot\mathbf{A}\,d\mathscr{V} = \int \mathbf{A}\cdot d\mathscr{A}$$

(The limits of the volume of integration on the left are fixed by the closed surface of integration on the right.)

(A-10) (Stokes's Theorem) $$\oint \mathbf{A}\cdot d\mathbf{l} = \int \nabla\times\mathbf{A}\cdot d\mathscr{A}$$

(The surface of integration on the right has the line of integration on the left as a perimeter.)

(A-11) $$\int \nabla\times\mathbf{A}\,d\mathscr{V} = -\int \mathbf{A}\times d\mathscr{A}$$

(The surface and volume integrals are related as in Gauss's Theorem.)

(A-12) $$\oint \psi\,d\mathbf{l} = -\int \nabla\psi\times d\mathscr{A}$$

(The line and surface integrals are related as in Stokes's Theorem.)

The foregoing identities are independent of the type of coordinate system, but the forms of the gradient, the divergence, the curl, and the Laplacian in terms of the derivatives with respect to the coordinate variables depend on the kind of coordinate system used. These functions are listed for the three most commonly used systems: Cartesian, cylindrical, and spherical.

In the Cartesian system, **i**, **j**, and **k** are unit vectors in the directions of increasing x, y, and z, respectively, where x, y, and z, in that order, form a right-handed system.

In cylindrical coordinates, $\mathbf{r}_u$, $\boldsymbol{\theta}_u$, and **k** are unit vectors in the directions of increasing r, θ, and z, respectively, where r, θ, and z, in that order, form a right-handed system.

In spherical coordinates, $\mathbf{r}_u$, $\boldsymbol{\theta}_u$, and $\boldsymbol{\varphi}_u$ are unit vectors in the directions of increasing r, θ, and φ, where θ is the co-latitude and φ the azimuth, and r, θ, and φ in that order form a right-handed system.

(A-13) The Gradient:

(Cart.)

$$\nabla \psi = \mathbf{i}\frac{\partial \psi}{\partial x} + \mathbf{j}\frac{\partial \psi}{\partial y} + \mathbf{k}\frac{\partial \psi}{\partial z}$$

(Cyl.)

$$\nabla \psi = \mathbf{r}_u \frac{\partial \psi}{\partial r} + \boldsymbol{\theta}_u \frac{1}{r}\frac{\partial \psi}{\partial \theta} + \mathbf{k}\frac{\partial \psi}{\partial z}$$

(Sph.)

$$\nabla \psi = \mathbf{r}_u \frac{\partial \psi}{\partial r} + \boldsymbol{\theta}_u \frac{1}{r}\frac{\partial \psi}{\partial \theta} + \boldsymbol{\varphi}_u \frac{1}{r \sin \theta}\frac{\partial \psi}{\partial \varphi}$$

(A-14) The Divergence:

(Cart.)

$$\nabla \cdot \mathbf{A} = \frac{\partial A_x}{\partial x} + \frac{\partial A_y}{\partial y} + \frac{\partial A_z}{\partial z}$$

(Cyl.)

$$\nabla \cdot \mathbf{A} = \frac{1}{r}\frac{\partial}{\partial r}(rA_r) + \frac{1}{r}\frac{\partial A_\theta}{\partial \theta} + \frac{\partial A_z}{\partial z}$$

(Sph.)

$$\nabla \cdot \mathbf{A} = \frac{1}{r^2}\frac{\partial}{\partial r}(r^2 A_r) + \frac{1}{r \sin \theta}\frac{\partial}{\partial \theta}(\sin \theta \, A_\theta) + \frac{1}{r \sin \theta}\frac{\partial A_\varphi}{\partial \varphi}$$

(A-15) The Curl:

(Cart.)

$$\nabla \times \mathbf{A} = \mathbf{i}\left(\frac{\partial A_z}{\partial y} - \frac{\partial A_y}{\partial z}\right) + \mathbf{j}\left(\frac{\partial A_x}{\partial z} - \frac{\partial A_z}{\partial x}\right) + \mathbf{k}\left(\frac{\partial A_y}{\partial x} - \frac{\partial A_x}{\partial y}\right)$$

(Cyl.)

$$\nabla \times \mathbf{A} = \mathbf{r}_u\left(\frac{1}{r}\frac{\partial A_z}{\partial \theta} - \frac{\partial A_\theta}{\partial z}\right) + \boldsymbol{\theta}_u\left(\frac{\partial A_r}{\partial z} - \frac{\partial A_z}{\partial r}\right) + \mathbf{k}\left[\frac{1}{r}\frac{\partial}{\partial r}(rA_\theta) - \frac{1}{r}\frac{\partial A_r}{\partial \theta}\right]$$

(Sph.)

$$\nabla \times \mathbf{A} = \mathbf{r}_u \frac{1}{r \sin \theta}\left[\frac{\partial}{\partial \theta}(A_\varphi \sin \theta) - \frac{\partial A_\theta}{\partial \varphi}\right] + \boldsymbol{\theta}_u \frac{1}{r}\left[\frac{1}{\sin \theta}\frac{\partial A_r}{\partial \varphi} - \frac{\partial}{\partial r}(rA_\varphi)\right] + \boldsymbol{\varphi}_u \frac{1}{r}\left[\frac{\partial}{\partial r}(rA_\theta) - \frac{\partial A_r}{\partial \theta}\right]$$

(A-16) The Laplacian:

(Cart.)

$$\nabla^2\psi = \frac{\partial^2\psi}{\partial x^2} + \frac{\partial^2\psi}{\partial y^2} + \frac{\partial^2\psi}{\partial z^2}$$

(Cyl.)

$$\nabla^2\psi = \frac{1}{r}\frac{\partial}{\partial r}\left(r\frac{\partial\psi}{\partial r}\right) + \frac{1}{r^2}\frac{\partial^2\psi}{\partial\theta^2} + \frac{\partial^2\psi}{\partial z^2}$$

(Sph.)

$$\nabla^2\psi = \frac{1}{r^2}\frac{\partial}{\partial r}\left(r^2\frac{\partial\psi}{\partial r}\right) + \frac{1}{r^2\sin\theta}\frac{\partial}{\partial\theta}\left(\sin\theta\frac{\partial\psi}{\partial\theta}\right) + \frac{1}{r^2\sin^2\theta}\frac{\partial^2\psi}{\partial\varphi^2}$$

Appendix B

The algebra of complex numbers

A complex number is defined as a number in the form

(B-1) $$z = x + jy$$

where $j = \sqrt{-1}$ and x and y are real quantities. x is called the real part of the complex number z and y the imaginary part. The real part is also on occasion indicated by $R(z)$ or $Re(z)$, and the imaginary part by $I(z)$ or $Im(z)$. Complex numbers need not, of course, always explicitly be in the form B-1, but any complex number can be manipulated into this form.

The ordinary algebraic operations of adding, subtracting, multiplying, and dividing apply to complex numbers. In such operations use is made of the special properties of j:

(B-2) $$j^2 = -1, j^3 = -j, j^4 = 1, \text{etc}$$

The conjugate of a complex number, often identified with a star, such as z^*, is formed by changing the sign of j, wherever it appears. By multiplying numerator and denominator of a complex quotient such as

(B-3) $$z = \frac{x + jy}{u + jv}$$

by the conjugate of the denominator, such a quotient is put into the form of B-1, a sum of a real and an imaginary part. Thus

(B-4) $$\frac{(x + jy)(u - jv)}{(u + jv)(u - jv)} = \frac{(xu + yv) + j(yu - xv)}{u^2 + v^2}$$

$$= \frac{(xu + yv)}{u^2 + v^2} + j\frac{(yu - xv)}{u^2 + v^2}$$

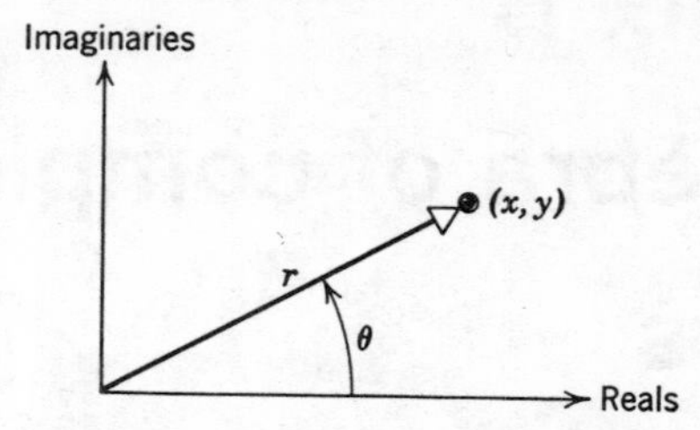

Figure B1. Argand diagram for the complex number $z = x + jy$.

In general the product of a complex number with its conjugate is a real number, whose square root is called the magnitude, absolute value, or modulus of the complex number.

Complex numbers are represented graphically on a pair of orthogonal axes, along one of which, conventionally the horizontal axis, is plotted the real part and along the other the imaginary part. Such a representation is called an Argand diagram. The length of the directed line (or "vector") from the origin to the point whose coordinates are the real and imaginary parts is the magnitude or modulus of the complex number, given by

$$r = \sqrt{x^2 + y^2} \qquad \text{(B-5)}$$

The angle which the "vector" makes with the axis of reals is given by

$$\tan \theta = \frac{y}{x} \qquad \text{(B-6)}$$

where θ is called the argument of z. Another form in which the complex number z may be written is

$$z = r(\cos \theta + j \sin \theta) \qquad \text{(B-7)}$$

The sum of two complex numbers is obtained by the parallelogram method of adding vectors (Figure B2) thus:

$$\begin{aligned} &(x + jy) + (u + jv) \\ &\quad = (x + u) + j(y + v) \\ &\quad = \sqrt{(x + u)^2 + (y + v)^2}\,(\cos \psi + j \sin \psi) \end{aligned} \qquad \text{(B-8)}$$

where $\tan \psi = (y + v)/(x + u)$.

The trigonometric form, or Euler form, B-7, of a complex number is on occasion abbreviated by the following notation:

$$r(\cos \theta + j \sin \theta) = r \angle \theta \qquad \text{(B-9)}$$

The Euler form is often convenient in obtaining the product or quotient of complex numbers. Thus, if

(B-10) $$z_1 = x + jy = r_1(\cos\theta_1 + j\sin\theta_1)$$
$$z_2 = u + jv = r_2(\cos\theta_2 + j\sin\theta_2)$$

then

(B-11) $$z_1 z_2 = r_1 r_2[\cos(\theta_1 + \theta_2) + j\sin(\theta_1 + \theta_2)]$$
$$= r_1 r_2 \underline{/(\theta_1 + \theta_2)}$$

and

(B-12) $$\frac{z_1}{z_2} = \frac{r_1}{r_2}[\cos(\theta_1 - \theta_2) + j\sin(\theta_1 - \theta_2)]$$
$$= \frac{r_1}{r_2}\underline{/(\theta_1 - \theta_2)}$$

These results are readily demonstrated with the aid of the trigonometric identities for the cosine and sine of the sum and difference of two angles.

The series expansions for the sine and the cosine are:

(B-13) $$\sin\theta = \theta - \frac{\theta^3}{3!} + \frac{\theta^5}{5!} - \frac{\theta^7}{7!} + \cdots$$

(B-14) $$\cos\theta = 1 - \frac{\theta^2}{2!} + \frac{\theta^4}{4!} - \frac{\theta^6}{6!} + \cdots$$

so that

(B-15) $$\cos\theta + j\sin\theta = 1 + j\theta - \frac{\theta^2}{2!} - j\frac{\theta^3}{3!} + \frac{\theta^4}{4!} + j\frac{\theta^5}{5!} + \cdots$$

The series expansion of e^w is

(B-16) $$e^w = 1 + w + \frac{w^2}{2!} + \frac{w^3}{3!} + \frac{w^4}{4!} + \cdots$$

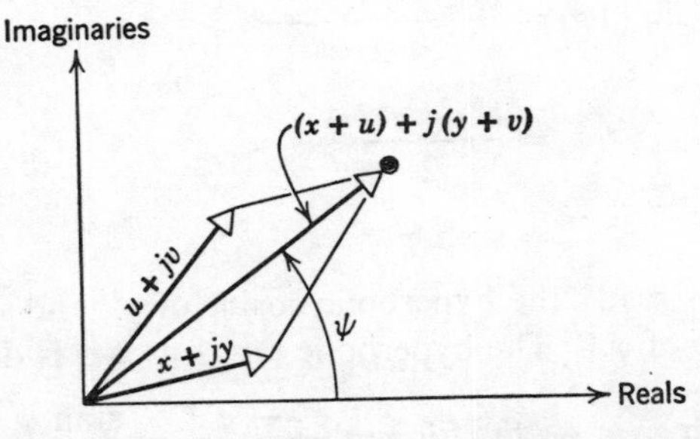

Figure B2. Adding two complex numbers on the Argand diagram.

When $w = j\theta$, B-16 becomes identical with B-15, so that we can also write

(B-17) $$z = x + jy = re^{j\theta}$$

where $\tan\theta = y/x$ and $r^2 = x^2 + y^2$, and consequently B-11 and B-12 can also be put into the form:

(B-18) $$z_1 z_2 = r_1 r_2 e^{j(\theta_1+\theta_2)}$$

$$\frac{z_1}{z_2} = \frac{r_1}{r_2} e^{j(\theta_1-\theta_2)}$$

From the sum of $\cos\theta + j\sin\theta = e^{j\theta}$ and its conjugate, we find that

(B-19) $$\cos\theta = \frac{e^{j\theta} + e^{-j\theta}}{2}$$

and from the difference of $\cos\theta + j\sin\theta = e^{j\theta}$ and its conjugate, we find that

(B-20) $$\sin\theta = \frac{e^{j\theta} - e^{-j\theta}}{2j}$$

and hence

(B-21) $$\tan\theta = \frac{e^{j\theta} - e^{-j\theta}}{j(e^{j\theta} + e^{-j\theta})} = -j\frac{e^{j\theta} - e^{-j\theta}}{e^{j\theta} + e^{-j\theta}}$$

The trigonometric functions of an imaginary angle occur on occasion and may be expressed in terms of the above forms. Where $\theta = j\varphi$, B-20 and B-21 become

(B-22) $$\cos(j\varphi) = \frac{e^{j(j\varphi)} + e^{-j(j\varphi)}}{2}$$

$$= \frac{e^{\varphi} + e^{-\varphi}}{2}$$

$$= \cosh\varphi$$

$$-j\sin(j\varphi) = \frac{e^{-j(j\varphi)} - e^{j(j\varphi)}}{2}$$

$$= \frac{e^{\varphi} - e^{-\varphi}}{2}$$

$$= \sinh\varphi$$

where $\cosh\varphi$ is read "the hyperbolic cosine of φ" and $\sinh\varphi$ is read "the hyperbolic sine of φ." The hyperbolic tangent of φ is defined as

(B-23) $$\tanh\varphi = -j\tan(j\varphi) = \frac{e^{\varphi} - e^{-\varphi}}{e^{\varphi} + e^{-\varphi}} = \frac{\sinh\varphi}{\cosh\varphi}$$

Values of $\sinh \varphi$, $\cosh \varphi$, and $\tanh \varphi$ are found in tables similar to those for the natural trigonometric functions.

From B-22 and B-23 the following identities are readily demonstrated:

(B-24)
$$\cosh^2 \varphi - \sinh^2 \varphi = 1$$
$$\sinh 2\varphi = 2 \sinh \varphi \cosh \varphi$$
$$\cosh 2\varphi = \cosh^2 \varphi + \sinh^2 \varphi$$
$$2 \sinh^2 \varphi = \cosh 2\varphi - 1$$
$$2 \cosh^2 \varphi = \cosh 2\varphi + 1$$
$$\sinh (\theta + \varphi) = \sinh \theta \cosh \varphi + \cosh \theta \sinh \varphi$$
$$\cosh (\theta + \varphi) = \cosh \theta \cosh \varphi + \sinh \theta \sinh \varphi$$
$$\tanh^2 \varphi = \frac{\cosh 2\varphi - 1}{\cosh 2\varphi + 1}$$
$$\sinh (j\varphi) = j \sin \varphi$$
$$\cosh (j\varphi) = \cos \varphi$$
$$\tanh (j\varphi) = j \tan \varphi$$

Appendix C
Fourier analysis of periodic functions

Most periodic functions encountered in physics can be expanded as series of sine and cosine terms. For example,

$$f(x) = \frac{a_0}{2} + a_1 \cos x + a_2 \cos 2x + \cdots + a_n \cos nx + \cdots \\ + b_1 \sin x + b_2 \sin 2x + \cdots + b_n \cos nx + \cdots \tag{C-1}$$

where n is an integer. The periodicity of $f(x)$ is over a range 2π of the variable x. The conditions on $f(x)$ in order that it be represented by Equation C-1 are not stringent; it must be bounded and have no more than a finite number of discontinuities in any one period.

The coefficients, a_n and b_n, are the weights assigned to the successive harmonic terms. Multiplication of each term by $\cos nx$ and integration over the range of one period yields

$$\frac{1}{\pi}\int_{x_0}^{x_0+2\pi} f(x) \cos nx \, dx = a_n \tag{C-2}$$

since the integrals of all terms where there is a product of cosines of different n's and all terms where there is a product of sines and cosines are zero. Note that the factor $\frac{1}{2}$ in the constant term in the series C-1 makes it possible to include a_0 as a special case of C-2.

Multiplication of each term by $\sin nx$ and integration over a period yields

$$\frac{1}{\pi}\int_{x_0}^{x_0+2\pi} f(x) \sin nx \, dx = b_n \tag{C-3}$$

Where the periodicity in a function $f(u)$ is over a range P of the variable u, a change of variable to

$$u = \frac{Px}{2\pi} \tag{C-4}$$

converts the function to a periodicity of 2π in x so that C-2 and C-3 apply.

For an even function where

$$f(x) = f(-x) \tag{C-5}$$

the coefficients of the sine terms are all zero, whereas for an odd function where

$$f(x) = -f(-x) \tag{C-6}$$

the coefficients of the cosine terms (and the constant term) are all zero. Often a shift of the origin of coordinates can convert a function into an even or an odd function.

As was seen in the section on complex numbers, sine and cosine functions can be transformed into functions involving exponentials. In terms of exponentials, the series C-1 becomes

$$f(x) = \sum_{n=-\infty}^{n=\infty} c_n e^{jnx} \tag{C-7}$$

with the coefficients determined from

$$c_n = \frac{1}{2\pi} \int_{x_0}^{x_0+2\pi} f(x) e^{-jnx}\, dx \tag{C-8}$$

Appendix D

Systems of Units

The meter-kilogram-second system of units (mks) is used in the body of this book. A second system, called the Gaussian system, also finds considerable favor, especially in atomic and nuclear physics. Although necessarily self-consistent, it grew as a kind of hybrid system out of two earlier ones, the electromagnetic system and the electrostatic system. As a foundation for the Gaussian system of units, as well as to assist understanding of older references, we shall develop the electromagnetic system (emu) and the electrostatic system (esu) in some detail. Both the emu and the esu systems employ the centimeter, the gram, and the second as the fundamental units of length, mass, and time. The already familiar mks system of units provides a reference for the comparison of magnitudes. In what follows, the symbols for current, charge, field, etc., refer to magnitudes as measured in the units indicated by the subscripts. They do not stand for the size of the unit.

In any of the systems of units, certain definitions are common:

(D-1) Current = time rate of change of Charge

$$I = \frac{dQ}{dt}$$

(D-2) Electric field = Force per unit positive Charge

$$E = \frac{F}{Q}$$

(D-3) Potential = Work per unit positive Charge

$$V = \frac{W}{Q}$$

In addition to these definitions, in each system certain arbitrary choices are made. For the electromagnetic system these choices are made in the same equations as in the mks system, and we shall begin with the emu system.

The Electromagnetic System of Units

The emu system begins with the definition of the unit of current, the abampere, based on the force between parallel wires carrying equal currents. The force on a straight wire of length L, distant r from another and parallel wire of infinite length may be generalized as

$$F = k_1 \frac{2I^2L}{r} \tag{D-4}$$

where I is the current in either wire and k_1 is a constant whose value and dimensions depend on the system of units. In the mks system, F is in newtons, I in amperes, L and r in meters, and $k_1 = \mu_0/4\pi$ henrys/meter. In the emu system, on the other hand, F is in dynes, I in abamperes, L and r in centimeters, and k_1 is chosen as a dimensionless number whose value in free space is unity. Thus a given current is expressed either as

$$I^2_{\text{emu}} = \frac{F(\text{dynes})\, r(\text{cm})}{2L(\text{cm})} \tag{D-5}$$

or as

$$I^2_{\text{mks}} = \frac{4\pi F(\text{newtons})\, r(\text{meters})}{2\mu_0 L(\text{meters})} \tag{D-6}$$

With $\mu_0 = 4\pi \times 10^{-7}$ henry/meter and $F(\text{dynes}) = F(\text{newtons}) \times 10^5$, we find that

$$\frac{I_{\text{emu}}}{I_{\text{mks}}} = \frac{1}{10} \tag{D-7}$$

Since the magnitude of a current measured in abamperes is $\frac{1}{10}$ the value measured in amperes, the abamperes *unit* is 10 times the ampere.

With the second as a unit of time common to all systems, it follows from the definition D-1 that measurements of charge in the two systems are related by

$$\frac{Q_{\text{emu}}}{Q_{\text{mks}}} = \frac{1}{10} \tag{D-8}$$

The emu of charge is the abcoulomb.

From D-2 and D-8 we obtain for the electric field ratio:

$$\text{(D-9)} \qquad \frac{E_{\text{emu}}}{E_{\text{mks}}} = \frac{F(\text{dynes})\, Q(\text{coulomb})}{F(\text{newtons})\, Q(\text{abcoulombs})} = 10^6$$

and from D-3 and D-8, using $W(\text{ergs}) = W(\text{joules}) \times 10^7$, we find that the ratio of potentials is

$$\text{(D-10)} \qquad \frac{V_{\text{emu}}}{V_{\text{mks}}} = \frac{W(\text{ergs})\, Q(\text{coulombs})}{W(\text{joules})\, Q(\text{abcoulombs})} = 10^8$$

where the emu of potential is called the abvolt. The emu of capacitance is the abfarad and capacitance in abfarads is related to capacitance in farads by

$$\text{(D-11)} \qquad \frac{C_{\text{emu}}}{C_{\text{mks}}} = \frac{Q(\text{abcoulomb})\, V(\text{volts})}{Q(\text{coulomb})\, V(\text{abvolts})} = 10^{-9}$$

We come now to Coulomb's law, which we write in the form:

$$\text{(D-12)} \qquad F = k_2 \frac{QQ'}{r^2}$$

Since the units of force, distance, and charge are already fixed, the constant k_2 depends on these units. In the mks system where F is measured in newtons, r in meters, and Q in coulombs, $k_2 = 1/4\pi\epsilon_0$ meters/farad, with a magnitude very close to 9×10^9. The dimensions of k_2 in the emu system are obtained from D-1, D-5, and D-12, as (cm/sec)^2, and its magnitude can be found in the following way:

In the mks system, we write from D-12,

$$\text{(D-13)} \qquad k_2 = \frac{1}{4\pi\epsilon_0} = \frac{F(\text{newtons})\, [r(\text{meters})]^2}{[Q(\text{coulombs})]^2}$$

and in the emu system,

$$\text{(D-14)} \qquad k_2 = \frac{F(\text{dynes})\, [r(\text{cm})]^2}{[Q(\text{abcoulombs})]^2}$$

$$= \frac{F(\text{newtons}) \times 10^5\, [r(\text{meters}) \times 10^2]^2}{[Q(\text{coulombs}) \times 10^{-1}]^2}$$

$$= \frac{10^{11}}{4\pi\epsilon_0} = 9 \times 10^{20} = c^2$$

where c is the velocity of light in cgs units.

Magnitudes of Polarization, the electric dipole moment per unit volume, are related by

$$(\text{D-15})\qquad \frac{P_{\text{emu}}}{P_{\text{mks}}} = \frac{Q(\text{abcoulombs})\, r(\text{cm})}{[L(\text{cm})]^3} \times \frac{[L(\text{meters})]^3}{Q(\text{coulombs})\, r(\text{meters})} = 10^{-5}$$

Gauss's law (cf: Chapter 1, Section 8) for a point charge q within an arbitrary volume is written

$$(\text{D-16})\qquad \int \mathbf{E} \cdot d\mathscr{A} = \int \frac{c^2 q\, d\mathscr{A} \cos\theta}{r^2} = 4\pi c^2 q$$

which, after introducing polarization charge density explicitly, becomes

$$(\text{D-17})\qquad \int \nabla \cdot \mathbf{E}\, d\mathscr{V} = 4\pi c^2 \left[\int \rho_{\text{free}}\, d\mathscr{V} - \int \nabla \cdot \mathbf{P}\, d\mathscr{V} \right]$$

We choose to regroup D-17 as

$$(\text{D-18})\qquad \int \left[\nabla \cdot \left(\frac{\mathbf{E}}{c^2} + 4\pi \mathbf{P} \right) \right] d\mathscr{V} = 4\pi \int \rho_{\text{free}}\, d\mathscr{V}$$

from which we define the displacement as

$$(\text{D-19})\qquad \mathbf{D} = \frac{\mathbf{E}}{c^2} + 4\pi \mathbf{P}$$

Then it follows that

$$(\text{D-20})\qquad \nabla \cdot \mathbf{D} = 4\pi \rho_{\text{free}} \qquad \nabla \cdot \mathbf{E} = 4\pi c^2 \rho_{\text{total}}$$

A comparison of D-19 with the corresponding mks form shows that

$$(\text{D-21})\qquad \frac{D_{\text{emu}}}{D_{\text{mks}}} = 4\pi \times 10^{-5}$$

The form of the magnetic flux density, **B**, is the result of choosing to include in **B** all factors in such relations as D-4 that are not associated with the current at the point where B is measured (cf. Chapter 5, Section 6). Alternatively, we may choose to define **B** from the force on a moving point charge q:

$$(\text{D-22})\qquad \mathbf{F} = k_3 q(\mathbf{v} \times \mathbf{B})$$

In both mks and emu systems k_3 is chosen dimensionless and of magnitude unity. (In the Gaussian system $k_3 = 1/c$.) Then

(D-23) $$\frac{B_{\text{emu}}}{B_{\text{mks}}} = 10^4$$

B_{emu} is measured in gauss. The reasoning of Section 13 of Chapter 5 shows that in the emu system

(D-24) $$\oint \mathbf{B} \cdot d\mathbf{l} = 4\pi I \qquad \text{or} \qquad \oint \mathbf{B} \cdot d\mathbf{l} = 4\pi \int \mathbf{J} \cdot d\mathscr{A}$$

and from Sections 1 and 2 of Chapter 7:

(D-25) $$\int (\mathbf{B} - 4\pi\mathbf{M}) \cdot d\mathbf{l} = 4\pi \int \mathbf{J} \cdot d\mathscr{A}$$

where **J** is the "true" current density and **M** the magnetization. Thus in the emu system:

(D-26) $$\mathbf{H} = \mathbf{B} - 4\pi\mathbf{M}$$

and

(D-27) $$\frac{H_{\text{emu}}}{H_{\text{mks}}} = 4\pi \times 10^{-3}$$

H and **B** are dimensionally the same in the emu system, but it has become the custom to speak of **H** as measured in oersteds.

The charge continuity relation

(D-28) $$\nabla \cdot \mathbf{J} = -\dot{\rho}$$

must follow from D-1, so that

(D-29) $$\nabla \times \mathbf{H} = 4\pi\mathbf{J} + \dot{\mathbf{D}}$$

and the remaining Maxwell equations are

(D-30) $$\nabla \cdot \mathbf{B} = 0$$

(D-31) $$\nabla \times \mathbf{E} = -\dot{\mathbf{B}}$$

The Electrostatic System

This system starts by defining charge from Coulomb's law, D-12, after choosing k_2 to be dimensionless and of magnitude unity in free space. The unit of charge is the statcoulomb and has the dimensions of $(\text{dynes})^{1/2}$ centimeter. (Note that in emu, the dimensions of charge are $(\text{dynes})^{1/2}$

seconds.) By the method already used for the emu system, it is readily verified that

$$\text{(D-32)} \qquad \frac{Q^2_{\text{esu}}}{Q^2_{\text{mks}}} = \frac{1}{4\pi\epsilon_0} \frac{F(\text{dynes})\,[r(\text{cm})]^2}{F(\text{newtons})\,[r(\text{meters})]^2} = 9 \times 10^{18}$$

from which

$$\text{(D-33)} \qquad \frac{Q_{\text{esu}}}{Q_{\text{mks}}} = \frac{I_{\text{esu}}}{I_{\text{mks}}} = 3 \times 10^9$$

Together with D-7 and D-8, D-33 shows that

$$\text{(D-34)} \qquad \frac{Q_{\text{esu}}}{Q_{\text{emu}}} = \frac{I_{\text{esu}}}{I_{\text{emu}}} = c$$

The unit of current in esu is the statampere.

Field, potential, and capacitance ratios are:

$$\text{(D-35)} \qquad \frac{E_{\text{esu}}}{E_{\text{mks}}} = \frac{1}{3 \times 10^4} \qquad \frac{E_{\text{esu}}}{E_{\text{emu}}} = \frac{1}{c}$$

$$\text{(D-36)} \qquad \frac{V_{\text{esu}}}{V_{\text{mks}}} = \frac{1}{300} \qquad \frac{V_{\text{esu}}}{V_{\text{emu}}} = \frac{1}{c}$$

$$\text{(D-37)} \qquad \frac{C_{\text{esu}}}{C_{\text{mks}}} = 9 \times 10^{11} \qquad \frac{C_{\text{esu}}}{C_{\text{emu}}} = c^2$$

The esu of potential is the statvolt, and of capacitance is the statfarad.

The constant k_1 in D-4 is now fixed by D-34. Thus,

$$\text{(D-38)} \qquad k_1 = \frac{F(\text{dynes})\, r(\text{cm})}{2[I(\text{statamperes})]^2\, L(\text{cm})}$$

$$= \frac{F(\text{dynes})\, r(\text{cm})}{2[I(\text{abamperes}) \times c]^2\, L(\text{cm})}$$

but

$$\frac{F(\text{dynes})\, r(\text{cm})}{2[I(\text{abamperes})]^2\, L(\text{cm})} = 1$$

so that

$$\text{(D-39)} \qquad k_1 = \frac{1}{c^2}$$

from which it follows that

$$\text{(D-40)} \qquad \frac{B_{\text{esu}}}{B_{\text{emu}}} = \frac{1}{c} \quad \text{and} \quad \frac{B_{\text{esu}}}{B_{\text{mks}}} = \frac{1}{3 \times 10^6}$$

Using D-39 and the reasoning of Sections 1 and 2 of Chapter 7, we write:

$$\text{(D-41)} \qquad \oint \mathbf{B} \cdot d\mathbf{l} = \frac{4\pi}{c^2}\left[\int \mathbf{J} \cdot d\mathscr{A} + \int \nabla \times \mathbf{M} \cdot d\mathscr{A}\right]$$

$$= \frac{4\pi}{c^2}\left[\int \mathbf{J} \cdot d\mathscr{A} + \int \mathbf{M} \cdot d\mathbf{l}\right]$$

which we choose to regroup in the form

$$\text{(D-42)} \qquad \int (c^2\mathbf{B} - 4\pi\mathbf{M}) \cdot d\mathbf{l} = 4\pi \int \mathbf{J} \cdot d\mathscr{A}$$

from which the magnetic field is defined as

$$\text{(D-43)} \qquad \mathbf{H} = c^2\mathbf{B} - 4\pi\mathbf{M}$$

It follows that

$$\text{(D-44)} \qquad \frac{H_{\text{esu}}}{H_{\text{emu}}} = c \qquad \frac{H_{\text{esu}}}{H_{\text{mks}}} = 12\pi \times 10^7$$

Reasoning similar to D-16ff leads to

$$\text{(D-45)} \qquad \mathbf{D} = \mathbf{E} + 4\pi\mathbf{P}$$

and the Maxwell equations in the esu system are

$$\text{(D-46)} \qquad \begin{aligned} &\nabla \cdot \mathbf{D} = 4\pi\rho_{\text{free}} \qquad \nabla \cdot \mathbf{E} = 4\pi\rho_{\text{total}} \\ &\nabla \cdot \mathbf{B} = 0 \\ &\nabla \times \mathbf{E} = -\dot{\mathbf{B}} \\ &\nabla \times \mathbf{H} = 4\pi\mathbf{J} + \dot{\mathbf{D}} \end{aligned}$$

In both the emu system and in the esu system, the relative dielectric constant and the relative permeability are the same dimensionless ratio as in the mks system, so that

$$\text{(D-47)} \qquad \begin{aligned} &B_{\text{emu}} = \kappa_m H_{\text{emu}} \qquad \text{and} \qquad B_{\text{esu}} = \frac{\kappa_m}{c^2} H_{\text{esu}} \\ &D_{\text{emu}} = \frac{\kappa}{c^2} E_{\text{emu}} \qquad \text{and} \qquad D_{\text{esu}} = \kappa E_{\text{esu}} \end{aligned}$$

The following table summarizes the relations among quantities in the three systems of units. Here $c = 3 \times 10^{10}$ cm/sec (more precisely $c =$ 2.99793 cm/sec). Where the units for a quantity has been given a separate name, this name is indicated.

Charge	$\frac{Q_{emu}}{Q_{mks}} = \frac{1}{10}\frac{\text{abcoul}}{\text{coul}}$	$\frac{Q_{esu}}{Q_{mks}} = 3 \times 10^9 \frac{\text{statcoul}}{\text{coul}}$	$\frac{Q_{esu}}{Q_{emu}} = c$
Charge density	$\frac{\rho_{emu}}{\rho_{mks}} = 10^{-7}$	$\frac{\rho_{esu}}{\rho_{mks}} = 3 \times 10^3$	$\frac{\rho_{esu}}{\rho_{emu}} = c$
Current	$\frac{I_{emu}}{I_{mks}} = 10^{-1}\frac{\text{abamp}}{\text{amp}}$	$\frac{I_{esu}}{I_{mks}} = 3 \times 10^9 \frac{\text{statamp}}{\text{amp}}$	$\frac{I_{esu}}{I_{emu}} = c$
Current density	$\frac{J_{emu}}{J_{mks}} = 10^{-5}$	$\frac{J_{esu}}{J_{mks}} = 3 \times 10^5$	$\frac{J_{esu}}{J_{emu}} = c$
Electric field	$\frac{E_{emu}}{E_{mks}} = 10^6$	$\frac{E_{esu}}{E_{mks}} = \frac{1}{3 \times 10^4}$	$\frac{E_{esu}}{E_{emu}} = \frac{1}{c}$
Electric displacement	$\frac{D_{emu}}{D_{mks}} = 4\pi \times 10^{-5}$	$\frac{D_{esu}}{D_{mks}} = 12\pi \times 10^5$	$\frac{D_{esu}}{D_{emu}} = c$
Potential	$\frac{V_{emu}}{V_{mks}} = 10^8 \frac{\text{abvolt}}{\text{volt}}$	$\frac{V_{esu}}{V_{mks}} = \frac{1}{300}\frac{\text{statvolt}}{\text{volt}}$	$\frac{V_{esu}}{V_{emu}} = \frac{1}{c}$
Capacitance	$\frac{C_{emu}}{C_{mks}} = 10^{-9}\frac{\text{abfrd}}{\text{farad}}$	$\frac{C_{esu}}{C_{mks}} = 9 \times 10^{11}\frac{\text{statfd}}{\text{farad}}$	$\frac{C_{esu}}{C_{emu}} = c^2$
Resistance	$\frac{R_{emu}}{R_{mks}} = 10^9 \frac{\text{abohm}}{\text{ohm}}$	$\frac{R_{esu}}{R_{mks}} = \frac{1}{9 \times 10^{11}}\frac{\text{statohm}}{\text{ohm}}$	$\frac{R_{esu}}{R_{emu}} = \frac{1}{c^2}$
Power	$\frac{P_{emu}}{P_{mks}} = 10^7 \frac{\text{erg}}{\text{jou}}$	$\frac{P_{esu}}{P_{mks}} = 10^7 \frac{\text{erg}}{\text{joule}}$	$\frac{P_{esu}}{P_{emu}} = 1$
Polarization	$\frac{P_{emu}}{P_{mks}} = 10^{-5}$	$\frac{P_{esu}}{P_{mks}} = 3 \times 10^5$	$\frac{P_{esu}}{P_{emu}} = c$
Magnetic flux density	$\frac{B_{emu}}{B_{mks}} = 10^4 \frac{\text{gauss}}{\text{wbr/m}^2}$	$\frac{B_{esu}}{B_{mks}} = \frac{1}{3 \times 10^6}$	$\frac{B_{esu}}{B_{emu}} = \frac{1}{c}$
Magnetic field	$\frac{H_{emu}}{H_{mks}} = \frac{4\pi}{10^3}\frac{\text{oersted}}{\text{amp-trn/m}}$	$\frac{H_{esu}}{H_{mks}} = 12\pi \times 10^7$	$\frac{H_{esu}}{H_{emu}} = c$
Magnetization	$\frac{M_{emu}}{M_{mks}} = 10^{-3}$	$\frac{M_{esu}}{M_{mks}} = 3 \times 10^7$	$\frac{M_{esu}}{M_{emu}} = c$
Inductance	$\frac{L_{emu}}{L_{mks}} = 10^9 \frac{\text{abhenry}}{\text{henry}}$	$\frac{L_{esu}}{L_{mks}} = \frac{1}{9 \times 10^{11}}\frac{\text{stathry}}{\text{henry}}$	$\frac{L_{esu}}{L_{emu}} = \frac{1}{c^2}$

mks	emu	esu	Gaussian
$\nabla \cdot \mathbf{E} = \dfrac{\rho_{\text{total}}}{\varepsilon_0}$	$\nabla \cdot \mathbf{E} = 4\pi c^2 \rho_{\text{total}}$	$\nabla \cdot \mathbf{E} = 4\pi \rho_{\text{total}}$	$\nabla \cdot \mathbf{E} = 4\pi \rho_{\text{total}}$
$\nabla \cdot \mathbf{D} = \rho_{\text{free}}$	$\nabla \cdot \mathbf{D} = 4\pi \rho_{\text{free}}$	$\nabla \cdot \mathbf{D} = 4\pi \rho_{\text{free}}$	$\nabla \cdot \mathbf{D} = 4\pi \rho_{\text{free}}$
$\nabla \cdot \mathbf{B} = 0$	$\nabla \cdot \mathbf{B} = 0$	$\nabla \cdot \mathbf{B} = 0$	$\nabla \cdot \mathbf{B} = 0$
$\nabla \times \mathbf{E} = -\dot{\mathbf{B}}$	$\nabla \times \mathbf{E} = -\dot{\mathbf{B}}$	$\nabla \times \mathbf{E} = -\dot{\mathbf{B}}$	$\nabla \times \mathbf{E} = -\dfrac{1}{c}\dot{\mathbf{B}}$
$\nabla \times \mathbf{H} = \mathbf{J} + \dot{\mathbf{D}}$	$\nabla \times \mathbf{H} = 4\pi \mathbf{J} + \dot{\mathbf{D}}$	$\nabla \times \mathbf{H} = 4\pi \mathbf{J} + \dot{\mathbf{D}}$	$\nabla \times \mathbf{H} = \dfrac{4\pi}{c}\mathbf{J} + \dfrac{1}{c}\dot{\mathbf{D}}$
$\mathbf{D} = \varepsilon_0 \mathbf{E} + \mathbf{P}$	$\mathbf{D} = \dfrac{1}{c^2}\mathbf{E} + 4\pi \mathbf{P}$	$\mathbf{D} = \mathbf{E} + 4\pi \mathbf{P}$	$\mathbf{D} = \mathbf{E} + 4\pi \mathbf{P}$
$\mathbf{H} = \dfrac{\mathbf{B}}{\mu_0} - \mathbf{M}$	$\mathbf{H} = \mathbf{B} - 4\pi \mathbf{M}$	$\mathbf{H} = c^2\mathbf{B} - 4\pi \mathbf{M}$	$\mathbf{H} = \mathbf{B} - 4\pi \mathbf{M}$
$\mathbf{F} = Q(\mathbf{E} + \mathbf{v} \times \mathbf{B})$	$\mathbf{F} = Q(\mathbf{E} + \mathbf{v} \times \mathbf{B})$	$\mathbf{F} = Q(\mathbf{E} + \mathbf{v} \times \mathbf{B})$	$\mathbf{F} = Q\left(\mathbf{E} + \dfrac{\mathbf{v} \times \mathbf{B}}{c}\right)$
$\mathbf{D} = \varepsilon_0 \kappa \mathbf{E}$	$\mathbf{D} = \dfrac{\kappa}{c^2}\mathbf{E}$	$\mathbf{D} = \kappa \mathbf{E}$	$\mathbf{D} = \kappa \mathbf{E}$
$\mathbf{B} = \mu_0 \kappa_m \mathbf{H}$	$\mathbf{B} = \kappa_m \mathbf{H}$	$\mathbf{B} = \dfrac{\kappa_m}{c^2}\mathbf{H}$	$\mathbf{B} = \kappa_m \mathbf{H}$

The Gaussian System

For reasons of convenience in some fields of physics it has become the custom to use a mixed system of units, called the Gaussian system, in which electrical quantities—charge, current, potential, electric field, etc.—are explicitly displayed in esu, whereas magnetic quantities—flux density, field, etc.—are in emu. From the fourth column of the previous table it is evident that equations using the Gaussian units differ from similar equations in emu or esu by the introduction of factors involving powers of c at appropriate places. A sampling of electromagnetic equations in the four systems of units is shown in the table on page 428.

Index